Recent Results in Cancer Research 94

Predictive Drug Testing on Human Tumor Cells

Edited by
V. Hofmann M.E. Berens G. Martz

With 87 Figures and 107 Tables

Springer-Verlag
Berlin Heidelberg New York Tokyo 1984

Priv. Doz. Dr. Victor Hofmann
Universitätsspital Zürich
Departement für Innere Medizin
Abteilung für Onkologie
Rämistrasse 100, 8091 Zürich, Switzerland

Michael E. Berens, Ph. D.
Department of Obstetrics Gynecology
Bowman Gray, School of Medicine
300 S. Hawthorne, Winston Salem NC 27103, USA

Professor Dr. Georg Martz
Universitätsspital Zürich
Departement für Innere Medizin
Abteilung für Onkologie
Rämistrasse 100, 8091 Zürich, Switzerland

Sponsored by the Swiss League against Cancer

ISBN-13: 978-3-642-82297-1 e-ISBN-13: 978-3-642-82295-7
DOI: 10.1007/978-3-642-82295-7

Library of Congress Cataloging in Publication Data. Main entry under title: Predictive drug testing on human tumor cells. (Recent results in cancer research; 94) Includes bibliographical references and index. 1. Antineoplastic agents-Testing. 2. Cancer cells-Growth-Regulation. 3. Cloning. 4. Cancer-Chemotherapy-Evaluation. I. Hofmann, V. (Victor). 1948– . II. Berens, M. E. (Michael E.). 1953– . III. Martz, G. IV. Series: Recent results in cancer research. v. 94 [DNLM: I. Antineoplastic Agents-pharmacodynamics. 2. Colony-Forming Units Assay. 3. Neoplasms-drug therapy. 4. Tumor Stem Cell Assay. W1 RE 106P v. 94/ QZ 267 P923] RC261.R35 vol. 94 616.99'4s 84-5599 [RC271.C5] [616.99'2061]

2125/3140–5 4 3 2 1 0

Preface

Predictive drug testing on human tumor cells in order to define the appropriate chemotherapy will remain imperative as long as the anticancer agents available are few in number and show only limited activity. The advantages of an effective test would lie in obviating the need for testing antineoplastic agents on large cohorts of patients for assessment of drug activity (phase II studies) and in allowing determination of optimal use of anticancer agents (phase III trials). Such an in vitro test could help to better define dose and schedule of drugs preclinically. The additive value of individual drugs could be determined on tumor cells in vitro in order to define the best combination chemotherapy in vivo.

Test-directed therapy would avoid unnecessary drug-related morbidity in patients with refractory tumors. Chemotherapy treatment would be more than justified even with side effects if palliation or even prolonged survival could be anticipated as a result. The benefits of predictive drug testing on human tumor cells would extend beyond improvement of individual patient treatment if the testing helped to identify new active agents. This spectrum of benefits to the entire field of oncology provides tremendous motivation for the development of such testing. Although a number of chemosensitivity tests have been proposed since the advent of modern anticancer chemotherapy, interest has been renewed by the possibility of cloning human tumor cells on agar plates, with a view to testing drug activity on cells with high proliferation capacity.

Other tests, referred to as nonclonogenic assays, measure the effects of the drugs on all the cells comprising the tumor.

After the initial optimism concerning immediate application of such tests, this book, arising from an international conference held in Zurich in July 1983, provides the reader with a more realistic perspective on the achievements to date in this rapidly moving field. It is fair to say that test-directed treatment of individual patients is not yet a reality, although some initial studies suggest that patients treated according to in vitro results may fare better. Now that the number of chemosensitivity-testing laboratories has dramatically expanded throughout the world, it is hoped that predictive tests will improve step by step and find their place in clinical oncology practice.

Zurich, July 1984 The Editors

Contents

*List of Contributors**

* The address of the principal author is given on the first page of each contribution

1 Page on which contribution begins

Clonogenic Assays for Solid Tumors

In Vitro Growth of Human Malignancies in a Cloning Assay

M. Rozencweig, V. Hofmann, C Sanders, W. Rombaut, U. Früh, and G. Martz*

Bristol-Myers Company, Pharmaceutical Research and Development Division, P.O. Box 4755, Syracuse, NY 13221–4755, USA

Introduction

The human tumor cloning assay developed by Hamburger and Salmon (1977) has great potential for cell biology and drug studies. A recent prospective study by Von Hoff et al. (1983) further supports its relevance as a chemosensitivity test. Among 246 in vitro-in vivo trials, the assay identified 43 tumors sensitive to chemotherapy. Clinical responses were achieved in 61% of these 43 tumors vs 23% of the total number of trials. Clinical failures were noted in 77% of the 246 trials, whereas 85% of the negative tests were confirmed clinically. In fact, many more trials had been conducted but correlations could be established in only 41% of these because of inadequate tumor growth and a number of other problems. The relatively small proportion of patients in whom these correlations could be made illustrates some of the limitations extensively described by Selby et al. (1983).

This paper summarizes the experience in two different laboratories with a soft agar cloning assay based on that developed by Hamburger and Salmon (1977). In both laboratories, the assay was essentially utilized for in vitro drug testing. Our report will focus on in vitro growth results with a number of human malignancies.

Materials and Methods

Origin of Tumor Samples

Tumor cells were obtained from various sources. Ascitic fluids, pleural effusions, and bone marrow aspirates were collected with preservative-free heparin. Biopsy specimens from primary and/or metastatic solid tumors and lymphomas were obtained from different surgeons. Most prostate and bladder samples were removed through transurethral resection. All specimens were rapidly brought to the laboratory. Solid samples were processed immediately. Effusions and bone marrow aspirates were occasionally kept at 4° C for 12–18 h, apparently without any loss of cell viability.

* The authors acknowledge the assistance of Mrs. Geneviève Decoster in the preparation of this manuscript. This work was supported in part by the Fonds National de la Recherche Scientifique Médicale (FRSM-3.4535.79, Belgium), the National Cancer Institute (NCI-NIH NO1/CM 53840, Bethesda, Maryland), and l'Association Sportive contre le Cancer (Belgium)

Isolation of Tumor Cells

Bone marrow cells were separated over a Ficoll gradient. Effusions were centrifuged at 1,200 rpm for 10 min, or 2,000 rpm for 5 min, at room temperature. Red cells were eliminated over a Ficoll gradient or by a short exposure to hypotonic solutions. Solid tumors were first washed in Hank's balanced salt solution or McCoy's 5A with 10% fetal calf serum (FCS). Then the samples were mechanically disrupted using scissors. Cells were passed through raspers or needles of decreasing diameters (18–23 gauge) to disrupt small aggregates. Cells were counted in a hemocytometer and cell viability was measured by the trypan blue exclusion method.

Culture System

Cultures were performed in 35 × 10 mm Petri dishes with a 2-mm grid. The double agar layer was prepared as recommended by Hamburger and Salmon (1977) with some variations. The underlayer or feeder layer contained 40 ml enriched McCoy's 5A, 10 ml tryptic soy broth, 0.6 ml asparagine (6.6 mg/ml), and 0.3 ml DEAE dextran (50 mg/ml). The enriched McCoy's 5A was prepared as follows: 500 ml McCoy's 5A, 25 ml horse serum, 50 ml heat-inactivated FCS, 5 ml sodium pyruvate (2.2%), 1 ml L-serine (21 mg/ml), 5 ml glutamine (200 m*M*), and 5 ml penicillin-streptomycin (10,000 U/ml). In Brussels, horse serum was replaced by FCS. To prepare the 0.5 agar underlayer, 1.7 ml of a 3% agar solution kept liquid in a warm water bath (about 45°–50° C) was added to 8.7 ml prewarmed (37° C) enriched McCoy's. This material was sufficient for nine plates. The upper layer was prepared with enriched CMRL 1066: 100 ml CMRL 1066, 15 ml horse serum (see above), 200 U insulin in 2 or 5 ml, 1 ml vitamin C (30 m*M*), 2 ml glutamine and penicillin-streptomycin. Immediately before plating, asparaginase (0.6 mg/40 ml medium), DEAE dextran (0.3 ml/40 ml medium), and mercaptoethanol (1 : 100) were added. The cell concentration was adjusted to 3.0×10^6/ml. The 0.3% upper layer was prepared by adding 0.3 ml 3% agar to 2.2 ml of the enriched CMRL and 0.5 ml of the cell suspension corresponding to 1.5×10^6 cells. This material was sufficient for three plates and was layered over the 0.5 agar, which was solid at that point.

Colony Formation

Tumor colony formation was monitored using an inverted microscope. Immediately after plating, all dishes were checked for the presence of preformed cell aggregates. Thereafter, plates were incubated at 37° C in a 5%–7% humidified atmosphere and examined two or three times a week until final counting. Small clusters of 3–20 cells appeared within 10 days and colonies consisting of ≥ 40 cells could be observed within 7–28 days. Final scoring of the number of colonies was generally possible during the 3rd or the 4th week.

Results

Of 204 tumor samples processed in Zurich, 46 which failed to grow were excluded from the study for a variety of reasons. In 15 tumor specimens, too few cells were obtained after mechanical disaggregation (in all cases $< 1.0 \times 10^6$). Among these samples, there were four superficial malignant melanomas and six biopsy specimens from bladder and prostatic cancers obtained by transurethral resection.

Nine effusions from nine patients with documented disseminated neoplastic disease were also excluded because cytological examination revealed no malignant cells. In nine other effusions, malignant cells were found but chemotherapy given 6–27 days before plating the cells could have been responsible for the lack of subsequent growth in vitro. Six specimens were obtained from patients in whom malignancy was clinically suspected, but further workup failed to disclose any malignant tissues.

Of the 289 samples received in Brussels 24% were excluded, mainly for insufficient cell collection (10%) and contamination upon arrival or in culture (9%). Exudates in which no malignant cells were identified were not excluded from this analysis.

In both laboratories, cell viability as determined by the trypan blue exclusion method was generally superior to 80% in cells isolated from effusions and bone marrow aspirates. In contrast, most solid tumors had median cell viability values below 50%. No correlations could be demonstrated between cell viability and likelihood of subsequent in vitro colony formation. A total of 377 samples plated were evaluable (Table 1). A large variety of tumor types were accrued, especially breast cancer (20%), ovarian cancer (16%), melanoma

Table 1. In vitro growth

Tumor type	Samples with ≥ 5 col per plate/total plated		
	Total	Zurich	Brussels
Breast	58/77 (75%)	18/22	40/55
Ovary	56/62 (90%)	24/26	32/36
Melanoma	29/44 (66%)	4/4	25/40
Lung (non-small cell)	19/28 (69%)	8/11	11/17
Unknown	14/25 (56%)	2/2	12/23
Bladder	14/14	13/13	1/1
Colon	8/14	3/5	5/9
Testicular	9/13	9/13	
Myeloma	9/11	7/8	2/3
Head and neck	6/11	4/9	2/2
Prostate	8/10	8/10	
Stomach	6/9	2/4	4/5
Sarcoma	6/9	4/5	2/4
Lung (small cell)	7/8	1/1	6/7
Renal	5/6	5/6	
Cervix	4/5	2/2	2/3
Hodgkin's	1/5	1/4	0/1
Mesothelioma	2/4		2/4
Endometrium	2/3	1/2	1/1
Lymphoma	1/3		1/3
Neuroblastoma	1/2		1/2
Esophagus	1/1	1/1	
Hepatoma	1/1		1/1
Gallbladder	1/1		1/1
Miscellaneous	0/11	0/10	0/1
Total	268/377 (71%)	117/158 (74%)	151/219 (69%)

(12%), non-small cell lung cancer (7%), and malignancies of unknown origin (7%). In vitro growth with formation of a minimum of five colonies per plate was seen in 71% of the samples. This occurred in 90% of the ovarian cancer specimens and 56%–75% of the other major tumor types represented in these series.

Colonies were seen regardless of the source of specimen, i.e., solid or liquid (Table 2). A minimum of 30 colonies per control plate is commonly used as a criterion of suitability for in vitro drug testing. Accordingly, in our experience 49% of the effusions but only 27% of the solid specimens would qualify. Results with bone marrow aspirates were even poorer, but these samples were mostly obtained from patients with hematologic disorders for which culture media were not optimal. Whether growth suitable for drug testing actually depends upon the source of specimen, the tumor type, or both is not apparent from our data. Different distributions of tumor types according to source of specimen, as well as small number of samples in most disease categories, preclude definite conclusions. Thus the majority of breast cancer cells had been obtained from pleural effusions or ascites, whereas most melanoma samples had been obtained by surgical excision (Tables 3 and 4). It is

Table 2. Tumor colony formation and suitability for in vitro drug testing according to source of specimens (Zurich series)

Source of specimen	No. of samples plated	Colony formation (%)	Suitable samples (%)
Solid	90	73	27
Effusion	52	77	49
Bone marrow	16	56	13
Total	158	74	32

Table 3. Liquid specimens suitable for drug testing

Tumor type	Samples with ≥ 30 col per plate/total plated		
	Total	Zurich	Brussels
Breast	30/65 (46%)	5/16	25/49 (51%)
Ovary	27/39 (69%)	10/17	17/22
Unknown	8/20	1/2	7/18
Lung (non-small cell)	10/19	4/7	6/12
Myeloma	1/11	1/8	0/3
Stomach	4/7	1/2	3/5
Melanoma	2/5	1/1	1/4
Lung (small cell)	3/4	1/1	2/3
Colon	1/4	1/1	0/3
Endometrium	2/3	1/2	1/1
Cervix	1/3	0/1	1/2
Head and neck	1/1		1/1
Esophagus	1/1	1/1	
Miscellaneous	0/15	0/9	0/6
Total	91/197 (46%)	27/68 (40%)	64/129 (50%)

Table 4. Solid specimens suitable for drug testing

Tumor type	Samples with ≥ 30 col per plate/total plated		
	Total	Zurich	Brussels
Melanoma	11/39 (28%)	2/3	9/36 (25%)
Ovary	16/23	6/9	10/14
Bladder	5/14	4/13	1/1
Testis	2/13	2/13	
Breast	4/12	2/6	2/6
Colon	2/10	0/4	2/6
Head and neck	2/10	2/9	0/1
Prostate	1/10	1/10	
Lung (non-small cell)	2/9	0/4	2/5
Sarcoma	2/8	1/5	1/3
Unknown	2/5		2/5
Renal	2/5	2/5	
Lung (small cell)	4/4		4/4
Mesothelioma	2/4		2/4
Hodgkin's	1/3	1/3	
Cervix	2/2	1/1	1/1
Gallbladder	1/1		1/1
Miscellaneous	0/8	0/5	0/3
Total	61/180 (34%)	24/90 (27%)	37/90 (41%)

Table 5. Growth of liquid specimens

Tumor type	Zurich		Brussels	
	Successful growth/ total plated	No. of col/plate: median (range)	Successful growth/ total plated	No. of col/plate: median (range)
Breast	12/14	25 (5–504)	35/49	49 (10–1,806)
	2/2	18 (5–30)		
Ovary	15/17	45 (6–824)	19/22	140 (21–700)
Unknown	2/2	37 (13–60)	8/18	108 (24–1,050)
Lung (non-small cell)	6/7	35 (7–180)	8/12	58 (11–1,027)
Myeloma	7/8	8 (5–27)	2/3	8 (7–9)
Stomach	1/2	30	4/5	303 (26–933)
Melanoma	1/1	103	2/4	30 (14–45)
Lung (small cell)	1/1	75	2/3	261 (72–450)
Colon	1/1	54	0/3	0
Endometrium	1/2	25	1/1	96
Cervix	1/1	7	1/2	407
Lymphoma			1/2	28
Neuroblastoma			1/2	12
Head and neck			1/1	2,675
Esophagus	1/1	130		
Miscellaneous	0/9		0/2	
Total	51/68		85/130	

Table 6. Growth of solid specimens

Tumor type	Zurich		Brussels	
	Successful growth/ total plated	No. of col/plate: median (range)	Successful growth/ total plated	No. of col/plate: median (range)
Melanoma	3/3	28 (20–136)	23/36	23 (5–1,049)
Ovary	9/9	32 (8–183)	13/14	183 (15–1,037)
Bladder	13/13	18 (6–139)	1/1	36
Testis	9/13	7 (5–34)		
Breast	4/6	10 (6–35)	5/6	19 (8–205)
Colon	2/4	15 (6–23)	5/6	28 (15–133)
Head and neck	4/9	9 (6–41)	1/1	9
Prostate	8/10	18 (5–42)		
Lung (non-small cell)	2/4	10 (7–12)	3/5	68 (6–163)
Sarcoma	4/5	5 (5–105)	2/3	25 (5–44)
Unknown			4/5	54 (7–196)
Renal	5/5	20 (6–40)		
Lung (small cell)			4/4	321 (34–2,315)
Mesothelioma			2/4	295 (73–516)
Hodgkin's	1/3	150		
Cervix	1/1	51	1/1	111
Stomach	1/2	20		
Gallbladder			1/1	49
Miscellaneous	0/3		0/3	
Total	66/90		65/90	

interesting to note that, in ovarian cancer, the percentage of samples that yielded a minimum of 30 colonies per plate seemed independent of the source of the specimen.

Consierable variations in the numbers of colonies were observed with each tumor type (Tables 5 and 6). In both laboratories, the highest plating efficiencies were consistently seen in ovarian carcinoma, with median values of 0.006%–0.04%. Results with small-cell carcinoma of the lung were also of interest, but small numbers of samples were studied. The highest number of colonies per plate was 2,315 in a specimen of small-cell lung cancer.

Conclusion

Our results are consistent with those reported by Salmon et al. (1978) and Von Hoff et al. (1981). In the latter series, a minimum of 30 colonies was produced by 27%–80% of the samples, depending on the type of malignancy. Among solid tumors, a noticeably lower propensity to form colonies was also noted with biopsy or surgical samples as compared to specimens from other sources.

The findings confirm that this assay may be used in various different laboraties with reproducible growth rates. A large variety of tumors grow and form colonies in this system.

However, as abundantly discussed elsewhere and in other chapters of this volume, plating efficiency is still too low for its routine use as a chemosensitivity test for individual patients. A number of methods have been proposed to improve this plating efficiency, with variable results. These methods include enzymatic cell disaggregation (Kern et al. 1982; Pavelic et al. 1980; Tveit et al. 1981); incorporation of other media, such as Chee's modified Eagle's nedium (Kern et al. 1982) or cell-free ascites (Uitendaal et al. 1983); and use of red blood cells and low oxygen concentration (Tveit et al. 1981). Current data are encouraging for further extensive work in this most important area of research.

References

Hamburger AW, Salmon SE (1977) Primary bioassay of human tumor stem cells. Science 197: 461–463

Kern DH, Campbell MA, Cochran AJ, Burk MW, Morton DL (1982) Cloning of human solid tumors in soft agar. Int J Cancer 30: 725–729

Pavelic ZP, Slocum HK, Rustum YM, Creaven PJ, Nowak NJ, Karakousis C, Takita H, Mittelman A (1980) Growth of cell colonies in soft agar from biopsies of different human solid tumors. Cancer Res 40: 4151–4158

Salmon SE, Hamburger AW, Soehnlen B, Durie BGM, Alberts DS, Moon TE (1978) Quantitation of differential sensitivity of human-tumor stem cells to anticancer drugs. N Engl J Med 298: 1321–1327

Selby P, Buick RN, Tannock I (1983) A critical appraisal of the "human tumor stem-cell assay". N Engl J Med 308: 129–134

Tveit KM, Fodstad O, Pihl A (1981) Cultivation of human melanomas in soft agar. Factors influencing plating efficiency and chemosensitivity. Int J Cancer 28: 329–334

Uitendaal MP, Hubers HAJM, McVie JG, Pinedo HM (1983) Human tumour clonogenicity in agar is improved by cell-free ascites. Br J Cancer 48: 55–59

Von Hoff DD, Cowan J, Harris G, Reisdorf G (1981) Human tumor cloning: feasibility and clinical correlations. Cancer Chemother Pharmacol 6: 265–271

Von Hoff DD, Clark GM, Stogdill BJ, Sarosdy MF, O'Brien MT, Casper JT, Mallox DE, Page CP, Cruz AB, Sandbach JF (1983) Prospective clinical trial of a human tumor cloning system. Cancer Res 43: 1926–1931

*Development and Applications of a Human Tumor Colony Assay for Chemosensitivity Testing**

S.E. Salmon

The University of Arizona, Health Sciences Center, Section of Hematology and Oncology, Tucson, AZ 85724, USA

Introduction

A major objective in cancer research has been to develop simple techniques to predict drug sensitivity of human cancers. Studies carried out with murine tumor models had suggested that in vitro tissue culture assays performed in semisolid media such as agar could be used to assess growth and chemosensitivity of clonogenic tumor cells (which are considered to be closely related to tumor stem cells in vivo) (Park et al. 1971). Tumor stem cells are a key subpopulation of cells within a tumor, which are responsible for self-renewal of the tumor cell population, as well as recurrence and metastasis after subcurative local or systemic treatment (Bruce et al. 1966; Steele 1977; Ogawa et al. 1973). In the mid-1970s, Hamburger and I developed an assay system which proved capable of supporting tumor colony formation from fresh biopsies of human cancers (Hamburger and Salmon 1977a). Our initial emphasis was on multiple myeloma and ovarian cancer (Hamburger and Salmon 1977b; Hamburger et al. 1978). An analogous assay procedure was developed independently by Courtenay and Mills (1978) and applied preclinically to human tumor xenografts. Our intent has been to develop a reproducible, pharmacologically based colony assay system which would prove useful for studies of cancer biology and prediction of response and survival in cancer patients, and to aid in screening and assessment of new anticancer drugs We described the assay as a "tumor stem cell assay" with the anticipation that clonogenic tumor cells as measured in vitro would have some correspondence to tumor stem cells in vivo. This paper briefly reviews progress in these areas.

Methods

The methods described briefly below are as carried out at the University of Arizona Cancer Center and detailed elsewhere. A flow diagram of the procedure appears in Fig. 1. The soft agar procedure of Hamburger and Salmon (1977a, b) was utilized for studies of multiple myeloma and the various solid tumors tested. Solid tumor biopsies were disaggregated into single-cell suspensions using mechanical (Hamburger et al. 1978) or enzymatic methods (Pavelic et al. 1980; Hamburger et al. 1981). Drug exposures were performed either with a

* Studies supported by grants CA-21839, CA-17094, and CA-23074 and contract NO-1-CM-17497 from the National Institutes of Health, Bethesda, MD 20205, USA

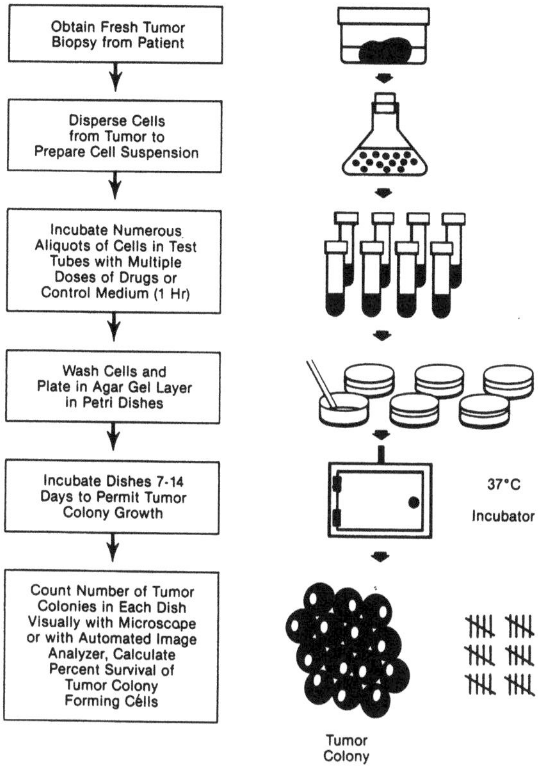

Fig. 1. Steps in drug-sensitivity assay (Salmon et al. 1980)

1-h exposure prior to plating or by continuous exposure by incorporating the agent into the agar (Salmon et al. 1980; Alberts et al. 1981a). The exposure mode was selected to simulate the likely pharmacologically achievable plasma concentration-time product for the specific drug tested. At least two concentrations of the drug were tested so that a dose-response curve could be constructed.

Plates were examined by inverted microscopy on the day of plating to assess the quality of the cell suspension. Our current positive control is with the plant lectin abrin, added at 10 µg/ml by continuous exposure (Salmon et al. 1983a). Control plates are scanned every few days by inverted microscopy and colonies counted on days 10–16 using a Bausch and Lomb FAS II image analysis system which has been optimized for counting tumor colonies and excluding artifacts (Kressner et al. 1980). Use of an initial day 1 FAS II count for background subtraction and the iodophenyl-nitrophenyl-phenyltetrazolium (INT) stain for selective counting of viable colonies further enhances the specificity of the assay (Salmon 1980b, pp 135–152; Alley et al. 1982). In our laboratory, the definition of adequate in vitro growth for evaluation of drug effect is a minimum of 30 colonies per 35-mm petri dish in the controls. The median number of tumor colonies from a large series of experiments was 108 colonies per control dish (500,000 nucleated cells plated). In the initial studies, the in vitro survival-concentration curves were ranked for sensitivity using an "area under the curve technique" (Salmon et al. 1978; Moon 1980). More recently, a mathematically simpler ranking of the percentage survival data up to empirically derived "cuttoff" drug concentrations has been used (Moon et al. 1981). In each instance, correlations are drawn between the in vitro and in vivo data in order to form a "training set" (for drugs as well as tumor types) for prospective application thereafter (Moon 1980; Moon et al. 1981). At the cutoff concentration, we classify tumors as sensitive to specific drugs if survival of tumor-colony-forming units (TCFU) is reduced to 30% of control or less, and as intermediate in the range of 30%–50%. When survival of TCFUs is greater than 50% of control at the cutoff concentration, the tumor is considered to be resistant to the specific drug tested.

Results and Discussion

Tumor Colony Growth

Although not all tumor specimens grow in the human tumor colony assay (HTCA), most tumor types have been grown with the assay system (Salmon 1980b, pp 3–14). Some of the recognized limitations in HTCA in its current stage of development include difficulties in obtaining true single-cell suspensions, and very low cloning efficiencies observed for many tumor types (e.g., 0.01%–0.1%). Cloning efficiencies vary widely and neoplasms such as lymphoma, multiple myeloma, and osteosarcoma have relatively poor growth rates. The best growth rates have been observed in ovarian and lung cancer and melanoma. About 50% of specimens for various types of cancers give rise to at least 30 colonies per plate. A linear relationship is generally observed between the number of cells plated and the number of tumor colonies formed. Evidence that the colonies are comprised of tumor cells has been obtained with morphological, cytogenetic, and biomarker studies (Salmon 1980b, pp 135–152). Colony formation by normal fibroblasts is inhibited by the agar in the culture system, whereas growth of normal hemopoietic precursors is limited by the lack of conditioning factors required for their growth. Tumor colonies arising from single tumor cells by definition arise from clonogenic tumor cells. Obtaining true single-cell suspensions remains a key objective. Evidence that tumor colonies contain cells with stem cell characteristics has been obtained by growth of tumors in nude mice from tumor colonies (e.g., in lung cancer) (Carney et al. 1981), and by "self-renewal" studies in vitro. In vitro self-renewal has been achieved in melanoma (Thomson and Meyskens 1982), and ovarian cancer (Buick and MacKillop 1981). Thus it is apparent that at least some of the cells giving rise to colonies in HTCA are closely related to tumor stem cells. The low cloning efficiencies observed may in fact reflect tumor stem cell characteristics (MacKillop et al. 1983).

Drug Sensitivity

Survival curves for tumors exposed to cycle-nonspecific drugs for 1 h prior to plating generally show progressively increasing lethality with increasing drug dosage. However, substantial heterogeneity in sensitivity is observed from patient to patient. The use of 1 h as a routine incubation time for most drugs may be viewed as arbitrary (Selby et al. 1983), but was selected to encompass the major pharmacologic exposure time in vivo after intravenous push injection of typical short-acting anticancer drugs (Alberts et al. 1981a). Many of the apparently flat survival curves observed in early experiments appear to have been artifactual and were likely to be due to the presence of aggregates in the cell suspensions. Use of day 1 subtraction counts, the viability stain, and the addition of a positive control have substantially reduced interpretative problems. Drugs which are schedule dependent often nanifest flat survival concentration curves with 1-h exposure, but steep inhibition curves at substantially lower doses when the drug is added to the agar for continuous exposure studies. Schedule-dependent curves have been observed with bleomycin, VP-16, and vinblastine, but not with doxorubicin, *cis*-platinum, or alkylating agents (Ludwig et al. to be published). A common phenomenon observed with many of the standard anticancer drugs is that a majority of solid tumor specimens exhibit resistance to the agent up to dose levels which exceed pharmacologically achievable plasma concentrations. Whenever possible we obtain large tumor specimens and attempt to test at least seven or eight drugs (at two concentrations) to increase the probability of identifying at least one active drug. In ovarian cancer, the frequency of in vitro sensitivity to common agents such as *cis*-platinum, doxorubicin, and bleomycin was greater in patients who had received no prior chemotherapy than in those who were in relapse (Alberts et al. 1981b)

Serial in vitro studies have also been evaluated in a series of patients before and after initiation of therapy (Salmon et al. 1980). Tumors which initially exhibited sensitivity to a specific drug in vitro prior to therapy frequently showed resistance in vitro on rebiopsy after the patient had relapsed ($p = 0.03$) (Salmon 1980b).

Correlation of In Vitro Drug Sensitivity and Clinical Response

Correlation between in vitro chemosensitivity in HTCA and clinical response was first evaluated at the University of Arizona Cancer Center in a series of patients with multiple myeloma or ovarian cancer (Salmon et al. 1978). Thirty-one clinical trials were assessed. A highly significant correlation was observed between in vitro tumor resistance to specific drugs and failure of the patient to respond to the same drugs clinically. On the other hand, in vitro sensitivity was also noted in most instances where the patient responded to the drug; however, there was a subset of patients whose TCFUs exhibited in vitro sensitivity but failed to show objective clinical response (false positives) (Salmon et al. 1978). This initial report also detailed practical and logistic problems needing resolution before broad clinical application of HTCA could become a reality. A number of subsequent studies have independently confirmed and extended the initial report (Alberts et al. 1980, 1981a; Salmon et al. 1978; Meyskens et al. 1981, 1983; Von Hoff et al. 1981a, 1983; Ozols et al. 1982; Carney et al. 1983; Mann et al. 1982; Durie et al. 1983). Overall, HTCA had a 71% true positive rate and a 91% true negative rate (Table 1). For most tumor types, the limitations in growth rate and cloning efficiency, the lack of effective drugs (as reflected by a high frequency of in vitro resistance) and clinical decisions for selection of alternative chemotherapy irrespective of assay results currently preclude routine application of HTCA. Additionally, correlations between in vitro sensitivity and clinical response to combination chemotherapy are difficult to evaluate. Clinical studies of ovarian cancer may advance rapidly as adequate growth is observed frequently, and a number of active drugs are available for testing in vitro and in the clinic. Major efforts at present are directed toward enhancing in vitro growth conditions by the use of specific growth or conditioning factors for specific tumor types such as lung (Carney et al. 1983) or breast cancer (Hug et al. 1983), or by use of an altered microenvironment (von Hoff and Hoang 1983). It is of importance to realize that tumor cells in a petri dish are not a tumor in the patient, and that many major differences exist including the lack of inherent means to activate or inactivate drugs which undergo metabolism.

Table 1. Published series of clinical correlations with human tumor colony assays

Year of publication	Tumor types	No. trials	True positive in vivo/ total positive in vitro (%)	True negative in vivo/ total negative in vitro (%)	Cloning method	Reference
1978	Ovarian, myeloma	32	11/12 (92)	20/20 (100)	H & S	Salmon et al. (1978)
1981	Ovarian	44	8/11 (73)	33/33 (100)	H & S	Alberts et al. (1981a, b)
1981	Melanoma	48	12/19 (63)[a]	25/29 (86)	H & S	Meyskens et al. (1981)
1981	Multiple	123	15/21 (71)	100/102 (98)	H & S	Von Hoff et al. (1981a, b, c)
1982	Melanoma	49	10/11 (91)[b]	38/38 (100)	C & M	Tveit et al. (1982)
1982	Multiple	36	9/11 (82)	24/25 (96)	H & S	Mann et al. (1982)
1983	Lung	24	8/12 (67)	12/12 (100)	H & S	Carney et al. (1983)
1983	Myeloma	33	11/15 (73)	15/18 (83)	H & S	Durie et al. (1983)
1983	Multiple	246	26/43 (60)	172/203 (85)	H & S	Von Hoff et al. (1983)
Total		635	110/155 (71)	439/480 (91)		

H & S, method of Hamburger and Salmon (1977a); *C & M*, method of Courtenay and Mills (1978)

[a] Mixed responses included as responses

[b] Mixed responses and stable disease included as responses

In Vitro Phase II Trials

The correlations of in vitro drug sensitivity and clinical response have provided evidence for validity of HTCA and have also provided an additional scientific basis for its application to new drug development. One application has been described as the "in vitro phase II trial" (Salmon 1980b, pp 291–312; Salmon et al. 1981), which has two key objectives. The first is to determine concentrations of a new agent which exhibit significant cytotoxic activity against TCFUs. The second objective of the in vitro phase II trial is to obtain a preliminary projection of the antitumor spectrum and "response rate" by tumor type for specific new agents. Such information can be of great value in focusing clinical resources with respect to tumor types which may be anticipated to exhibit clinical sensitivity to a given new agent. Optimally, such data should be obtained against a large battery of human tumors of major histological types, and published prior to the initiation of phase II clinical trials. In vitro phase II studies published on mitoxantrone (Von Hoff et al. 1981b), bisantrene (Von Hoff et al. 1981c) and recombinant leukocyte interferons (Salmon et al. 1983b) illustrate this application.

Preclinical Screening

Another important application of HTCA for new drug development is in the initial search for new anticancer drugs (Salmon 1980a, b). Routinely, the National Cancer Institute (NCI) has used an in vivo P388 leukemia screen for antitumor activity. The NCI has recently launched a major effort to test the HTCA as a means for identifying novel antitumor agents which may have been missed with the leukemia prescreen. Initial results from this study have recently been reported (Shoemaker et al. 1983). They indicate that HTCA can detect most clinically active agents (with the exception of a few requiring bioactivation and a few antimetabolites). It also correctly detected as negatives a variety of simple sugars, amino acids, and salts. Screening of P388 negative compounds in HTCA was recently initiated. On preliminary assessment, a small proportion of P388 negative compounds appear to exert antitumor activity in vitro in HTCA (Shoemaker et al. 1983). A major long-term goal of the NCI study is to select novel agents identified with HTCA screening for advancement to clinical trial to establish whether an in vitro screen can identify unique and effective compounds.

Concluding Remarks

The human tumor colony assay provides a useful new tool with which to assess in vitro chemosensitivity in relation to cancer chemotherapy, as well as serving as an aid in drug development. The relative infrequency of marked in vitro sensitivity to pharmacologically achievable concentrations of many of the currently available anticancer drugs is consistent with clinical experience in a number of solid tumors. The published clinical correlations provide strong evidence that HTCA is able to predict clinical chemosensitivity with reasonable accuracy and drug resistance with a high degree of accuracy. While these findings are analogous to those obtained with colony assays for antibiotic sensitivity testing in bacterial infections, there are a number of difficulties in routinely applying HTCA for chemosensitivity testing of human tumors. Critical review (Selby et al. 1983) and efforts to improve HTCA procedures are therefore clearly indicated. It would be desirable to have a short-term assay which could yield similar results to those obtained with HTCA. The use of thymidine labeling (e.g., Sanfillipo et al. 1981) has shown encouraging results. However, as pointed out by Rupniak et al. (1983) on systematic evaluation, short-term assays with either dye exclusion or thymidine labeling still appear inadequate as they lack the time scale required to allow both recovery from sublethal damage and expression of potentially lethal damage by various agents. Applications of HTCA to clinical chemosensitivity testing and new drug development are obviously still in their infancy and likely to undergo continued evolution. Improved methods are also needed to prepare better single-cell suspensions from solid tumors and to enhance tumor colony growth. Criteria for drug sensitivity are empirical and subject to change. After growth is improved, randomized clinical studies will be needed to assess the overall clinical impact of chemosensitivity testing. Effective use of HTCA is also dependent on the availability of effective anticancer drugs. Currently available agents are "first generation" drugs, and lack efficacy for a number of common tumor types. Therefore, the use of HTCA for new drug screening and development may prove to be its most useful application.

Summary

Using the human tumor colony assay, the growth and chemosensitivity of clonogenic tumor cells present in fresh biopsies of human tumors can be investigated. Excellent evidence has been obtained that colonies grown in HTCA are comprised of tumor cells, and that clonogenic cells within tumor colonies have self-renewal properties (the defining feature of tumor stem cells). Clinical correlations have been made between in vitro chemosensitivity and the response of patients with metastatic cancer to chemotherapy. In a series of trials, HTCA has had a 71% true positive rate and a 91% true negative rate for predicting drug sensitivity and resistance respectively of cancer patients to specific chemotherapeutic agents. HTCA has also had several areas of application to new drug development and screening. Ongoing developmental research is needed to improve growth rates for many tumor types and to further improve assay methodology and thereby enhance its applicability to predictive drug sensitivity testing.

References

Alberts DS, Salmon SE, Chen HSG, Surwit EA, Soehnlen B, Young L, Moon TE (1980) Predictive chemotherapy of ovarian cancer using an in vitro clonogenic assay. Lancet 2: 340–342

Alberts DS, Salmon SE, Chen HSG, Moon TE, Young L, Surwit EA (1981a) Pharmacologic studies of anticancer drugs using the human tumor stem cell assay. Cancer Chemother Pharmacol 6: 253–264

Alberts DS, Chen HSG, Salmon SE, Surwit EA, Young L, Moon TE, Meyskens FL et al. (1981b) Chemotherapy of ovarian cancer directed by the human tumor stem cell assay. Cancer Chemother Pharmacol 6: 279–285

Alley MC, Uhl CB, Lieber MM (1982) Improved detection of drug cytotoxicity in the soft agar colony formation assay through use of a metabolizable tetrazolium salt. Life Sci 31: 3071–3078

Buick RN, MacKillop WJ (1981) Measurement of self-renwal in culture of clonogenic cells from human ovarian carcinoma. Br J Cancer 44: 349–355

Bruce WR, Meeker BE, Valeriote FA (1966) Comparison of the sensitivity of normal hematopoietic and transplanted lymphoma colony-forming cells to chemotherapeutic agents administered in vivo. J Natl Cancer Inst 37: 233–245

Carney DN, Gazdar AF, Bunn PA Jr, Guccion JG (1981) Demonstration of the stem cell nature of clonogenic tumor cells from lung cancer patients. Stem Cells 1: 149–164

Carney DN, Broder L, Edelstein M, Gazdar AF, Hansen M, Havemann K, Matthews MJ, Sorenson GD, Videlov L (1983) Experimental studies of the biology of human small cell lung cancer. Cancer Treat Rep 67: 21–26

Courtenay FD, Mills J (1978) An in vitro colony assay for human tumors grown in immune-suppressed mice and treated in vivo with cytotoxic agents. Br J Cancer 37: 261–268

Durie BGM, Young L, Salmon SE (1983) Human myeloma stem cell culture: interrelationships between drug sensitivity, cell kinetics and patient survival duration. Blood 61: 929–934

Hamburger AW, Salmon SE (1977a) Primary bioassay of human tumor stem cells. Science 197: 461–463

Hamburger AW, Salmon SE (1977b) Primary bioassay of human myeloma stem cells. J Clin Invest 60: 846–854

Hamburger AW, Salmon SE, Kim MB, Trent JM, Soehnlen B, Alberts DS, Schmidt HJ (1978) Direct cloning of human ovarian carcinoma cells in agar. Cancer Res 38: 3438–3443

Hamburger AW, White CP, Tencer K (1981) Effect of enzymatic disaggregation on proliferation of human tumor cells in soft agar. J Natl Cancer Inst 68: 945–948

Hug V, Spitzer G, Drewinko B, Blumenschein G, Haidle C (1983) Improved culture conditions for the in vitro growth of human breast tumors. Proc Am Assoc Cancer Res 24:35 [abstract no. 138]

Kressner BE, Morton RRA, Martens AE, Salmon SE, Von Hoff DD, Soehnlen B (1980) Use of an image analysis system to count colonies in stem cell assays of human tumors. In: Salmon SE (ed) Cloning of human tumor stem cells. Liss, New York, pp 179–193

Ludwig R, Alberts DS, Miller TP, Salmon SE (to be published) Schedule dependency of anticancer drugs in the human tumor stem cell assay. Cancer Chemother Pharmacol

MacKillop WJ, Ciampi A, Till JE, Buick RN (1983) A stem cell model of human tumor growth: implications for tumor cell clonogenic assays. J Natl Cancer Inst 70:9–16

Mann BD, Kern DH, Giuliano AE, Burk MW, Campbell MA, Kaiser LR, Morton DL (1982) Clinical correlations with drug sensitivities in the clonogenic assay. Arch Surg 117:33–36

Meyskens FL, Moon TE, Dana B, Gilmartin E, Casey WJ, Chen HSG, Franks DH, Young L, Salmon SE (1981) Quantitation of drug sensitivity by human metastatic melanoma colony forming units. Br J Cancer 44:787–797

Meyskens FL, Loescher L, Moon TE, Salmon SE (1983) A prospective trial of single agent chemotherapy for metastatic malignant melanoma directed by in vitro colony survival in a clonogenic assay. Proc Am Assoc Cancer Res 24:143 [abstract no. 567]

Moon TE (1980) Quantitative and statistical analysis of the association between in vitro and in vivo studies. In: Salmon SE (ed) Cloning of human tumor stem cells. Liss, New York, pp 209–221

Moon TE, Salmon SE, White CS, Chen HSG, Meyskens FL, Durie BGM, Alberts DS (1981) Quantitative association between the in vitro human tumor stem cell assay and clinical response to cancer chemotherapy. Cancer Chemother Pharmacol 6:211–218

Ogawa M, Bergsagel DE, McCulloch EA (1973) Chemotherapy of mouse myeloma: quantitative cell culture predictive of response in vivo. Blood 41:7–15

Ozols RF, Young RC, Speyer JL, Sugarbaker PH, Greene R, Jenkins J, Myers CE (1982) Phase I and pharmacological studies of Adriamycin adninistered intraperitoneally to patients with ovarian cancer. Cancer Res 42:4265–4269

Park CH, Bergsagel DE, McCulloch EA (1971) Mouse myeloma tumor stem cells: a primary cell culture assay. J Natl Cancer Inst 46:411–422

Pavelic ZP, Slocum HK, Rustum YM, Creavin PJ, Karakousis C, Takita H (1980) Colony growth in soft agar of human melanoma, sarcoma and lung carcinoma cells disaggregated by mechanical and enzymatic methods. Cancer Res 40:2160–2164

Rupniak HT, Dennis LY, Hill BT (1983) An intercomparison of in vitro assays for assessing cytotoxicity after a 24-hour exposure to anticancer drugs. Tumori 69:37–42

Salmon SE (1980a) Application of the human tumor stem cell assay in the development of anticancer therapy. In: Burchenal JH, Oettgen HF (eds) Cancer: achievements, challenges and prospects for the 1980's. Grune & Stratton, New York, pp 2:33–43

Salmon SE (ed) (1980b) Cloning of human tumor stem cells. Liss, New York

Salmon SE, Hamburger AW, Soehnlen B, Durie BGM, Alberts DS, Moon TE (1978) Quantitation of differential sensitivity of human tumor stem cells to anticancer drugs. N Engl J Med 298:1321–1327

Salmon SE, Alberts DS, Meyskens FL, Durie BGM, Jones SE, Soehnlen B, Young L, Chen HSG, Moon TE (1980) Clinical correlations of in vitro drug sensitivity. In: Salmon SE (ed) Cloning of human tumor stem cells. Liss, New York, pp 223–245

Salmon SE, Meyskens FL, Alberts DS, Soehnlen B, Young L (1981) New drugs in ovarian cancer and malignant melanoma: in vitro phase II screening with the human tumor stem cell assay. Cancer Treat Rep 65:1–12

Salmon SE, Durie BGM, Young L, Liu R, Trown PW, Stebbing N (1983b) Effects of cloned human leukocyte interferons in the human tumor stem cell assay. J Clin Oncol 1:217–225

Salmon SE, Liu R, Hayes C, Persaud J, Roberts R (1983a) Usefulness of abrin as a positive control for the human tumor clonogenic assay. IND 1:277–281

Sanfilippo O, Daidone MG, Costa A, Canetta R, Silvestrini R (1981) Estimation of differential in vitro sensitivity of non-Hodgkin's lymphomas to anticancer drugs. Eur J Cancer 17: 217–226
Selby P, Buick RN, Tannock I (1983) A critical appraisal of the human tumor stem cell assay. N Engl J Med 308: 129–134
Shoemaker RH, Wolpert-DeFilippes MK, Makuch RW, Venditti JM (1983) Use of the human tumor clonogenic assay for new drug screening. Proc Am Assoc Cancer Res 24: 311 [abstract no. 1231]
Steele GG (1977) Growth and survival of tumour stem cells. In: Steele GG (ed) Growth kinetics of tumors, Clarendon, Oxford, pp 217–262
Thomson SP, Meyskens FL Jr (1982) method for measurement of self-renewal capacity of clonogenic cells from biopsies of metastatic human malignant melanoma. Cancer Res 42: 4606–4613
Tveit KM, Fodstad O, Lotsberg J, Vaage S, Pihl A (1982) Clonoy growth and chemosensitivity in vitro of human melanoma biopsies. Relationship to clinical parameters. Int J Cancer 29: 533–538
Von Hoff DD, Hoang M (1983) A new perfused capillary cloning system to improve cloning of human tumors. Proc Am Assoc Cancer Res 24: 310 [abstract no. 1225]
Von Hoff DD, Casper J, Bradley E, Sandbach J, Jones D, Makuch R (1981a) Association between human tumor colony forming assay results and response of an individual patient's tumor to chemotherapy. Am J Med 70: 1027–1032
Von Hoff DD, Coltman CA, Forseth B (1981b) Activity of mitoxantrone in a human tumor cloning system. Cancer Res 41: 1853–1855
Von Hoff DD, Coltman CA, Forseth B (1981c) Activity 9-10 anthracenedicarboxaldehyde bis [(4,5-dihydro-1*H*-N-imidazol-2yl)hydrazone] dihydrochloride (CL-216,942) in a human tumor cloning system. Cancer Chemother Pharmacol 6: 141–144
Von Hoff DD, Clark GM, Stogdill BJ, Sarosdy MF, O'Brien MT, Casper JT, Mattox DE, Page CP, Cruz AB, Sandbach JF (1983) Prospective clinical trial of a human tumor cloning system. Cancer Res 43: 1926–1931

A Replenishable Soft Agar Colony Assay for Human Tumour Sensitivity Testing

V.D. Courtenay

Radiotherapy Research Unit, Institute of Cancer Research, Belmont, Sutton, Surrey, United Kingdom

Introduction

The replenishable soft agar colony technique in test-tubes, first used in 1976 by Courtenay et al., is based on observations on the growth requirements of cells taken directly from excised tumours. Probably the most important feature of the method is the use of low oxygen concentrations in the gas phase, together with the addition of rat red blood cells (RBC) and the replenishment of the nutrients in the agar by adding liquid medium.
Under standard culture conditions in air-CO_2 incubators it has been a general observation, both in monolayer and in agar, that primary cultures of tumour cells fail when plated out at low cell densities but may survive when the number of cells is increased. A density of between 2×10^5 and 5×10^5 cells per millilitre (Hamburger and Salmon 1977) has therefore commonly been used. However, for colony assays such high cells numbers may lead to difficulties due to early exhaustion of the medium, limiting the growth of potentially clonogenic cells. The apparent need for high cell densities has been thought to be due to the conditioning of the medium by the release of various normal metabolites from large numbers of cells, thus modifying the biochemical constitution of the medium. But another consequence of cellular metabolic activity is the uptake of O_2, and in static culture at high cell density the O_2 concentration within the medium is substantially reduced. CO_2 produced at the same time is more rapidly lost by diffusion but causes some reduction in pH.
The effect of subatmospheric O_2 concentrations in promoting the growth of normal mouse embryo cells was first reported in 1972 by Richter et al. Then, in 1975, Courtenay developed an agar colony method for Lewis Lung mouse tumour studies (Shipley et al. 1975), which was based on the use of low oxygen levels. It was found that primary cultures of the mouse tumour grown in monolayer could be maintained at lower cell densities (Courtenay 1976) if they were incubated with a gas phase containing 5% O_2 instead of air. Increasing CO_2 concentration from 5% to 10% or adding conditioned medium from confluent cultures did not improve survival. Subsequent work on many human tumours in primary culture has consistently shown improved colony formation in cultures gassed with mixtures containing 3%–5% O_2 (Courtenay 1983).
Low O_2 concentrations provide more physiological conditions for tumour cell growth, since the oxygen concentration in venous blood is equivalent to about 5% O_2, considerably lower than that in air (21%). The majority of tumour cells in a solid tumour are in a micro-environment with O_2 concentrations ranging down to levels below 0.009%, as indicated by changed radiosensitivity (Millar et al. 1982).

Recent Results in Cancer Research. Vol. 94

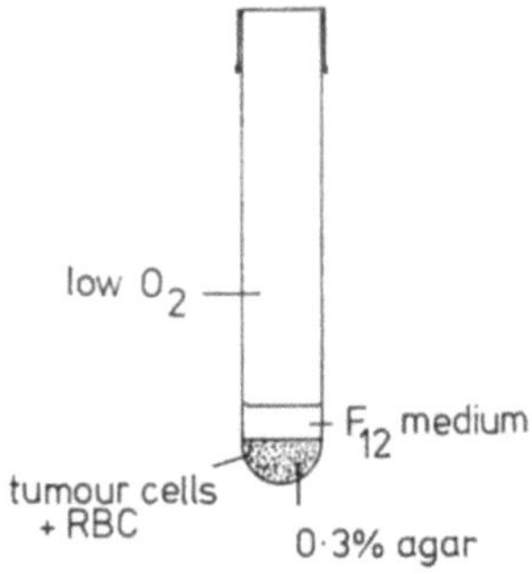

Fig. 1. The replenishable test-tube system for the growth of human tumour cell colonies in soft agar. *RBC,* red blood cells

Another feature of the Courtenay method is the addition of rat RBC, which lyse in agar to release labile growth factors lacking in standard culture media. RBC have consistently improved the colony-forming efficiency (CFE) of tumour cells in agar. The provision of a replenishable liquid phase renewing the supply of nutrients in the agar at weekly intervals has permitted the continued growth of colonies. This has been achieved by replacing the widely used double-layer technique in Petri dishes by a test-tube method. In test-tubes, the agar containing tumour cells and RBC forms a solid hemisphere in the bottom of the tube, as shown in Fig. 1. Culture medium can then be pipetted on top of the agar, and changed as necessary, without damaging the agar.

This method hsas now been extensively used, and has been found to have a number of other practical advantages. No underlayer is needed. The tubes are more rapidly and easily set up than Petri dishes, and the danger of mould infection, which can be a problem in Petri dishes incubated for several weeks, is virtually eliminated by the use of tightly fitting press-on caps. The important advantage of the greater depth of agar in the tubes in maintaining low oxygen tensions will be discussed later.

Experimental Methods

Preparation of Cell Suspensions

Cell suspensions are prepared from the chopped tumour by a variety of methods (Courtenay 1983), usually requiring incubation with enzymes. The cell suspension in Ham's F12m medium with 15% foetal calf serum is filtered through a 30-μm mesh to exclude clumps. In some cases, a further separation of clusters passing through the mesh may be obtained by keeping the vertical suspension at 4° C for 10 min. Cell clusters tend to settle out more rapidly than single cells and the top two-thirds of the suspension is then taken for use. The cells are stained with lissamine green to aid discrimination between live and dead cells, and the number of viable cells is counted in a haemocytometer, viewed under phase contrast. In our measurements, cell counts include stromal cells and the larger nucleated blood cells, as well as tumour cells. Small nucleated blood cells that can be clearly distinguished from tumour cells by their size are excluded from the counts.

Assay Procedure (Courtenay 1983)

The cell suspension to be assayed is diluted to a concentration of 5×10^4 cells per millilitre in culture medium. For melanomas and other tumours that may give a colony-forming

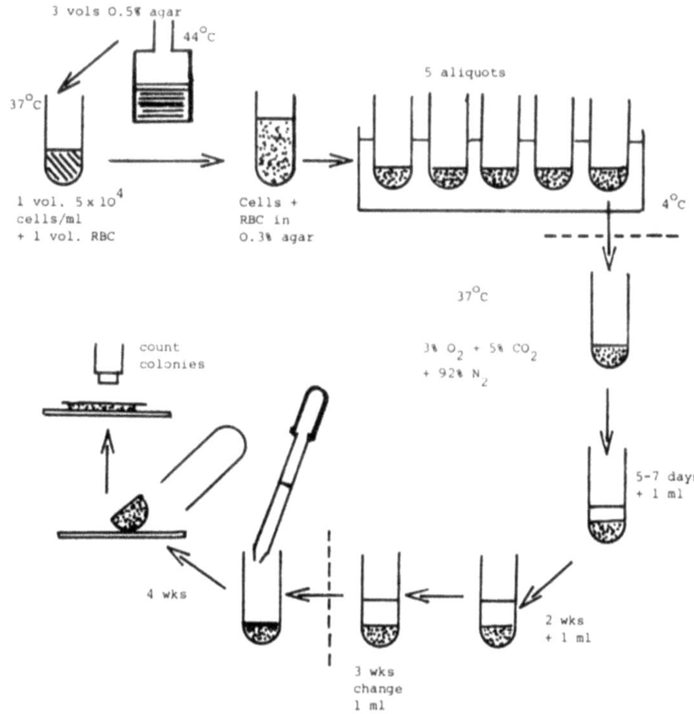

Fig. 2. Setting up and maintenance procedures for the replenishable test-tube system. *RBC,* red blood cells

efficiency (CFE) exceeding 2%, further dilutions are also set up. In this case the total number of tumour cells per millilitre is increased by adding up to 5×10^4 heavily irradiated (HR) cells pretreated on the same day with 40 Gy γ- or X-radiation. To set the cells up in agar, one volume of cell suspension is added to one volume of washed resuspended RBC, diluted 1 : 15. An 0.5% agar solution in Ham's F12m medium with 15% serum is made up immediately before use and held at 44° C. Three volumes of 0.5% agar medium are added to the cell/RBC mixture, it is inverted to mix and 1-ml aliquots are immediately pipetted into vertical test-tubes in ice water. When the agar is set, the tubes are gassed with a mixture of 3% O_2 + 5% CO_2 + 92% N_2 and incubated at 37° C in a three-gas incubator or suitable gas-tight box. The maintenance procedure is illustrated in Fig. 2. After 5–7 days, when the RBC are seen to be lysed, 1 ml of liquid medium is pipetted on to the agar in the tubes. A further 1 ml of medium is added 1 week later. Further additions of medium are made by removing and adding 1 ml of medium each week. When colonies are formed, they usually reach a size exceeding 50 cells after 3–4 weeks, but in some cases it may take rather longer.

For colony counting, the liquid medium is removed and the agar is decanted on to a microscope slide or Petri dish. A cover glass is placed on top, squashing the agar sufficiently to bring all the colonies into the same focal plane when examined under a dissecting microscope. With different tumours there are considerable differences in cell size, and therefore in the size of a 50-cell colony. For an unknown tumour it is therefore necessary first to determine the size of colony to be counted, using a $\times$ 20 objective to view the individual cells in a colony. By focussing up and down about 18 cells can be distinguished on the surface of a 50-cell colony, viewed from above or below. Colonies can then be counted under lower magnification, taking the size of the observed 50-cell colonies as a standard.

With the slide on a moveable stage, the whole area of the agar is covered by scanning back and forth.

CFE is measured as the number of colonies of 50 or more cells produced per 100 cells set up in agar. It has been shown that lethally irradiated tumour cells are capable of completing up to six divisions before dying, thus producing abortive colonies (Nias et al. 1965). The 50-cell colony is therefore a minimum size for accurate assessment of cell survival and a widely accepted criterion of survival in radiation studies. It is as yet uncertain how often abortive colony formation occurs after treatment with drugs which may have many different modes of action, but it may well apply to some categories of drug. Counting only colonies larger than 50 cells also helps to minimize errors from cell clusters introduced into the cell suspension.

CFE of Cells from Human Biopsy Specimens

The test-tube method has been applied to grow colonies from a wide range of tumour cells taken from human tumour biopsy material, as well as from xenografted human tumours not previously grown in culture. Table 1 shows CFEs obtained in our initial studies on human biopsy material in which the tubes were gassed with 5% O_2.

Altogether, 40 different tumours were tested and 22 gave colonies. Of these 19 gave a CFE greater than 0.1%. Colony formation appeared to be dependent on tumour type. No colonies were obtained from the six breast tumours and only two from five colon tumours, of which only one gave a CFE above 0.1%. Both of these tumours were found difficult to disaggregate and viable cell yields were low. In some cases the cells obtained may have been irreparably damaged in the process of disaggregation.

The best results were obtained with ovarian tumours (mainly from ascites) and with melanomas, and nine out of nine and seven out of 13 respectively gave colonies. The uterine tumour examined gave a CFE of 12%. The success with the melanomas has been amply confirmed by subsequent experience. From melanomas good single-cell suspensions with good cell yields can generally be obtained.

The biopsy studies were carried out in collaboration with Selby, who used an agar diffusion chamber (ADC) technique as an alternative to the in vitro method. In these studies, ADCs containing tumour cells in 0.3% agar (see Fig. 3) were implanted into the peritoneal cavity of mice immune suppressed by prior irradiation (Smith et al. 1976). The ADCs have the advantage that the cells are maintained under partially in vivo conditions, since there is an exchange of metabolites and nutrient substances from the peritoneal fluid of the mouse to the cells in agar in the ADCs. It had been thought that higher CFEs might be obtained. In fact, the cells tended to grow slightly more rapidly than in vitro, but the success rate (17 out of 29 tumours) and the average CFE were comparable with those from the in vitro system.

The failure to grow some categories of tumour, e.g., breast, could be related to a number of different factors, including a need for contact with stromal cells or for hormones and growth factors lacking in culture medium. However, the similar results obtained with the ADCs in which cells were maintained under partially in vivo nutritional conditions suggest that for many tumours (particularly the large proportion that can be grown as xenografts in mice), colony formation is unlikely to be improved by simple changes in culture medium.

The use of agar, which has the valuable property of preventing the growth of anchorage-dependent stromal cells, is common to both systems, and it may be that agar

Table 1. Colony-forming efficiency (CFE) of colonies grown in vitro from tumour biopsies

Tumour type	Form	Treatment of suspension	CFE in vitro
Melanoma	SC deposit	m	0
	SC deposit	m	15
	SC deposit	m	0.5
	SC deposit	m	3.0
	SC deposit	m	5.6
	SC deposit	m	0
	SC deposit	m	0.2
	Node deposit	m	0
	Node deposit	m	0
	Node deposit	m	0.5
	Node deposit	m	0.25
	Primary	m	0
	Ascites	u	0
Ovarian cancer	Ascites	u	2.7
	Ascites	u	0.25
	Ascites	u	0.02
	Ascites	u	0.2
	Ascites	u	1.0
	Ascites	u	1.0
	Primary	t	0.4
	Secondary	t	1.3
	Peritoneal deposit	m	4.5
Uterine cancer (body)	Secondary	m	12
Breast cancer	Primary	t	0
	Primary	t	0
	Primary	t	0
	Pleural effusion	u	0
	Pleural effusion	u	0
	Pleural effusion	u	0
Colorectal cancer	Primary	t + c	0
	Primary	t + c	0
	Primary	t	0.03
	Primary	t	0
	Secondary	t + c	0.3
Pancreatic carcinoma	Primary	t + c	0.17
Teratoma testis	Primary	t	0
Seminoma	Primary	t	0
Orchioblastoma	Primary	t	0.04
Hypernephroma	Secondary	t	0.1
Leiomyosarcoma	Primary	m	0

sc, subcutaneous; *m*, mechanical separation; *u*, untreated; *t*, trypsin; *c*, collagenase

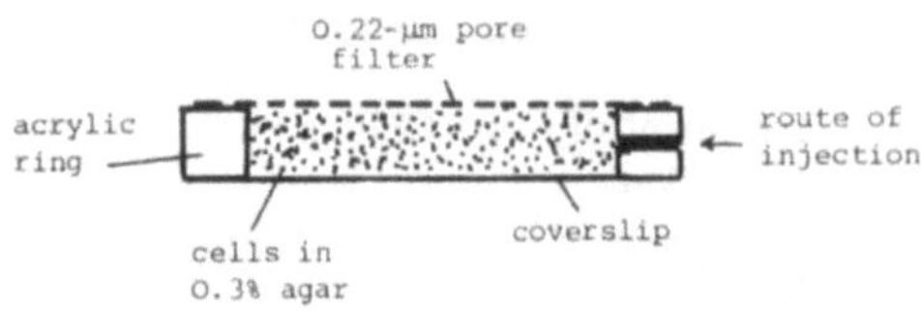

Fig. 3. Section of an agar diffusion chamber

imposes a limitation on the growth of some tumour cells. There is evidence from tumour cell lines normally grown in monolayer culture that not all are capable of growing in agar, and it could be that some tumour cells retain a degree of anchorage dependence. Hastings and Franks (1981), growing human bladder cell carcinomas, and Dodson et al. (1981), with a squamous cell carcinoma, have shown that the tumour cells were capable of forming xenografts in mice and could be grown in monolayer culture, but failed to give colonies in agar. From a recent series of 12 human sarcoma biopsy specimens obtained by Westbury at the Royal Marsden Hospital, none have given rise to compact colonies using the agar test-tube system, although six were successfully grown as xenografts and nine have been maintained through one to two subcultures in monolayer. Three of the monolayer cultures appeared to consist of a single cell type characteristic of the tumour of origin, and were grown up in monolayer culture both from the original biopsy specimen and from the xenografted tumour; but they did not form colonies in agar. The remaining six cultures were of mixed cell type. Attempts to grow these in agar resulted in two cases in producing spheroids of cells formed outside the agar. These results suggest that soft agar may not be a suitable medium for the growth of colonies from human sarcomas.

Experimental Basis of the Test-Tube Technique

Red Blood Cells. The growth-promoting action of lysed RBC was first demonstrated by Bradley et al. in 1971, who showed that the growth of colonies from mouse marrow in agar was enhanced by the addition of washed RBC from rats and mice but not sheep. Using RBC from the August rat strain, we have found a consistent enhancement of colony formation by human and animal tumour cells in primary culture (Courtenay et al. 1978; Tveit et al. 1981). The growth-promoting substances in RBC are evidently partly labile, since suspensions of whole RBC, which lyse after addition to the agar, are more effective than lysates. The nature of the growth factors involved has been studied by Bertoncello and Bradley (1977) and Kriegler et al. (1981), and it is apparent that two or more different growth factors may be released.

From our observations, the RBC factors aid in the initiation of colony formation by promoting cell division. To be effective, it is necessary that RBC lysis should be complete within the first 5–7 days of setting up. In an experiment to compare agarose as an alternative to agar, RBC failed to lyse, and there was no promotion of growth. The action of RBC does not appear to be related to the ability of haemoglobin to take up oxygen,

Fig. 4a–d. HX34 human melanoma cells grown in agar. Cells were plated out either without the addition of liquid medium at cell densities of **a** 1×10^4/ml or **b** 1×10^3/ml or with liquid medium added at cell densities of **c** 1×10^4/ml or **d** 1×10^3/ml

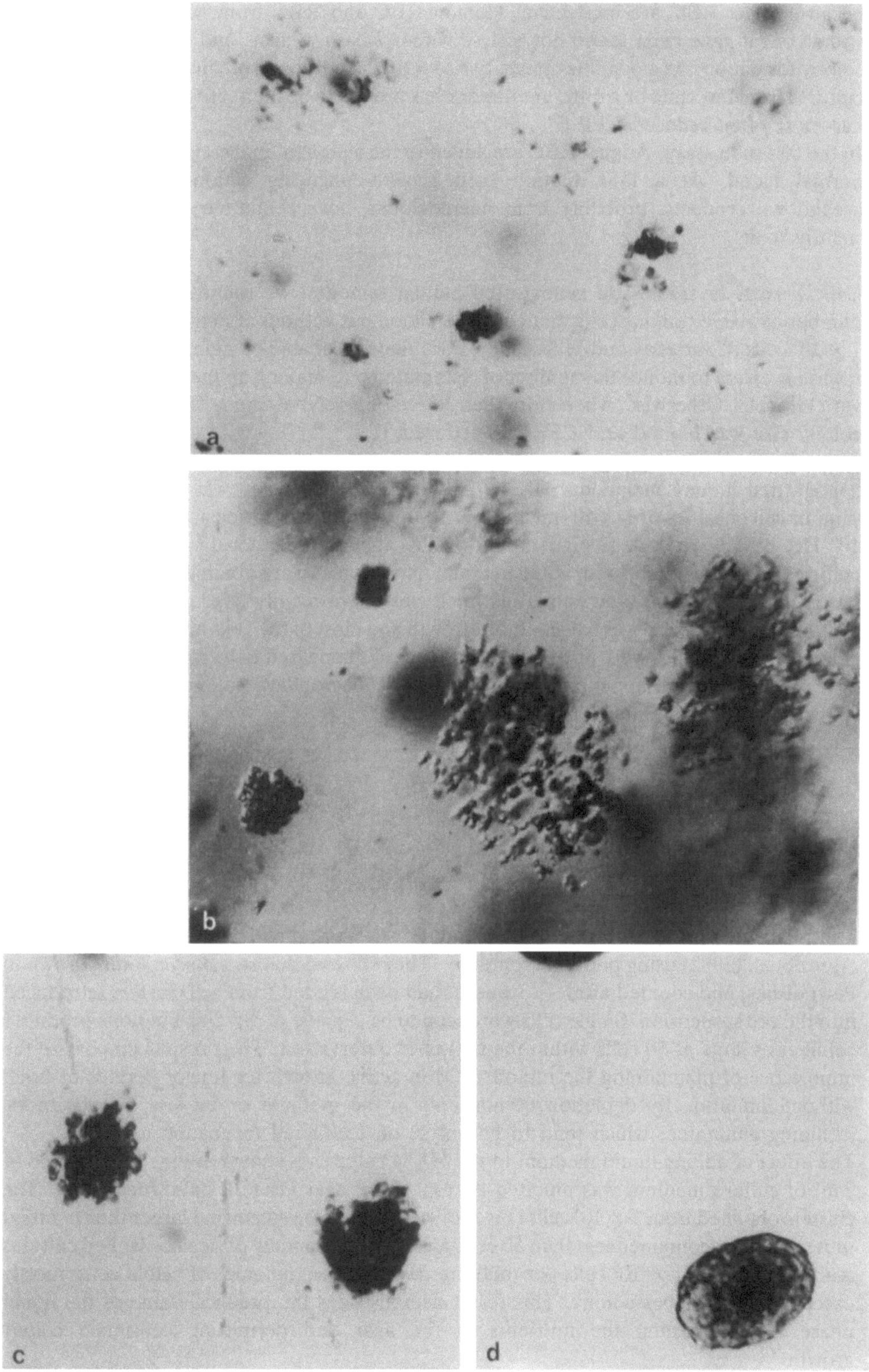
a
b
c
d

because intact RBC are ineffective. Human RBC and RBC from Marshall and Wistar strains of rat have been found not to lyse within 7 days in agar, and have not promoted colony formation. As a possible alternative to RBC, conditioned medium from tumour cell cultures from the same or a different tumour has proved to be ineffective and, in our hands, has most often reduced CFE.

In the test-tube assay, August RBC are added to the agar at a final concentration of 1/80 in normal blood. At a 1/40 dilution their growth-enhancing action is increased, but breakdown products, probably from haemoglobin, have sometimes been found to be slightly toxic.

Cell Density. In studies on xenografted human tumours, we found that the maximum number of viable tumour cells that could be plated out without affecting CFE was around 1×10^4 cells (Courtenay and Mills 1978). For tumours capable of giving a CFE above 2%, it was necessary to reduce the number of cells plated out, making up the total number to 10^4 with HR cells. Otherwise, when more than 300 colonies containing > 50 cells were formed, colony size was limited and CFE was reduced.

The effect of cell density on colony growth is illustrated in an experiment on the xenografted human melanoma HX34. A single-cell suspension was set up as usual at tumour cell densities of 1×10^3/ml and 1×10^4/ml. At the lower density of 1×10^3/ml, 1×10^4 HR cells were added. Figure 4 shows the appearance after 17 days incubation of cultures that were not fed with liquid medium. No colonies were obtained from 1×10^4 cells per millilitre (Fig. 4a), only numerous small clusters of mainly dead cells. From 1×10^3 cells per millilitre there was considerable growth and most of the cells appeared to be viable (Fig. 4b), but colonies were not compact and the cells appeared to be migrating out into the agar, giving a diffuse appearance more typical of macrophage colonies. Experience has shown that cells continue to migrate out from such diffuse colonies and colonies cannot then be distinguished from each other.

Liquid Phase

The relatively slow growth rate of human tumour cells in agar has been demonstrated by Agrez et al. (1982) using photomicrography. They showed that so-called colonies grown in Petri dishes, and counted after 2–3 weeks, had been seeded from cell clusters introduced into the cell suspension. Single cells were seen to be capable of dividing but none produced colonies as large as 50 cells within the period of observation. Their results emphasize the importance of maintaining the tumour cells in active growth for longer periods of time, without limitation by depletion of nutrients in the medium or by loss of substances, including glutamine, which tend to hydrolyse on continued incubation at 37° C.

The effect of adding liquid medium to the HX34 cultures is shown in Fig. 4. A volume of 2 ml of culture medium was pipetted on top of the agar after 10 days' incubation. The clusters obtained from 1×10^4 cells (Fig. 4c) were seen to be somewhat larger than in unfed tubes, but they contained less than 50 cells and consisted mainly of dead cells. Fed cultures seeded from only 1×10^3 cells per millilitre gave compact colonies of viable cells, mostly exceeding 50 cells per colony. This result demonstrates the practical value of the liquid phase in replenishing the nutrients in the agar and permitting continued colony growth.

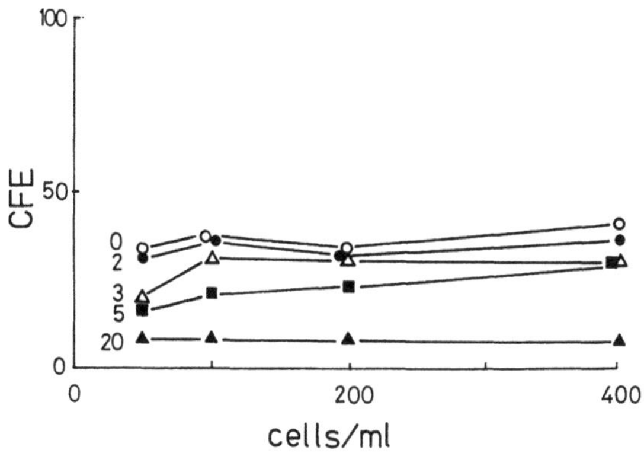

Fig. 5. Effect of O_2 concentration on colony-forning efficiency *(CFE)* of HX34 melanoma cells. Tubes were gassed with O_2 concentrations from 0 to 20% in mixtures containing 5% CO_2 + 95 to 75% N_2. The tubes in gassed boxes were maintained in an air incubator

Oxygen Concentration

The need for low oxygen tensions in the gas phase has consistently been found. In early studies a 5% oxygen concentration was used as the standard, but subsequent experience showed an advantage in using lower concentrations. Figure 5 shows the result of an experiment to examine the effect of different oxygen concentrations on colony formation by HX34 cells. Between 100 and 500 tumour cells were set up in test-tubes with 10^4 HR cells and gassed at weekly intervals with gas mixtures containing 0%, 2%, 3%, 5% or 20% O_2 together with 5% CO_2 and N_2. The gassed and tightly capped tubes were incubated in sealed plastic boxes gassed with the same mixtures. CFE was highest from boxes gassed with between 0 and 3% O_2. The fact that colonies were obtained with a nominal absence of O_2 illustrates the problem of maintaining low O_2 levels in plastic boxes and tubes, which may adsorb and release oxygen and from which there is always a small gas leak so that the actual O_2 level wou have been slightly higher. With the use of a three-gas incubator, O_2 concentration can be more accurately controlled than in boxes, as well as saving time in gassing individual boxes.

In any static culture system it can be shown (McLimans et al. 1968) that a gradient of O_2 concentration is established within the culture medium and the O_2 concentration adjacent to the cells is dependent not only on the O_2 concentration in the gas phase, but also on the rate at which O_2 is metabolized by the cells.

The rate of flow of O_2 in the agar medium is given by Fick's law, which states that the rate of flow of a gas is directly proportional to the gradient of gas concentration. For O_2 diffusing through agar medium

$$\text{flow rate } (Q) = -D\frac{d\Omega}{dx}$$

where D = diffusion coefficient of O_2
Ω = O_2 concentration
x = distance.

The O_2 concentration at the bottom of the agar is given by

$$Q = -D\frac{(\Omega_h - \Omega_s)}{h} \qquad (1)$$

where h = depth of agar
Ω_s = O_2 concentration at surface of agar
Ω_h = O_2 concentration at bottom of agar

In the steady state, O_2 flow rate equals the rate of consumption of O_2 by the cells:

$$Q = nc \tag{2}$$

where n = number of cells per tube
c = O_2 consumption per cell.

Substituting from Eq. (1),

$$nc = -D\frac{(\Omega_h - \Omega_s)}{h}$$

and

$$\Omega_h - \Omega_s = -\frac{nch}{D},$$

hence

$$\Omega_h = \Omega_s - \frac{nch}{D}. \tag{3}$$

Thus the O_2 concentration at the bottom of the agar depends not only on the cell density, but also on the depth of the agar.

A practical demonstration of this is seen in Fig. 6, which shows colony formation by HX34 melanoma cells in three open 6 × 90 mm tubes, made by sealing the ends of Pasteur pipettes. The cells were set up in agar with RBC as usual at a range of cell concentrations and gassed with 5% O_2. At cell densities of 4×10^5/ml and 4×10^4/ml, cells survived and divided only in the top 1.5 mm and 2.3 mm respectively. Cells below these levels soon died due to hypoxia produced as O_2 was used up by cell metabolism. The diffusion of O_2 downwards from the surface was prevented by O_2 uptake by the overlying cells. That conditions were hypoxic was shown by radiation experiments. Good colony formation was seen in tubes with only 4×10^3 cells per millilitre but it was restricted to a zone between 4.5 mm and 9.0 mm from the surface. The poor colony growth at the top of the agar was of interest, since it indicated that 5% O_2 is too high a concentration for colony growth, which occurred further down the O_2 gradient. As with the higher cell concentrations, the lower cut-off in colony formation is taken to correspond to the hypoxic region. This result indicates that any increase in the standard agar depth of 10 mm in the tubes might be expected to decrease the CFE of HX34 cells.

Another experiment that has demonstrated the importance of agar depth for optimum cell growth was originally devised to investigate the use of multidishes as an alternative to test-tubes. Into 17-mm and 24-mm well multidishes, 100 HX34 cells per well with RBC and with or without HR cells were plated out in volumes of 0.5 ml and 1.0 ml 0.3% agar over an underlayer of 0.5 ml 0.5% agar. At the same time, 100 HX34 cells were set up as usual in test-tubes. All were maintained with humidification in an incubator gassed with 3% oxygen. After 1 and 2 weeks, 1 ml of medium was added to wells and tubes. Colonies were counted at 18 days. Wide differences in CFE were obtained, and the CFE was always highest in the tubes and lowest in the 24-mm wells. The results are shown in Fig. 7, in which

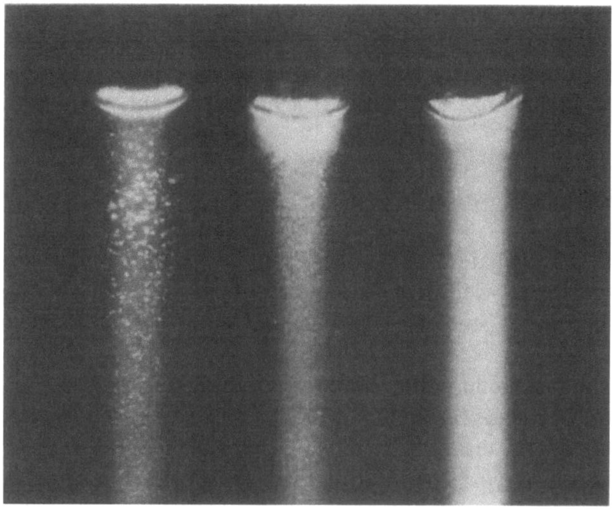

Fig. 6. The effect of agar depth on colony formation by HX34 melanoma cells plated out at densities of 4×10^3, 4×10^4, and 4×10^5/ml in open 6×90 mm tubes gassed with 5% O_2

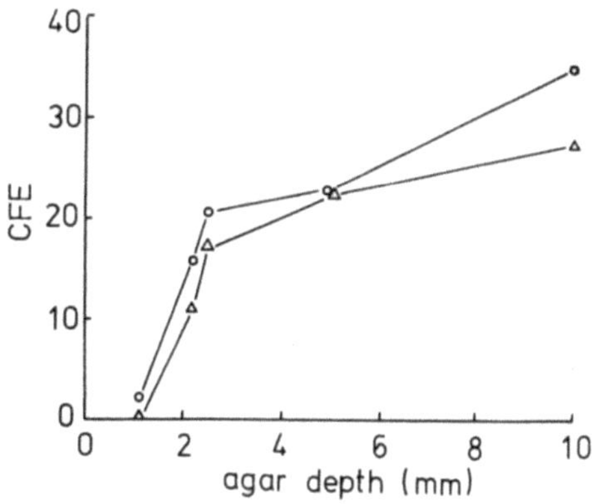

Fig. 7. The effect on colony-forming efficiency *(CFE)* of different agar depths obtained in multidishes or test-tubes. Depths of 1.1 and 2.2 mm were obtained in 24-mm diameter multidishes, 2.5 and 5.0 mm in 17-mm diameter multidishes, and 10 mm in test-tubes. Each container received 100 cells + red blood cells with (○) or without (△) 1×10^4 HR cells

CFE is plotted against agar depth in the different containers. A clear correlation was obtained between agar depth and CFE. Without HR cells, CFE ranged from 0.16% with a 1-mm depth of agar to 27.5% in test-tubes. The addition of HR cells consistently increased CFE, possibly due to their action in taking up O_2. These results show that the test-tubes gave the best conditions for colony formation using a gas phase with 3% O_2. The low CFE obtained with the 24-mm dishes in which the average O_2 concentration would be closer to 3% indicates that the optimum O_2 concentration for cell growth is considerably lower than 3% O_2.

With shallow agar, the use of gas mixtures with lower O_2 concentrations could be expected to give better results. However, as yet it is uncertain what is the optimum range of O_2 concentration for cell growth, and whether it is the same for all cells. In practice, it is not easy to maintain low O_2 concentrations at a constant level. Deep agar in test-tubes provides an O_2 gradient with a wide range of O_2 concentrations and permits colony formation from unknown tumours, with possibly different O_2 requirements, although the maximum possible CFE may not necessarily be attained.

Sensitivity Testing

Sensitivity studies on biopsy material using the test-tube assay have been carried out mainly on melanomas and also, to a limited extent, on ovarian ascites.

Radiation Sensitivity

In vitro radiation studies using γ-radiation were undertaken on a series of melanoma biopsy specimens. Cells from these tumours gassed with 3% O_2 gave rather higher CFEs than in the earlier studies (Table 1), and eight out of 12 gave a CFE of between 1.0% and 12.5%.

The melanomas differed in ease of disaggregation, and enzyme methods using collagenase with pronase (Howell and Koch 1980) were used to release cells from tumours that were not readily disaggregated by mechanical means. With two of the tumours, which were very fibrous, trypsin was also used in the final stage of preparation of the suspension. After filtration through a 30-µm mesh, 1-ml aliquots of the cell suspension, diluted to 2×10^5 cells per millilitre in plastic test-tubes, were exposed to γ-radiation from a ^{60}Co source. The cells were irradiated at room temperature under oxic conditions. A range of doses was given between 0.5 and 6.0 Gy covering the magnitude of dose fraction commonly given in radiotherapy schedules. Radiation was delivered in 1 min, the tubes being fixed on a board at calibrated distances from the source.

Survival measurements were obtained from five of the tumours and the results are shown in Fig. 8. Surviving fraction was calculated from the number of colonies per 100 cells set up in agar, divided by the CFE. The results from a sixth tumour were discarded because the number of colonies in untreated controls exceeded 300. In Fig. 8a, results from the individual tumours are shown on a semi-log plot. In Fig. 8b the same data are fitted to a single survival curve calculated from the equation giving survival $S = e^{-\alpha D - \beta D^2}$ (where D is radiation dose). The results from the biopsies fall within a narrow range and are comparable to those obtained from an unselected series of eight different xenografted

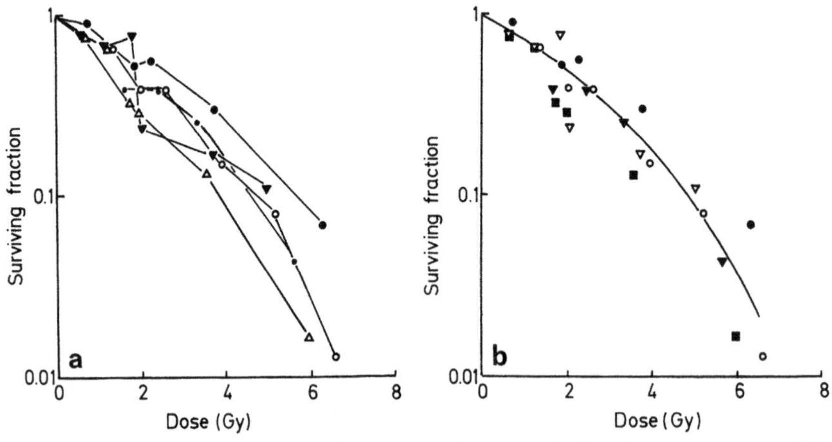

Fig. 8a, b. Cell survival measured from colony formation in five human melanoma biopsy specimens exposed to γ-radiation in vitro. **a** Separate survival curves are shown for the individual tumours. **b** A single survival curve of the form $S = e^{-\alpha D - \beta D^2}$ is fitted to the same data

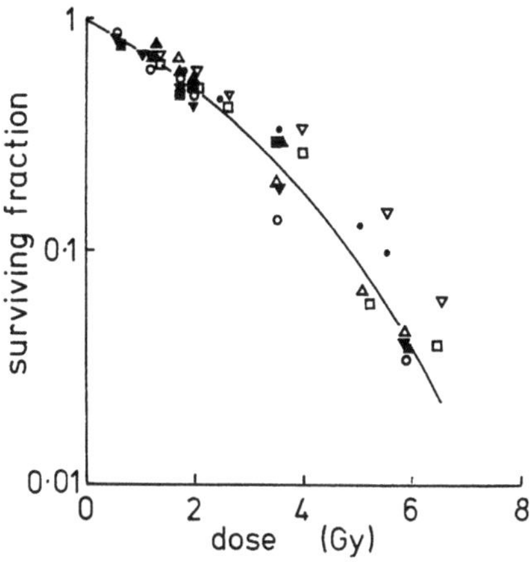

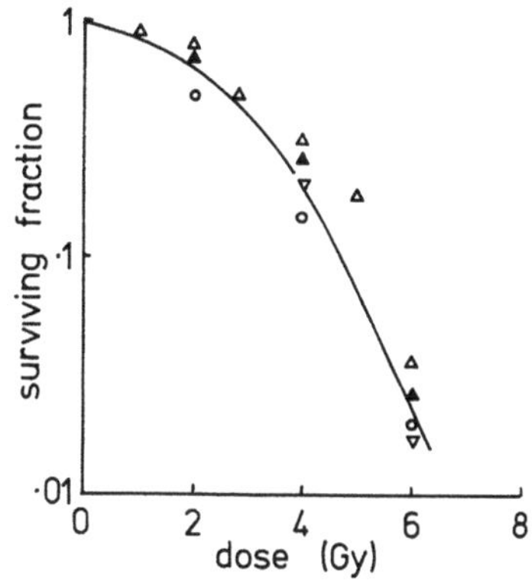

Fig. 9 (Left). Cell survival data from eight different melanoma xenografts exposed to γ-radiation in vitro. The survival curve shown was fitted to the biopsy data and is redrawn from Fig. 8b

Fig. 10 (Right). Cell survival from an ovarian ascites tumour exposed to γ-radiation in vitro. Results obtained from fresh material are shown as (○) and from material stored in liquid N_2 for different times by (△▲▽)

melanomas in early passage. Figure 9 shows surviving fractions obtained from the xenografts, fittet to the same curve as is shown in Fig. 8b. These results show the uniformity of radiation sensitivity of oxygenated melanoma cells. They also indicate the degree of accuracy that can be obtained from the assay.

One of the main problems in sensitivity testing is the inability to repeat measurements. Often this is due to the limited amount of tissue available. However, in some cases it may be possible to store the tissue in liquid N_2 and repeat experiments later on. Figure 10 shows the results of measurements on an ovarian ascites tumour in which radiation survival was measured from the fresh specimen and then repeated on three occasions from frozen material. Good agreement was obtained between cell survival from the stored and that from the fresh material. Similar results have been reported by Selby and Steel (1981) using ADC.

Drug Sensitivity

Studies undertaken on xenografted tumours have shown some of the problems involved in relating tumour sensitivity in vitro to response in vivo.

The response of two human pancreatic adenocarcinoma xenografts, HX32 and HX58, treated in situ in the mouse (Courtenay et al. 1982) was assayed in vitro. The tumours were implanted in the hind legs of immune-suppressed CBA_{lac} mice and treated at a size of 5–8 mm diameter by one of a spectrum of drugs, injected intraperitoneally, at doses up to the LD_{10} value (lethal dose for 10% of group). The tumours were excised and assayed 18 h after treatment, to allow time for the completion of drug action. For the five effective drugs hexamethylmelamine, cyclophosphamide, *cis*-dichloro-diammino platinum II (*cis*-Pt II), methyl cyclohexyl-chloroethyl-nitrosourea (MeCCNU) and melphalan, the survival curves (Fig. 11) were exponential, giving a straight line on a semi-log plot. There was no sign of a

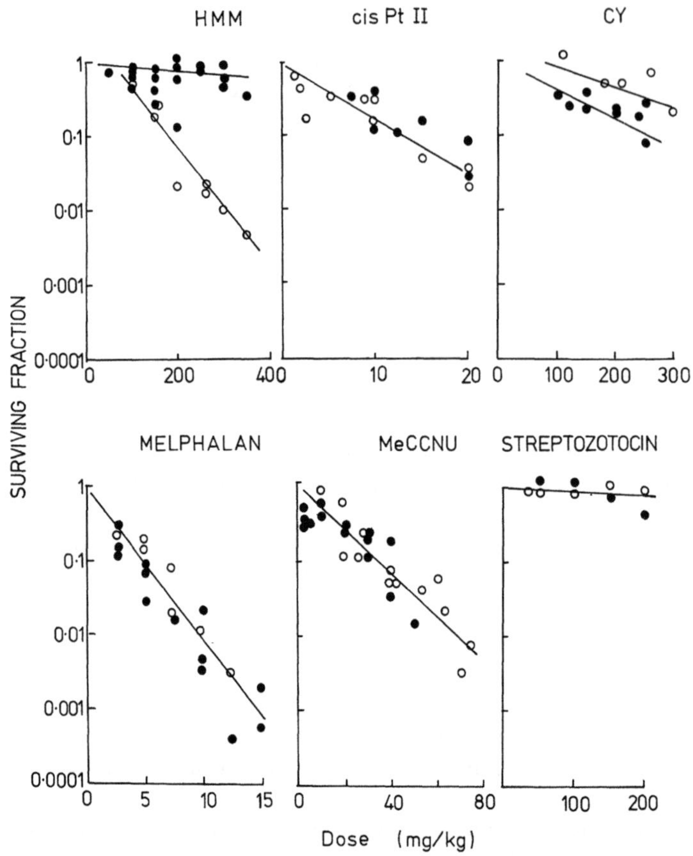

Fig. 11. Cell survival curves for two human pancreatic adenocarcinoma xenografts – HX32 (○) and HX58 (●) – treated in vivo with six different cytotoxic drugs. For each drug, the dose scale extends up to the approximate LD_{10} dose level

resistant tail. The shape of these in vivo curves differs from those commonly obtained from in vitro measurements.

The response of one of the tumours, HX32, to in vitro treatment with melphalan was measured, and it was found that the survival curves obtained differed, depending on the exact methodology used in the test. HX32 cells were treated by a commonly used method in which the drug was added to 10^6 tumour cells in 1 ml of medium containing 15% special bobby calf serum in a vertical test-tube, and incubated at 37° C for 1 h. During treatment the cells were seen to settle out on the bottom of the test-tube. After rinsing twice by centrifugation, the cells were set up in the agar assay system. The upper survival curve, shown in Fig. 12, was obtained when similarly treated cells were maintained in suspension, using a blood suspension mixer. Survival was lower, as shown by the lower curve. Keeping the tube horizontal during treatment gave the same result as the mixer. These results suggest that, in vertical tubes, drug access to cells towards the bottom of the cell pellet formed may be hindered by drug uptake by the overlying cells.

Other factors found to affect survival after treatment in vitro were the concentration of serum in the culture medium during melphalan treatment (Fig. 13) and the type of serum in

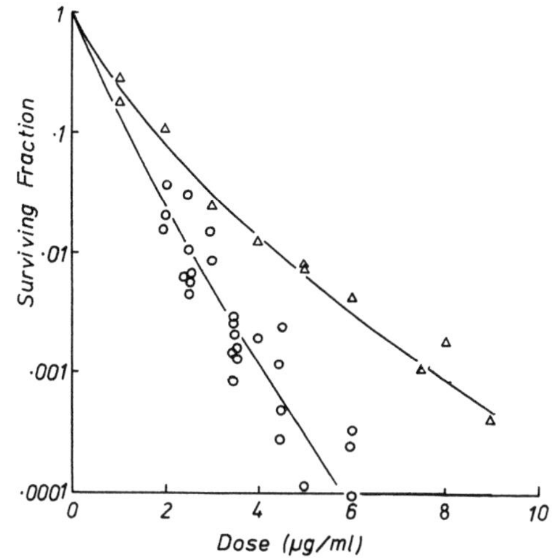

Fig. 12. Cell survival curves for HX32 cells incubated in vitro with melphalan for 1 h, cells being allowed to settle out in static vertical tubes (△) or kept in suspension in tubes on a blood suspension mixer (○)

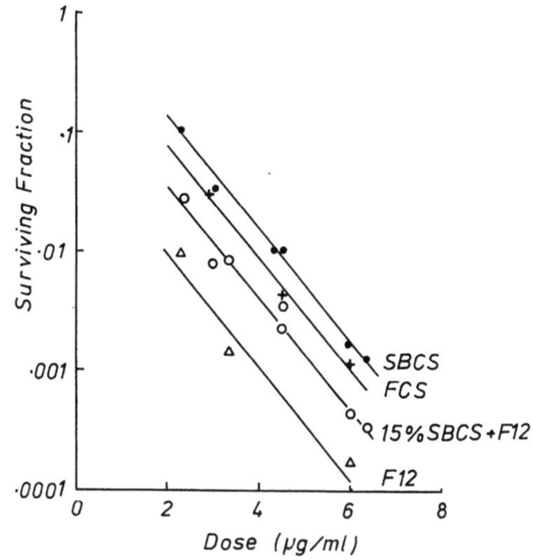

Fig. 13. Cell survival curves for HX32 cells treated in horizontal tubes for 1 h with melphalan in serum [special bobby calf serum (*SBCS*) (●) or foetal calf serum (*FCS*) (+)], in medium with 15% SBCS (○) or in medium without serum (△). All were subsequently cultured in SBCS. *F12,* Ham's F12m medium

which the cells were grown. Pooled results from a large number of measurements (Fig. 14) show different survival curves depending on whether the cells were cultured throughout the experiments using special bobby calf serum or faetal calf serum in the medium.

These results show the need to maintain standard conditions in in vitro sensitivity testing and the importance of detailed methodology. They also provide further evidence of the difficulties involved in relating in vitro response to drug concentration.

Drug sensitivity testing is most likely to be of immediate value when used to predict patient response to a single drug. In an ongoing study we have been supplied by Mr. Meirion Thomas at Westminster Hospital with biopsy material from melanoma patients undergoing isolated limb perfusion with melphalan. Sensitivity testing has been carried out by Miss Judith Mills.

Tumours were disaggregated and treated within 24 h of receipt. Treatment with melphalan at concentrations of 1 and 2 µg/ml was carried out on 5×10^5 cells in 1 ml of medium with

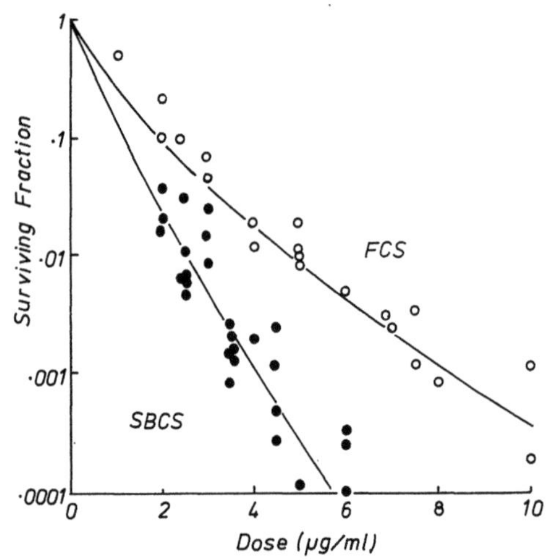

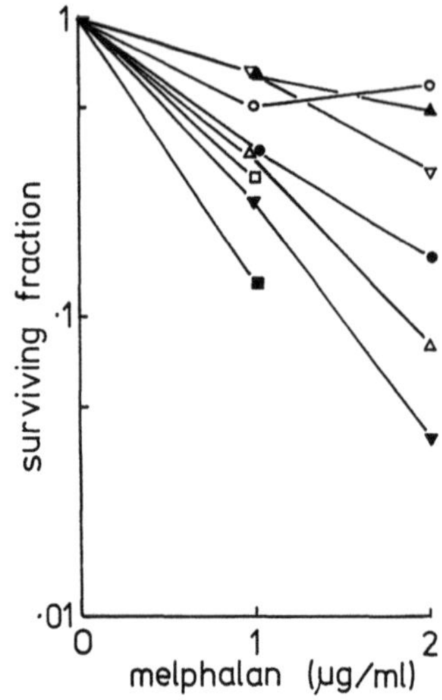

Fig. 14 (Left). Cell survival curves for HX32 cells treated with melphalan for 1 h in horizontal tubes. Cells were treated and subsequently cultured either in foetal calf serum *(FCS)* (○) or special bobby calf serum *(SBCS)* (●)

Fig. 15 (Right). Drug sensitivity measurements on melanoma biopsies. Cell survival after treatment in vitro with melphalan

15% foetal calf serum in 12 × 75 mm tubes (Falcon 2058) laid in a horizontal position and incubated at 37° C for 1 h.

Measurements were obtained from eight of the nine biopsy specimens received. The results shown in Fig. 15 indicate a wide range of sensitivity to the drug. As yet the final clinical assessment of the response of these patients has not been made.

Although much larger numbers of patients are required, a retrospective trial of this type, evaluating in vitro sensitivity to a single drug to which clinical response may be practically assessed, is most likely to give evidence of the predictive value of sensitivity testing.

Conclusions

The primary objective in these studies has been to obtain improved colony growth in soft agar from a range of human tumours, and some of the requirements for colony formation have been identified. It has been shown that growth factors, released from rat RBC in agar, aid the initiation of colony formation and that continued colony growth may be maintained by providing a replenishable liquid phase. The use of low oxygen concentrations in the gas phase has been found to be critically important for colony formation, and it has been shown that CFE can vary more than 100-fold depending only on oxygen concentration. By providing an oxygen gradient with a wide range of O_2 concentration, the test-tube method provides suitable growth conditions for tumours with potentially different tolerated O_2 levels.

Although further studies be needed to adapt the method to the specific growth requirements of some categories of tumour, the method may represent a step towards raising the CFE of a wide range of tumours.

The radiation survival measurements on melanoma biopsy specimens agreed well with those obtained from melanoma xenografts and suggest that the assay is capable of giving an accurate measure of tumour cell survival.
The drug sensitivity studies on a xenografted pancreatic tumour have yielded evidence of the complex problems involved in relating in vitro results to drug action in vivo. After treatment in vitro with melphalan, survival curves were clearly influenced by differences in the exact methodology used in conducting the test. This must introduce an element of uncertainty into the attempt to compare the effectiveness of different drugs.
The results from a small study ranking the sensitivity of melanoma biopsies to melphalan suggest that the test may be usefully used to rank sensitivity to a single drug.

References

Agrez MV, Kovacs JS, Lieber MM (1982) Cell aggregates in the soft agar "human tumour stem-cell assay". Br J Cancer 46: 880–887

Bertoncello I, Bradley TR (1977) The physicochemical properties of erythrocyte derived activity which enhances murine bone marrow colony growth in agar culture. Aust J Exp Biol Med Sci 55: 281–292

Bradley TR, Telfer PA, Fry P (1971) The effect of erythrocytes on mouse bone marrow colony development in vitro. Blood 38: 353–359

Courtenay VD (1976) A soft agar colony assay for Lewis lung tumour and B16 melanoma taken directly from the mouse. Br J Cancer 34: 39–45

Courtenay VD (1983) The Courtenay clonogenic assay. In: Dendy PP, Hill BT (eds) Human tumour drug sensitivity testing in vitro: techniques and clinical applications. Academic Press

Courtenay VD, Mills J (1978) An in vitro colony assay for human tumours grown in immune-suppressed mice and treated in vivo with cytotoxic agents. Br J Cancer 37: 261–268

Courtenay VD, Smith IE, Peckham MJ, Steel GG (1976) In vitro and in vivo radiosensitivity of human tumour cells obtained from a pancreatic carcinoma xenograft. Nature 263: 771–772

Courtenay VD, Selby PJ, Smith IE, Mills J, Peckham MJ (1978) Growth of human tumour cell colonies from biopsies using two soft-agar techniques. Br J Cancer 38: 77–81

Courtenay VD, Mills J, Steel GG (1982) The spectrum of chemosensitivity of two human pancreatic tumour xenografts. Br J Cancer 46: 436–439

Dodson MG, Slota J, Lange C, Major E (1981) Distinction of the phenotypes of in vitro anchorage-independent soft-agar growth and in vivo tumorigenicity in the nude mouse. Cancer Res 41: 1441–1446

Hamburger AW, Salmon SE (1977) Primary bioassay of human tumour stem cells. Science 197: 461–463

Hastings RJ, Franks KM (1981) Chromosome pattern, growth in agar and tumorigenicity in nude mice of four human bladder carcinoma cell lines. Int J Cancer 27: 15–21

Howell RL, Koch CJ (1980) The disaggregation, separation and identification of cells from irradiated and unirradiated EMT6 mouse tumours. Int J Radiat Oncol Biol Phys 6: 311–318

Kriegler AB, Bradley TR, Hodgson GS, McNeice IK (1981) Identification of the "factor" in erythrocyte lysates which enhances colony growth in agar cultures. Exp Haematol 9: 11–21

McLimans WF, Blumenson LE, Tunnah KV (1968) Kinetics of gas diffusion in mammalian cell culture systems, II. Theory. Biotechnol Bioeng 10: 741–763

Millar BC, Fielden EM, Jinks HS (1982) Further studies on the nature of the biphasic radiation survival response of Chinese hamster cells V-79-753B to molecular oxygen. In: Bicher HI, Bruley DF (eds) Hyperthermia. Plenum, New York

Nias AHW, Gilbert CW, Lajtha LG, Lange CS (1965) Clone size analysis in the study of cell growth following single or during continuous irradiation. Int J Radiat Biol 9: 275–290

Richter A, Sanford KK, Evans VJ (1972) Influence of oxygen and culture media on plating efficiency of some mammalian tissue cells. J Natl Cancer Inst 49: 1705–1712

Selby PJ, Steel GG (1981) Clonogenic cell survival in cryopreserved human tumour cells. Br J Cancer 43: 143–148

Shipley WU, Stanley JA, Courtenay VD, Field SB (1975) Repair of radiation damage in Lewis lung carcinoma cells following in situ treatment with fast neutrons and γ rays. Cancer Res 35: 932–938

Smith IE, Courtenay VD, Gordon M-Y (1976) A colony forming assay for human tumour xenografts using agar in diffusion chambers. Br J Cancer 34: 476–483

Tveit KM, Fodstad O, Pihl A (1981) Cultivation of human melanomas in soft agar. Factors influencing plating efficiency and chemosensitivity. Int J Cancer 28: 329–334

Analysis of Malignant Effusions by Cellular Composition, Proliferation Kinetics, and In Vitro Clonogenicity

V. Hofmann, M.E. Berens, and U. Früh

Universitätsspital Zürich, Department für Innere Medizin, Abteilung für Onkologie, Rämistrasse 100, 8091 Zürich, Switzerland

Introduction

The human tumor clonogenic assay (HTCA) has received wide international attention and has been introduced in numerous laboratories since its original description by Hamburger and Salmon (1977). The dominant adavantage of the HTCA is its specificity for selecting neoplastic clones to the exclusion of other cellular elements. The neoplastic origin of cells comprising colonies has been confirmed by light and electron microscopy (Salmon and Liu 1979; Harris et al. 1982), production of tumor markers (Von Hoff et al. 1980), cytogenetics (Trent and Salmon 1980), and colony transplants into nude mice. Furthermore, it is believed that tumor clones growing in semisolid agar represent the cell fraction responsible for in vivo tumor propagation.

Human epithelial tumors clone with varying and unpredictable degrees of success in the HTCA (Hofmann et al., to be published). Despite considerable investigation of fresh tumor specimens, little is known of the characteristics which favor successful colony formation in vitro. Tumor kinetics may affect colony formation, as observed in multiple myeloma (Durie and Salmon 1980). Macrophages can influence clonal growth of ovarian adenocarcinoma (Hamburger et al. 1978) and malignant melanoma (Meyskens 1980). While these findings have suggest ed a role of some inherent characteristics of tumor specimens in subsequent cloning, no systematic analysis of a large cohort of samples has been done. In this report, we have investigated the association of tumor cell number, lymphoid and macrophage infiltration, and tumor cell proliferation kinetics with subsequent clonogenic growth.

Material and Methods

Patients

Included in this study are ascites and pleural effusions from 84 patients with various epithelial neoplasias. Patients were either untreated or had received their last treatment course more than 4 weeks prior to paracentesis.

Recent Results in Cancer Research. Vol. 94

Cytology

Cell differentials were done on cytospin preparations stained with May-Grünwald-Giemsa or according to Papanicolaou.

Tumor Proliferation Kinetics

The labeling index was determined by the high-speed scintillation autoradiography technique as described by Durie and Salmon (1975). Briefly 2×10^6 cells were exposed to 10 μCi ^{3}H-thymidine (specific activity 77 Ci/mmol) for 1 h; unincorporated nucleotide was removed by two washes. Cells were cytocentrifuged, dipped in emulsion (Kodak, NTB 3), and developed. Labeling index was calculated as the percentage of tumor cells with ≥ 5 granules located over the nuclei.

Tumor Cloning Assay

Cells were isolated by centrifugation from effusions anticoagulated with 10 IU preservative-free heparin per milliliter (Novo). Thereafter, cells were washed twice, adjusted to appropriate concentrations, and aliquoted for differentials, labeling index, and tumor cloning.
The double-layer agar system was prepared exactly as described by Hamburger and Salmon (1978), with the exception that no conditioned medium was included.
Growth was monitored every other day using an inverted microscope, and the final scoring was usually performed between days 10 and 28. A colony was defined as any new, round cell aggregate of at least ≥ 40 cells and/or greater than 80 μm in diameter. Successful growth was defined as ≥ 5 colonies arising from 0.5×10^5 cells plated.

Results

Of 84 effusions, 72 (86%) were positive for tumor cells. None of the 12 cytologically negative specimens developed colonies in vitro. Colony growth was observed in 34 of 72 cytologically positive samples (47%). Colony growth ranged from five to 307 (median: 22 colonies). The effusions (Table 1) were composed of 0.1%–93% tumor cells with a median number of 7.9; the number of lymphocytes ranged from 3%–98% (median: 38.5%) and

Table 1. Characteristics of 72 malignant effusions

	Median (%)	Range (%)
Cellular composition		
Tumor cells	7.9	0.1–93.4
Lymphocytes	38.5	2.8–97.8
Monocytes/macrophages	24.0	1.0–86.8
Labeling index	3.3	0.0–20.8
Cloning success	47	5 –307 colonies

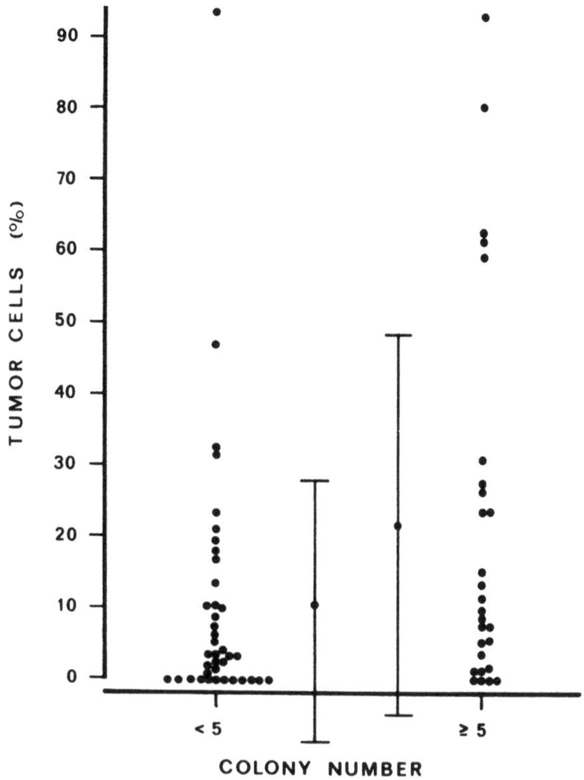

Fig. 1. Association of tumor cells with colony number

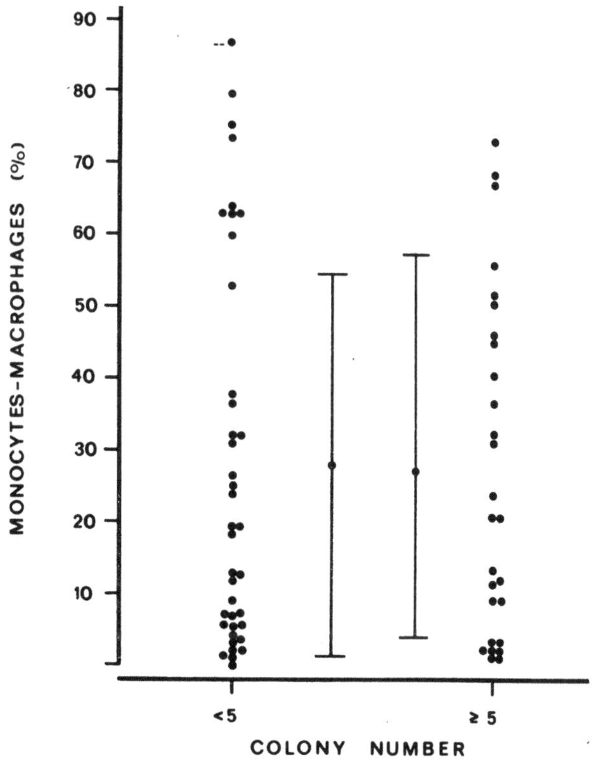

Fig. 2. Association of monocytes/macrophages with colony number

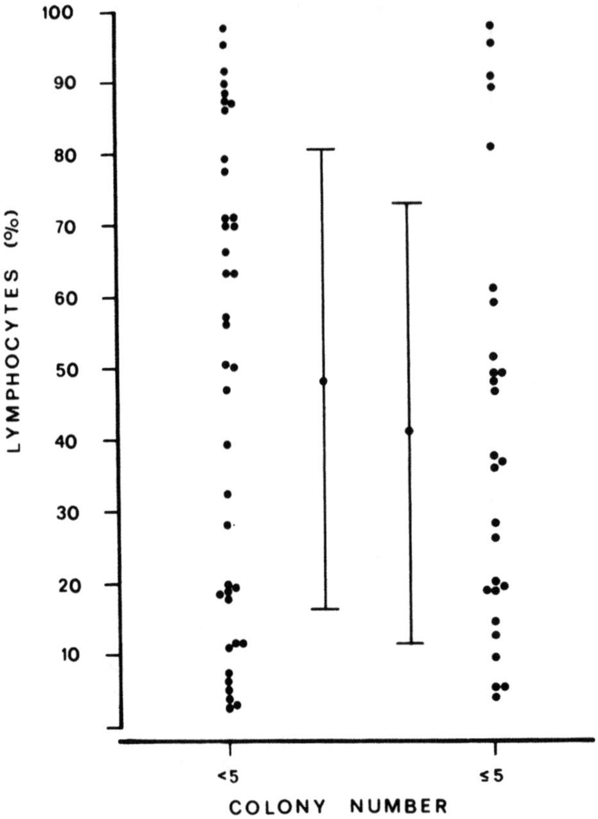

Fig. 3. Association of lymphocytes with colony number

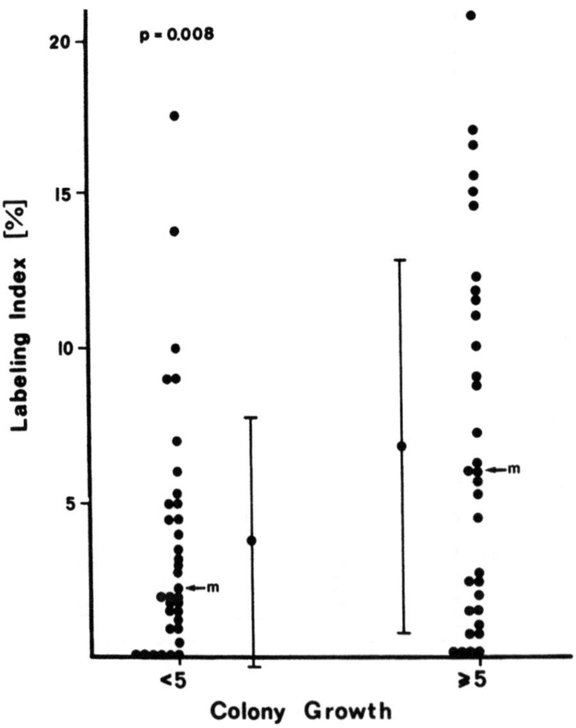

Fig. 4. Association of preplating labeling index with subsequent tumor colony formation

macrophages from 1.0%–87.0% (median: 24%). The labeling index ranged from 0% to 21% (median: 3.23%). Linear regressions of in vitro colony formation with number of tumor cells, lymphocytes, macrophages or labeling indices were not significant.

Samples were grouped into those that grew ⩾ 5 colonies and those that failed to grow (< 5 colonies). The number of tumor cells, lymphocytes, or monocytes/macrophages in these two groups did not show any significant difference (Figs. 1–3). We next examined whether samples containing tumor cells with a high proliferative capacity (high labeling index) had a subsequent growth advantage. Indeed, the labeling index of samples that grew was significantly higher than in those that failed to grow (Fig. 4).

Discussion

The in vitro characteristics that allow tumor colony formation are still poorly understood. Whereas parameters can be defined and repetitively studied for established tumor cell lines, each fresh human tumor specimen is unique. This is the most common reason for the failure of routine cloning.

In this study, we have looked at four readily measurable parameters in 84 fresh human tumors. We found that neither the number of tumor cells nor lymphocytes nor macrophages indicated which samples would produce tumor clones in a defined culture system.

Previous studies had suggested that macrophages could influence tumor colony formation. Macrophages have been shown to stimulate ovarian tumor-colony-forming units (Hamburger et al. 1978) and normal myeloid progenitor cells (CFU_c) (Kurland et al. 1979). However, autologous macrophages inhibited myeloma colony growth (Hamburger and Salmon 1980). These examples emphasize that host cells can affect in vitro tumor colony growth in divergent ways.

The influence of T-lymphocytes on the myeloma cell line RPMI 8226 has recently been studied, showing that OKT-8 positive cells from patients with myeloma and normal donors can suppress clonal growth (Ludwig et al. 1983). While these data demonstrate host cell modulation of clonal growth for particular tumor types, the importance of these effects on a wide range of diverse tumors is unknown.

Our results show that cellular composition by itself does not reveal a set of tumors likely to grow. In contrast to the above mentioned experiments describing the effects of macrophages, this study did not employ any depletion/reconstitution techniques. In addition, the influence of host cells may be specific within certain cytological subtypes, but this has not been considered in this analysis. In this study, the only feature of malignant effusions which correlated with subsequent clonal growth was the preplating labeling index of the tumor cells. Among the growers 50% (17 of 34) of the samples had a labeling index ⩾ 6%, while this value was found in only 20% (seven of 35) of the nongrowers. Since the determination of the labeling index is a simple and rapid procedure, it could be useful for selecting samples with higher likelihood of growing. For the samples with low proliferative capacity better growth conditions need to be tested.

References

Durie BGM, Salmon SE (1975) High speed scintillation autoradiography. Science 190: 1093–1095

Durie BGM, Salmon SE (1980) Cell kinetic analysis of human tumor stem cells. In: Salmon SE (ed) Cloning of human tumor stem cells. Liss, New York, pp 153–163

Hamburger AW, Salmon SE (1977) Primary bioassay of human myeloma stem cells. J Clin Invest 60: 846–854

Hamburger AW, Salmon SE, Kim MB, Trent JM, Soehnlen BJ, Alberts DS, Schmidt HJ (1978) Direct cloning of human ovarian carcinoma cells in agar. Cancer Res 38: 3438–3444

Harris GJ, Zeagler J, Hodach A, Casper J, Harb J, Von Hoff DD (1982) Ultrastructural analysis of colonies growing in a human tumor cloning system. Cancer 50: 722–726

Hofmann V, Berens ME, Martz G (to be published) Drug selection for perioperative chemotherapy. In: Metzger U, Senn HJ, Largiader F (eds) Recent results in cancer research.

Kurland JI, Broxmeyer HE, Bockman RS, Moore MAS (1979) Role of monocyte-macrop age-derived colony stimulating factor and prostaglandin E in the positive and negative feedback control of myeloid stem cell proliferation. Blood 52: 388–407

Ludwig C, Hicks MJ, Pena B, Durie BGM (1983) OKT-8$^+$-Suppressor-T-Lymphozyten sind bei Patienten mit stabilem multiplem Myelom vermehrt. Schweiz Med Wochenschr 113: 1451–1454

Meyskens FL (1980) Human melanoma colony formation in soft agar. In: Salmon SE (ed) Cloning of human tumor stem cells. Liss, New York, pp 85–99

Salmon SE, Liu R (1979) Direct "wet" staining of tumor or hematopoetic colonies in agar culture. Br J Cancer 39: 779–781

Trent JM, Salmon SE (1980) Human tumor karyology: marked analytic improvement by short-term agar culture. Br J Cancer 41: 867–874

Von Hoff DD, Harris GJ, Johnson G, Glaubiger D (1980) Initial experience with the human tumor stem cell assay system: potential and problems. In: Salmon SE (ed) Cloning of human tumor stem cells. Liss, New York, pp 113–124

Direct Cloning of Human Ovarian Cancer in Soft Agar: Clinical Limitations and Pharmacologic Applications

R.F. Ozols, W.M. Hogan, and R.C. Young

Department of Health and Human Services, Public Health Service, National Institutes of Health, Building 10, Room 12N226, 9000 Rockville Pike, Bethesda, MD 20205, USA

Introduction

Epithelial ovarian cancer is a disease which is well suited to the evaluation of in vitro tests for the individual selection of a patient's chemotherapy, since a variety of antineoplastic drugs have significant activity in this disease (Young 1975). A predictive in vitro test could therefore potentially identify those drugs active for a particular patient and thereby spare her the unnecessary toxic effects of inactive drugs. In addition, many patients with advanced ovarian cancer have malignant effusions (ascites or pleural fluid) which provide an easily accessible source of cells for in vitro analysis. Additional malignant cells can also be obtained frequently by peritoneal lavage from patients who do not have clinically evident ascites (Ozols et al. 1981). The observation (Hamburger et al. 1978) that human ovarian cancer cells can form tumor colonies in soft agar along with the promising preliminary results of in vitro/in vivo correlations of drug resistance using this assay (Salmon et al. 1978; Alberts et al. 1980) led to our study on the growth of ovarian cancer tumor colonies from peritoneal washings as well as malignant effusions (Ozols et al. 1980a). In addition to evaluating the assays' potential for individualization of chemotherapy, we have also used the human tumor stem cell assay (HTSCA) to investigate patterns of drug resistance and cross-resistance in ovarian cancer. In this report we will summarize our results with 215 ovarian cancer specimens, which demonstrate that, while the HTSCA has major limitations for the individualization of chemotherapy, the patterns of drug resistance and cross-resistance have provided clinically relevant information for the design of novel therapeutic approaches in ovarian cancer, such as intraperitoneal chemotherapy and high-dose cisplatin.

Materials and Methods

Patient Specimens

Effusion specimens and peritoneal washings were obtained from patients undergoing treatment for pathologically confirmed advanced epithelial carcinoma of the ovary. All specimens were collected aseptically, heparinized (10 units of preservative-free heparin per milliliter of fluid), and transported to the laboratory at ambient temperature. One-half of the effusion specimens (78 samples) were obtained from patients who were being cared for in the Medicine Branch of the National Cancer Institute (MB-NCI) and half were collected at other institutions. Abdominal washings were obtained from patients either at staging or restaging laparotomy or laparoscopy or through a surgically implanted Tenckhoff dialysis catheter prior to intraperitoneal chemotherapy. Sterile 0.85% NaCl solution was used to lavage the abdomen at laparotomy or laparoscopy, while a 1.5% Inpersol solution (Abbott Laboratories, Chicago) was used in patients with a Tenckhoff catheter. All peritoneal washings were obtained at the MB-NCI and were processed within 2 h of collection. For malignant effusions, the interval from collection to plating varied with the geographic origin of the specimen; however, the majority (74%) were plated 18–30 h after collection.

Specimen Processing

The cellular fraction was first harvested by centrifugation at 150 g for 10 min. Contaminating RBCs were lysed and the remaining cells were washed twice by centrifugation in McCoy's 5A medium + 10% heat-inactivated fetal calf serum (HI-FCS). The total number of nucleated cells was determined by direct count in a hemocytometer, with no attempt being made to determine a differential count, and at the same time the specimen was evaluated for the presence of tumor cell clumping. Cell aggregates containing two to five tumor cells were invariably present, and efforts to eliminate them completely generally proved unsuccessful. Clumps large enough subsequently to be confused with colonies (i.e., aggregates of more than 30 cells) were also frequently seen. Attempts were made to eliminate these clumps by passage of the cell suspension through progressively smaller needles (21–25 gauge), gravity sedimentation, and/or filtration through 44-μm monofilament nylon mesh (Small Parts Inc., Miami). The concentration of the optimized cell suspension was adjusted to 6×10^6 nucleated cells per milliliter and sample agar plates were then prepared as a final check to assure that large, colony-sized aggregates had been eliminated. Cytocentrifuge preparations of the final suspension were made and stained with hematoxylin-eosin or be the Papanicolau method (American Histolabs, Silver Spring).

Soft Agar Cloning System

Cells were cultured in 35-mm Petri dishes using the double-layer agar system as previously described (Hamburger et al. 1978; Ozols et al. 1980a). The concentration of cells plated from malignant effusions varied from 0.5 to 1.0×10^6 nucleated cells per plate, depending in part upon the percentage of viable cells (as determined by trypan blue dye exclusion) in the specimen. The number of cells plated from peritoneal washings was dependent upon the cellularity of the specimens and ranged from 1.0×10^5 nucleated cells per plate in some laparoscopy specimens to 1.0×10^6 cells per plate in the majority of Tenckhoff washings. Cultures were incubated at 37° C in a humidified atmosphere of 6% CO_2 in air.

In Vitro Drug Sensitivity Tests

In vitro drug sensitivity tests were performed as previously described (Ozols et al. 1980b). When the cell yield was adequate, drug testing was performed over a 3-log dose range which included 0.1, 1.0, and 10× the clinically achievable peak plasma concentration. In many cases, however, cellular material was limited and, in an effort to increase the number of drugs tested, only the lowest drug concentration was employed. Cytotoxicity was evaluated using either a 1-h or continuous drug exposure.

1. One-hour exposure: Either 1.5 or 3.0 × 10^6 nucleated cells were suspended in 1.5 ml McCoy's 5A medium + 10% HI-FCS with and without the appropriate concentration of drug. Following a 1-h incubation at 37° C, the cells were washed twice in drug-free medium, resuspended in medium containing molten agar, plated in triplicate, allowed to solidify at room temperature, and incubated as described above.
2. Continuous drug exposure: Cells were suspended in 2.4 ml of enriched medium to which 0.3 ml of drug and 0.3 ml of molten agar were added. After thorough mixing, the suspension was plated in triplicate, allowed to solidify, and incubated.

Colony Scoring and Assessment of Drug Effect

Day 0 control plates were examined under an inverted microscope at × 40 and × 100 magnification and the size and number of cell aggregates recorded. In our experience, while large clumps could generally be eliminated during the preparation of the cell suspension (see above), an absolute single-cell suspension could generally not be obtained, and as a result the clonal origin of a colony from a single malignant ovarian cell could not be assured. Plates were evaluated for growth at days 7 and 10 and final colony counts were made at day 14 or 21. Only cell aggregates containing more than 30 cells were scored as colonies. A mean control colony count of 30 or greater was required for an in vitro drug trial to be considered evaluable.

In order to assess drug-induced cytotoxicity, the ratio between the number of colonies formed after a 1-h or continuous drug exposure (mean of triplicate plates) and the mean number of colonies formed in the control plates was determined. In vitro sensitivity was defined as a 70% or greater reduction in colony formation at a drug concentration which approximated one-tenth the clinically achievable peak serum concentration following a standard intravenous dose.

Results

Characterization of Colonies Grown in Soft Agar

We have previously described the morphology of the human ovarian cancer colonies grown in our laboratory (Ozols et al. 1980). The cytologic characteristics of individual colonies plucked from control plates or fixed in situ (Salmon and Buick 1979) and subsequently stained with hematoxylin-eosin were similar to those of the ovarian cancer cells seen in the cytocentrifuge preparations of the original cell suspensions. Colonies grown from peritoneal washings did not differ morphologically from those grown from malignant effusions.

Growth of Ovarian Cancer Colonies from Malignant Effusions and Peritoneal Washings

From October 1978 through May 1981, we studied a total of 215 cytologically positive specimens from 133 patients with epithelial carcinoma of the ovary. Our overall results with

Table 1. Growth of epithelial ovarian cancer in soft agar

	No. cytologically positive	No. with growth	Success rate
Effusions			
Ascitic	136	88	65%
Pleural	20	13	65%
Washings			
Peritoneoscopy	22	16	73%
Tenckhoff	33	26	79%
Solid	4	1	25%
Total	215	144	67%

Table 2. Growth of epithelial ovarian cells from malignant effusions

	Ascites	Pleural	Total
No. of specimens	140	20	160
No. cytologically positive	136	20	156
No. (%) of specimens with growth	88 (65%)	13 (65%)	101 (65%)
Colony count			
Mean	81	106	
Median	36	50	
Range	3–712	3–323	
No. (%) with colony count ≥ 30	54 (40%)	10 (50%)	64 (41%)
No. (%) with colony count ≥ 100	19 (14%)	5 (25%)	24 (15%)

respect to colony formation in the double-layer soft agar system are summarized in Table 1. A mean colony count of three or greater was observed in 144 (67%) of the control cultures. Growth was documented in control plates seeded with cells derived from malignant effusions, peritoneal washings, and solid tumors.

In Table 2 and Fig. 1 our experience with colony growth from malignant effusions is presented in greater detail. Sixty-five percent of effusion specimens demonstrated in vitro colony formation; however, in only 41% of the specimens was growth adequate to enable the assessment of in vitro drug effects (i.e., ≥ 30 colonies). Mean control colony counts of 100 or greater were unusual, occurring in only 15% of specimens. There was no apparent difference in the cloning ability of cells derived from pleural or ascitic fluids and the median cloning efficiency (number of colonies × 100%/total number of cells plated) was 0.007%.

Eighty-four abdominal washings were obtained from 34 patients at second-look laparoscopy, at laparotomy, or through a Tenckhoff catheter. Colony formation was observed in 76% of the 55 specimens containing cytologically malignant cells (Table 3). In all 29 cytologically negative washings no colony growth was observed. While colony formation from abdominal washings occurred at a frequency similar to that noted with effusions, the number of colonies formed was generally less, and in only 25% of peritoneal washings was there sufficient growth to perform drug sensitivity studies.

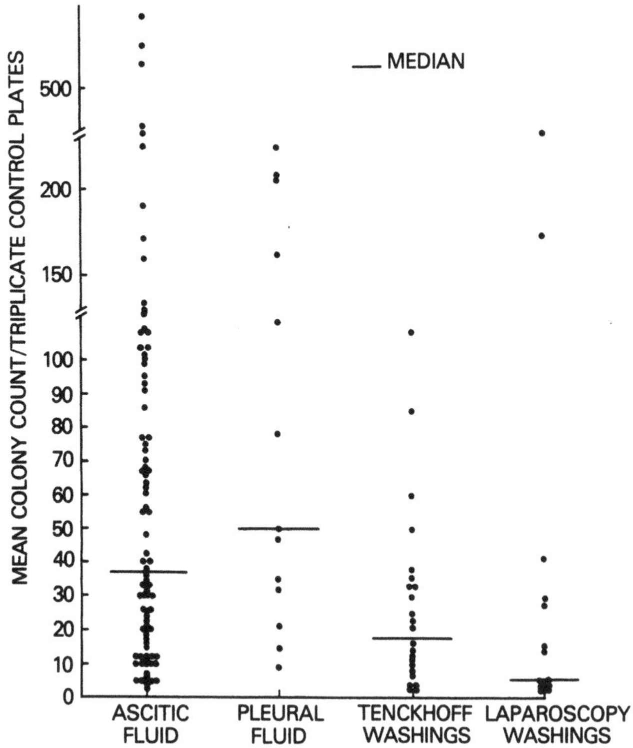

Fig. 1. Ovarian cancer specimens were obtained from a variety of sources and plated in the double-layer agar system. The mean colony count in triplicate plates (500,000 nucleated cells per plate) is plotted against the source of the malignant cells

Table 3. Growth of epithelial ovarian cells from peritoneal washings

	Peritoneoscopy	Tenckhoff	Total
No. of specimens	43	41	84
No. cytologically positive	22	33	55
No. (%) of specimens with growth	16 (73%)	26 (79%)	42 (76%)
Colony count			
Mean	45	28	
Median	6	18	
Range	3–369	3–117	
No. (%) with colony count ≥ 30	4 (18%)	10 (30%)	14 (25%)
No. (%) with colony count ≥ 100	2 (9%)	1 (3%)	3 (5%)

Patterns of Drug Resistance and Sensitivity

The broad drug resistance observed clinically in ovarian cancer patients who have relapsed after initial chemotherapy compared to previously untreated patients is reflected in the pattern of in vitro resistance seen in this study (Table 4). In previously untreated patients, in vitro sensitivity was observed in 41% of the assays, whereas in patients who had relapsed after primary chemotherapy, in vitro sensitivity in only 12%.

Table 4. Results of in vitro sensitivity assays in relation to previous drug treatment

Drug assayed	No. sensitive/total	
	Previously treated	Untreated
Doxorubicin	0/25 (12)	6/12
Melphalan	1/7 (7)	2/7
5-Fluorouracil	0/8 (6)	1/2
Cisplatin	2/15 (11)	2/6
Vinblastine	3/9 (0)	
Bleomycin	1/5 (0)	
Mitomycin C	1/3 (0)	
Methotrexate	1/4 (0)	
Total	9/76 (12%)	11/27 (41%)

All assays were performed using a 1-h drug exposure at one-tenth the clinically achievable peak plasma concentration. In vitro sensitivity is defined as a greater than 70% reduction in colony formation

Figures in parentheses indicate number of in vitro assays in which the drug tested had been proven clinically ineffective

In Vitro Dose-Response Relationships

In order to explore the potential therapeutic value of high-dose chemotherapeutic modalities such as intraperitoneal drug administration, high-dose cisplatin, and high-dose alkylating agent therapy with autologous bone marrow rescue, we examined the in vitro dose-response relationships to administration of doxorubicin, cisplatin, and melphalan over a 3-log dose range, including a dose ten times greater than that clinically achievable by standard systemic administration.

For doxorubicin, three distinct patterns of in vitro sensitivity were observed. The greatest degree of sensitivity was observed in cells obtained from previously untreated patients, 50% of whom demonstrated a greater than 70% reduction in colony formation following a 1-h exposure to a 0.1 μg/ml drug concentration (approximately one-tenth achievable peak plasma concentration). Among 12 patients who had relapsed after initial chemotherapy with a non-doxorubicin-containing combination, there were no instances of in vitro sensitivity at this dose level but 83% of this group did respond at a drug concentration of 10 μg/ml, a level which is not achievable by intravenous administration but can be achieved via the intraperitoneal route. In contrast, patients who had relapsed after initial therapy with a doxorubicin-containing combination demonstrated marked in vitro resistance even at the highest employed drug concentration; only one patient in the latter group showed in vitro sensitivity at the 10 μg/ml drug level. Four patients in whom secondary doxorubin-containing regimens failed showed an in vitro response pattern similar to that in patients in whom non-doxorubicin-containing primary chemotherapy had failed, perhaps because these patients received no more than three courses of doxorubicin, often at a reduced dose. Survival curves (Fig. 2) constructed by plotting the combined mean percentage colony survival for all specimens in each treatment group against doxorubicin concentration demonstrate a pattern of relative cross-resistance between doxorubin and non-doxorubicin drug regimens.

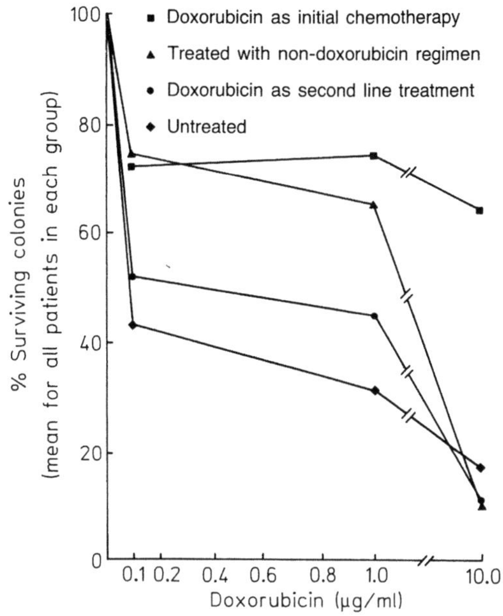

Fig. 2. Ovarian cancer cells were obtained from previously untreated ovarian cancer patients and from patients who had relapsed after therapy with a non-doxorubicin-containing combination or doxorubicin as a second-line treatment, or who had relapsed after primary therapy with a doxorubicin-containing combination. The in vitro effects of doxorubicin were examined and the results expressed as the mean percentage survival of colonies from patients in the four defined groups versus the concentration of doxorubicin

Table 5. In vitro dose response in relapsed ovarian cancer patients

Drug	Percentage (number) with in vitro sensitivity[a]			
	1 h exposure		Continuous exposure	
	Peak plasma level	Ten × peak plasma level	Peak plasma level	Ten × peak plasma level
Melphalan[b]	14% (1/7)	50% (3/6)	6% (1/16)	50% (3/6)
Cisplatin[c]	0 (0/6)	50% (3/6)	–	–

[a] Defined as < 30% colony survival
[b] Patients had received prior therapy with cyclophosphamide
[c] Patients had received prior therapy with cisplatin

Table 5 summarizes the dose-response relationships for melphalan and cisplatin in cells obtained from patients who were clinically refractory to cyclophosphamide and cisplatin. For both cisplatin and melphalan significant reduction in colony formation was observed in 50% of specimens following exposure to drug at ten times the peak achievable plasma level.

Discussion

The results of this study suggest that the stem cell assay is of limited value in the individualization of chemotherapy for patients with refractory ovarian cancer. A major reason which limits the routine application of the stem cell assay to individualization of

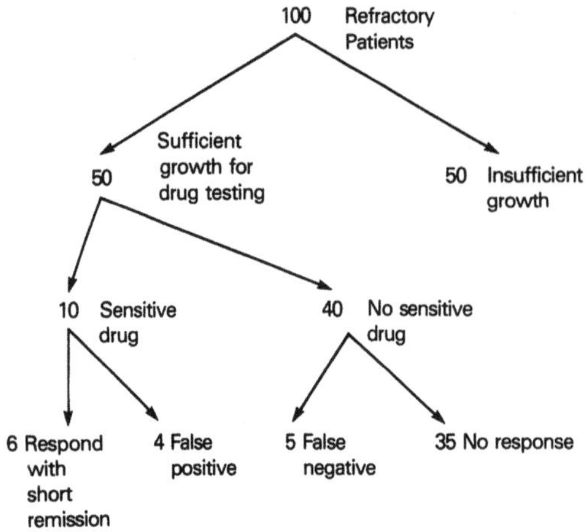

Fig. 3. Flow diagram of expected results of human tumor stem cell assay directed therapy in 100 relapsed ovarian cancer patients

chemotherapy is the frequent inability to obtain sufficient colonies for drug testing. In addition, it is clear that the theoretical goal of achieving a single-cell suspension from solid human tumors is not a practical reality and that the assay is usually not a "clonogenic" assay. Not only is the clinical application limited by frequent inability to obtain sufficient colonies, but in those instances where adequate colony formation is seen, it is usually not possible to identify an active drug in specimens from relapsed patients (Table 4).

The correlative accuracy for resistance in our series was 94% and for sensitivity 64% (Hogan et al. 1982). Recent larger studies have suggested that the predictive accuracy for sensitivity is approximately 50% and for resistance 85%–90% (von Hoff et al. 1983). In those patients where the stem cell assay is predicitve for response, the majority of the responses have been transient partial remissions (Alberts et al. 1980). The clinical limitations of the HTSCA are depicted in Fig. 3, which summarizes the expected outcone in 100 patients with refractory ovarian cancer if their therapy was directed solely on the basis of in vitro tests. The numbers are approximations derived from our own results and from those of Von Hoff et al. (1981, 1983). Approximately 50 patients would have sufficient in vitro colony formation to permit drug testing and in 10 of these an active drug is likely to be identified in vitro, while in the remaining 40 patients only drug resistance will be demonstrated. Of those ten patients in whom an active drug is identified, five will be likely to respond with a partial remission. In those 40 patients with in vitro resistance, the false negative rate may be 10%–15% [in Von Hoff's study of 203 resistant tumors of various histologic types there were 31 instances (15%) where a patient responded to a drug which was inactive in vitro]. Thus approximately as many patients may be denied an active drug as could potentially be benefited if all therapy was stem cell directed. It is clear that ovarian cancer patients should not be denied standard therapy on the basis of current in vitro tests. In addition, there is no justification for carrying out invasive surgical procedures (e.g., laparotomy) to obtain tissue solely for stem cell study. In contrast to the limitations of the assay for the individualization in refractory ovarian cancer patients, the nature of the in vitro dose response to cisplatin, doxorubicin, and melphalan using the HTSCA provides insight into the patterns of resistance and cross-resistance in ovarian cancer which are important in the design of new therapeutic modalities.

The in vitro dose response observed with doxrubicin has provide support for a phase I trial of intraperitoneally administered doxorubicin in patients who were refractory to a non-doxorubicin regimen. It is clear that cross-resistance to doxorubicin is present in cells obtained from patients treated with non-doxorubicin regimens (Fig. 2). The in vitro results have been confirmed by clinical studies demonstrating that intravenously administered doxorubicin is not effective second-line therapy. In contrast, intraperitoneal administration of doxorubicin, which allows for cytotoxic levels to be achieved in the peritoneal cavity, produced three responses in the phase I trial in ten patients (Ozols et al. 1982). Phase II studies are in progress at the MB-NCI to determine the rate of response to intraperitoneally administered doxorubicin.

The observation that cells from some patients who were clinically refractory to either alkylating agents or cisplatin demonstrated marked in vitro effects (less than 30% survival) after exposure to ten times peak achievable plasma levels of these same drugs suggests that high-dose chemotherapy with these agents may be beneficial in selected patients. It would be particularly important prospectively to identify those patients in whom ablative chemotherapy followed by autologous bone marrow infusion would have a high probability of producing a response, in order to avoid unecessary expense and severe toxicity. The in vitro results with cisplatin, together with previous clinical studies which demonstrated a dose response to cisplatin (see review by Hogan and Young 1982) and experimental studies in a human ovarian cancer cell line in which a steep dose response to cisplatin was observed (Ozols et al. 1984) have formed, in part, the rationale for an ongoing high-dose cisplatin study in refractory ovarian cancer patients (Ozols and Young 1983). In this study cisplatin (5×40 mg/m^2 daily) is administered in hypertonic saline with extensive hydration on the basis of experimental studies demonstrating that cisplatin nephrotoxicity can be reduced by the maintenance of a chloruresis (Litterst 1981).

Finally, the size of the colonies grown in the stem cell assay precludes most biochemical investigations required to analyze the molecular basis of drug resistance. We have therefore turned to human ovarian cancer lines as an alternative model in which to determine the mechanism of drug resistance and potential ways in which such drug resistance can be reversed (Hamilton et al. 1983).

References

Alberts DS, Chen HSG, Soehnlen B et al. (1980) In vitro clonogenic assay for predicting response of ovarian cancer to chemotherapy. Lancet 2: 340–342

Hamburger AW, Salmon SE, Kim MB, Trent JM, Soehnlen BJ, Alberts DS, Smith HJ (1978) Direct cloning of human ovarian cancer cells in agar. Cancer Res 38: 3438–3444

Hamilton TC, Foster BJ, Grotzinger KR et al. (1983) Development of drug sensitive and resistant human ovarian cancer cell lines: a model system for indentifying new drugs and procedures. Proc Am Assoc Cancer Res 24: 313

Hogan WM, Young RC (1982) Gynecologic malignancies. Cancer Chemotherapy 1982. Excerpta Medica, Amsterdam, pp 309–339

Hogan WM, Ozols RF, Willson JFV et al. (1982) Application of the human tumor stem cell assay to the study of epithelial ovarian cancer. (Abstract). Presented at Society of Gynecology and Oncology

Litterst CL (1981) Alterations in the toxicity of *cis*-dichlorodiammine platinum and in tissue localization as a function of NaCl concentrations in the vehicle of administration. Toxicol Appl Pharmacol 61: 99–108

Ozols RF, Young RC (1983) Phase II trial of high dose cisplatin in refractory ovarian cancer patients. Medicine Branch, NCI

Ozols RF, Willson JKV, Grotzinger KR, Young RC (1980) Cloning of human ovarian cancer cells in soft agar from malignant effusions and peritoneal washings. Cancer Res 40: 2743–2747

Ozols RF, Willson JKV, Weltz MD, Grotzinger KR, Myers CE, Young RC (1980b) Inhibition of human ovarian cancer colony formation by Adriamycin and its major metabolites. Cancer Res 40: 4109–4112

Ozols RF, Fisher RI, Anderson T et al. (1981) Peritoneoscopy in the management of ovarian cancer. Am J Obstet Gynecol 140: 611–619

Ozols RF, Young RC, Speyer JL et al. (1982) Phase I and pharmacologic studies of Adriamycin administered intraperitoneally to patients with ovarian cancer. Cancer Res 42: 4265–4269

Ozols RF, Corden BJ, Jacob J et al. (1984) High-dose cisplatin in hypertonic saline. Ann Int Med 100: 19–24

Salmon SE, Buick RN (1979) Preparation of permanent slides of intact soft-agar colony cultures of hematopoietic and tunor stem cells. Cancer Res 39: 1133–1136

Salmon SE, Hamburger AW, Soehnlen B, Durie BGM, Alberts DS, Moon TE (1978) Quantitation of differential sensitivity of human tumor stem cells to anticancer drugs. N Eng J Med 298: 1321–1327

Von Hoff DD, Casper J, Bradley E et al. (1981) Association between human tumor colony forming assay results and response of an individual patient's tumor to chemotherapy. Am J Med 70: 1027–1032

Von Hoff DD, Clark GM, Stogdill BJ, et al. (1983) Prospective clinical trial of a human tumor cloning system. Cancer Res 43: 1926–1931

Young RC (1975) Chemotherapy of ovarian cancer: past and present. Semin Oncol 2: 267–27

Technical Problems with Soft Agar Colony Formation Assays for In Vitro Chemotherapy Sensitivity Testing of Human Solid Tumors: Mayo Clinic Experience

M.M. Lieber

Department of Urology, Mayo Clinic, 200 First Street Southwest, Rochester, MN 55905, USA

Introduction

Since publication of the revolutionary papers by Salmon and colleagues from Tucson which about 5 years ago suggested that primary human tumor cells could be cloned in soft agar culture and in vitro chemotherapy sensitivity testing reliably performed (Hamburger and Salmon 1977; Salmon et al 1978; Salmon 1980), scores of cancer research groups around the world have looked into this testing technique. The number of papers already published (Johnson and Rossof 1983) and the breadth and intensity of the studies presented in this volume bear witness to the profound interest in soft agar colony formation assays for primary human tumors which the work of Salmon and his colleagues has generated. My own laboratory at the Mayo Clinic began performing the "human tumor stem cell assay" of Hamburger and Salmon for human solid tumor samples in 1979. In recent years we have endeavored to study by this technique all the suitable human solid tumor samples removed from the operating rooms at Rochester Methodist Hospital and St. Marys Hospital in Rochester, Minnesota, the two large general hospitals associated with the Mayo Clinic. To date, we have studied over 4,500 separate primary human tumor samples for in vitro chemotherapy sensitivity testing using soft agar colony formation assays similar to that described by Hamburger, Salmon, and colleagues. Our laboratory at the Mayo Clinic also has been one of four laboratories selected by the United States National Cancer Institute to perform soft agar colony formation assays with primary human tumor samples in a drug-screening mode, looking for new anticancer agents. This somewhat extensive practical experience in performing soft agar colony formation assays for in vitro anticancer drug sensitivity testing of primary human tumor cells generated the opportunities during which my colleagues and I observed a variety of technical problems.

Operational Axiom 1

Preparation of proliferative pure single-cell suspensions from primary human tumor samples is difficult and rare. Small cell aggregates are invariably seeded into soft agar cultures prepared by standard mechanical or enzymatic techniques. Proliferative cell colonies observed after culture incubation generally result from the growth of seeded small cell aggregates (Agrez et al. 1982).

The original publications of Salmon and colleagues and others suggested that primary human tumor samples could be readily dispersed into pure single-cell suspensions which could be seeded into bilayer soft agar cultures and then subsequently show proliferative clonal growth. This was the origin of the terms "human tumor stem cell assay" and "clonogenic assay," which became so popular. But from the beginning there was skepticism that clonal growth could be so commonly observed using preparations of primary human tumor cells (Editorial 1981).

Reliable studies of colony formation by primary human tumor cells in my own laboratory were greatly assisted by the availability of a computerized image analysis system designed expressly for soft agar colony formation assays (Bausch & Lomb FAS-II); this system enabled us objectively and quantitatively to assess the size, volume, and number of cell colonies present in cultures at multiple times. When the computerized image analysis system was used to carefully assess control culture plates the day after culture initiation, for primary human tumors disaggregated by common mechanical or enzymatic techniques, we were upset to find that approximately one-third of cultures studied contained more than 30 colonies (greater than 60 μm in diameter) the day after the cultures were made up. These cell aggregates appeared to result from incomplete disaggregation of the primary human tumors and not from reaggregation of cells during the culturing process (Agrez et al. 1982). Moreover, these cell aggregates had the identical appearance of proliferative tumor cell colonies and could not be distinguished from such cell colonies when the cultures were scored at 1–2 weeks after culture initiation.

Most important, if extensive filtration through fine-pore filters was carried out in order to make relatively more pure single-cell tumor suspensions and to eliminate larger cell aggregates, proliferative growth in agar to yield 30–50 cell colonies observable 1–2 weeks after culture initiation was virtually never observed (Agrez et al. 1982). When serial photomicrographic studies were carried out on a variety of soft agar cultures of primary human tumor cells, it was conclusively demonstrated for our laboratory that the larger colonies (greater than 60 μm in diameter) which arose by proliferation in 1–2 weeks always came from the growth of small cell aggregates seeded into the agar. We never observed colony formation (cell proliferation to form colonies of 60 μm) from seeded single cells photographed serially (Agrez et al. 1982). In this regard, soft agar cultures of cells from primary human solid tumors are quite different from similar cultures of cells from established tumor cell lines or xenografts.

These and continuing related observations of primary human tumor cultures in my own laboratory have led us to make the generalizations stated above as operational axiom 1: that is, that the standard mechanical, enzymatic, or combined mechanical and enzymatic methods which have been used to date to disaggregate primary human solid tumors, as well as tumor cell preparations from many malignant effusions, generally or invariably contain aggregates of tumor cells. Such preexisting cell aggregates are difficult to differentiate from proliferative cell colonies. In our experience, these small cell aggregates are often the origin of the larger cell colonies observed and counted 1–2 weeks after culture initiation. Because such cell aggregates are generally present in human tumor suspensions and are therefore invariably seeded into the culture, we believe that soft agar colony formation assays for primary human tumor cells must take this confounding problem into account and control for it.

In particular, control plates must be objectively counted immediately after culture initiation to document the presence or absence, size and size distribution of what cell aggregates are present initially. Such small cell aggregates are generally difficult to observe when freshly seeded into soft agar cultures. The optical density of the tumor cells and the agar culture medium is quite close. Therefore, at present, we believe such an objective assessment can only be made on soft agar control plates appropriately stained so that the cell aggregates are readily viable. In our own laboratory, we use the vital tetrazolium dye 2-(p-iodophenyl)-3-(p-nitrophenyl)-5-phenyl tetrazolium chloride (INT) (Alley et al. 1982) to stain such control plates and document the number of aggregates initially cultured.

We believe that older claims for high in vitro clonal growth rates for large numbers of primary human cancers studied in soft agar assays must be held in doubt. Moreover, claims for extensive clonal proliferative growth in soft agar cultures of cells derived from normal tissues and from benign tumors such as parathyroid adenomas, benign prostatic hyperplasia, and bladder papillomas must be regarded with some skepticisim. Certainly, claims that true clonal proliferation of single cells from these tissue sources took place in vitro requires more careful objective documentation than was used just several years ago.

In our laboratory, we believe that objective computerized colony counting must in fact demonstrate a substantial increase in colony size and number between the day of culture initiation and the day of counting for a valid proliferative soft agar colony formation assay to be considered to have been performed. Because the presence of small cell aggregates is generally present in cultures of most primary human tumors studied, operational axiom 2 has also become important to us.

Operational Axiom 2

Because tumor cell aggregates are so commonly present and can be confused with cell proliferation, (a) vital staining with an agent such as the tetrazolium dye INT is required to differentiate between living and nonviable cell colonies, (b) day 1 counting of control plates is necessary, and (c) all drug sensitivity experiments must contain one or more positive cytotoxic control compounds which should eliminate the presence of all detected viable cell aggregates.

None of these control measures were used when soft agar colony formation assays for primary human tumors were begun 5 or 6 years ago. Since smaller or larger cell aggregates are commonly present but difficult to detect when cultures are first initiated, their continued (but unknown) presence in the soft agar cultures led (in our experience) to false negative cytotoxic drug testing. That is, tumor cells (single cells and cell aggregates) were exposed to anticancer drugs in vitro. The small cell aggregates persisted in the soft agar cultures held in place by the semisolid medium, even though in many cases the cells in them were dead. But the cell aggregates ("ghosts") were eventually counted and scored as representing proliferative clonal colony growth in the presence of anticancer drugs. We believe that this class of culture artifact (the persistence of cell aggregates) was responsible for the unusual-looking dose-response curves seen in early years of studies of this type, when it often was reported that a subset of the tumor cell population was totally resistant to high doses of X-ray or anticancer drugs, seen as a flat plateau in the dose-response curves carried out to high doses of radiation or anticancer drugs.

In our own laboratory, Alley has demonstrated that the use of the tetrazolium vital dye INT allows differentiation between viable and nonviable cell colonies or aggregates in the soft agar assay system (Alley et al. 1982). When such a vital stain is used, it can be demonstrated that dose-response curves show a typical linear plot when charted in the usual semilog fashion. The plateaux showing in vitro resistance to high doses of drug are simply not seen when a vital stain is used.

During 1981, three laboratory/clinical research protocols, in which attempts were made prospectively to select single-agent therapy for patients with advanced ovarian, colorectal, and renal carcinomas based on results from soft agar colony formation assays, were carried out at the Mayo Clinic. Although single agents were tested in vitro at what we believed to be maximally achievable serum levels, we rarely identified a highly active drug in vitro even though a large number of drugs were tested per individual tumor (Williams et al. 1983). We now believe that the in vitro assay used was biased toward giving false negative in vitro sensitivities because we were unaware of the presence of small cell aggregates and the fact that such aggregates could persist in agar and be counted as viable colonies with the techniques then used. Only with vital staining of colonies have we begun to see a universal demonstration of cell killing with cytotoxic agents such as mercuric chloride and a much higher incidence of anticancer-drug-induced cytotoxicity in vitro (Alley et al. 1982).

Summary and Conclusions

If soft agar colony formation assays for primary human tumor cells are going to be performed and the results assessed by optical analysis of colony formation, then we believe it is mandatory in such assay techniques to (a) objectively count control plates on the day of culture initiation using stained plates, (b) use universal cytotoxic control compounds which should uniformly eliminate all viable cell proliferation, and (c) use a vital stain such as the tetrazolium dye INT to document the presence or absence of viable cell colonies when cultures are assessed.

There is no question, however, after careful observation of thousands of soft agar cultures of primary human tumor cells in my laboratory, that significant proliferation of small tumor cell aggregates often takes place in vitro and can in fact be used to assess cytotoxic drug effects in vitro. We do not believe that this is demonstrably stem cell or clonal growth. Nevertheless, it almost certainly is in vitro tumor cell proliferation. With very careful controls, we believe that optical methods can be used reliably to evaluate drug effects on this soft agar proliferative capacity of primary human tumor cells. However, it may eventually prove more useful to study human tumor cell proliferation in vitro (even soft agar colony growth) by other methods, such as incorporation of radiolabeled bases into newly synthesized DNA or other macromolecules, or by the simple use of vital dyes, as discussed elsewhere in this volume.

The National Cancer Institute Drug Screening Program using the human tumor stem cell assay, under the direction of Shoemaker and Wolpert, has recently required the use of day 1 control counts, positive cytotoxic control compounds, and vital staining with INT. We believe that such quality control techniques are useful and necessary to generate the maximum reproducibility and to observe the maximum cytotoxic drug sensitivity in assays of this type assessed by optical methods . Other investigators using optical methods for assessing tumor cell proliferation in vitro in other types of culture systems could also profitably learn from our difficulties.

Most common human cancers are clinically resistant to single-agent chemotherapy given empirically. If an in vitro assay method is also experimentally biased to give a picture of cytotoxic drug resistance, both because of technical artifacts and because of the use of relatively low concentrations of anticancer drugs studied, then it seems inevitable that favorable in vitro/in vivo correlations will be seen when the results of the laboratory assay and patients' clinical response to drug treatment are compared. That is, almost all in vitro assays will show pan-resistance and, similarly, almost all patients will fail to respond when treated with the drugs tested in vitro. Inevitably, however, as more clinical/laboratory in vitro/in vivo correlations are carried out in patients with chemotherapeutically sensitive tumors, false negative laboratory tests will be documented. In fact, examples of such false negative tests have recently been identified, for example in Von Hoff's recent study (Von Hoff et al. 1983).

It is hoped that this presentation makes it clear that there are a host of difficult technical problems in the reliable assessment of soft agar colony formation assays for in vitro chemotherapy sensitivity testing of cell suspensions derived from primary human tumors. For this reason alone I believe strongly that these tests are not suitable for widespread clinical use or commercial marketing at this time. Because of the technical problems in assessing such assays, the relatively low proliferation rate seen in vitro, and the absence of highly active drugs clinically to test in vitro, it seems evident that the expected clinical benefit for any given patient undergoing an assay of this type (which is quite expensive) is extremely limited. Soft agar colony formation assays for in vitro chemotherapy sensitivity testing of primary human tumor cells are still an area of cancer treatment which should be confined to cancer research laboratories. Carefully observed studies of tumor cell growth in vitro, such as those presented by Courtenay and others in this volume, will probably be far more useful in the long run than current massive clinical exploitation of an assay technique now fraught with technical difficulties in performance and interpretation.

References

Agrez MV, Kovach JS, Lieber MM (1982) Cell aggregates in the soft agar "human tumour stem-cell assay". Br J Cancer 46: 880–887

Alley MC, Uhl CB, Lieber MM (1982) Improved detection of drug cytotoxicity in the soft agar colony formation assay through use of a metabolizable tetrazolium salt. Life Sci 31: 3071–3078

Editorial (1981) Clonogenic assay for the chemotherapeutic sensitivity of human tumours. Lancet 1: 780–781

Hamburger AW, Salmon SE (1977) Primary bioassay of human tumor stem cells. Science 197: 461–463

Johnson PA, Rossof AH (1983) The role of the human tumor stem cell assay in medical oncology. Arch Intern Med 143: 111–114

Salmon SE (1980) Cloning of human tumor stem cells (Progress in clinical and biological research, vol 48). Liss, New York

Salmon SE, Hamburger AW, Soehnlen B, Durie BGM, Alberts DS, Moon TE (1978) Quantitation of differential sensitivity of human tumor stem cells to anticancer drugs. N Engl J Med 298: 1321–1327

Von Hoff DD, Clark GM, Stogdill BJ, Sarosdy MF, O'Brien MT, Casper JT, Mattox DE, Page CP, Cruz AB, Sandbach JF (1983) Prospective clinical trial of a human tumor cloning system. Cancer Res 43: 1926–1931

Williams TJ, Lieber MM, Podratz KC, Malkasian GD Jr (1983) Soft agar colony formation assay for in vitro testing of sensitivity to chemotherapy of gynecologic malignancies. Am J Obstet Gynecol 145: 940–947

*Cloning of Human Tumor Cells in Methylcellulose-Containing Medium**

C. Cillo and N. Odartchenko**

Schweizerisches Institut für Experimentelle Krebsforschung,
1066 Epalingess/Lausanne, Switzerland

Introduction

A colony assay for cells from human solid tumors has attracted a great deal of interest due to its use in studies comparing in vitro and in vivo sensitivities to anticancer drugs for clinical cancer chemotherapy (Salmon et al. 1978; Von Hoff et al. 1981). In addition, potential benefit for more basic studies on the biological characteristics of human tumor cells can be expected, since it allows on the growth of primary tumor cells at a clonal level.

In their original paper, Hamburger and Salmon (1977) have described a double-layer agar system for cloning human tumor cells using mouse spleen cells as a feeder layer in order to stimulate growth of the tumor cells. Although this system has since been considerably improved (Courtenay et al. 1978; Kern et al. 1982), primarily through changes in nutritional and chemical constituents, it remains an undefined system. Agar itself is not chemically well defined, is weakly mitogenic (Kincade et al. 1976), and has a promoting activity (Hoang et al. 1981) for the release of stimulating factors by cultivated cells.

In the present work, we show that the methylcellulose clonal assay compares favorably with the agar assay in terms of cloning efficiencies for several types of human solid tumors. We also show that colonies removed from the methylcellulose are capable of further growth.

Among the methods available for cultivating cells in semisolid or viscous media, methylcellulose offers several advantages over agar (Iscove and Schreier 1979). It is water soluble at room and incubator temperature. It immobilizes cells by increasing the viscosity of the medium rather than by gel formation, making it easier to recover cells or colonies for subculturing or for morphological and cytogenetic studies. Finally, it is chemically better defined ad does not contain mitogenic or promoting activities, allowing more definitive studies on the nutritional requirements of the cultivated cells.

* Supported by the Swiss National Foundation and the Swiss Cancer League

** We thank Drs. R. Abele, I. Alberto, L. Barrelet, A. Curchod, D. Dialdas, G. W. Locher, G. Losa, G. Maestroni, J. Pettavel, and P. Siegenthaler for providing us with tumor biopsy samples. We are grateful to S. Fakan for his help with electron microscopy, to M. Schreyer for the histological work

Materials and Methods

Tumor Samples

Biopsy fragments, mostly from breast and lung, were obtained from cancer-bearing patients undergoing routine care in various regional clinics. Tissue samples were immersed immediately in Iscove's modified Dulbecco's medium (IMDM; Gibco, Grand Island, New York) containing 10% fetal calf serum (FCS, Gibco). Pleural effusion samples were collected by paracentesis into heparinized tubes without medium. All tubes containing samples of malignant tissue or cells were kept at room temperature for periods varying from a few hours to no more than 18 h before processing.
Solid tumor samples were fragmented using scalpel blades, then passed through needles of decreasing bore size down to 23 gauge. The recovered suspension was centrifuged and treated with 1% trypsin (Gibco) in IMDM at 37° C for 10 min, washed twice with fresh IMDM + 10% FCS, and again passed through needles of decreasing diameter until a Romogeneous suspension of separated cells was obtained.
Malignant effusions were centrifuged at 400 *g* for 10 min and suspended in IMDM with 10% FCS. Dead cells and debris were eliminated by centrifugation at 400 *g* for 20 min over a Ficoll-Paque later (density 1,077 g/cm^3) (Pharmacia Fine Chemicals, Zurich, Switzerland) and collection of the interface cell layer. The cells were then washed twice with IMDM + 10% FCS and passed through needles of decreasing diameter as above, in order to obtain cell suspensions.
A portion of each cell suspension was cultured and, when sufficient numbers of cells were obtained, the remaining cells were immediately frozen in 1-ml aliquots at −80° C using 10% dimethyl sulfoxide (DMSO; Fluka, Buchs, Switzerland) as a protective agent (Selby and Steel 1981). A recovery of frozen cells of 60%−80% was usually obtained.

Methylcellulose Cultures

Tumor cell colonies were obtained by plating 10^5 viable cells suspended in 1 ml IMDM containing 0.8% methylcellulose (Methocel, 4,000 cp, premium grade; Dow Chemical, Midland, Michigan, or Fluka) and 15% FCS, in 35-mm petri dishes (Greiner, Nürtigen, Germany). These were bacteriological dishes, not pretreated for tissue culture. In addition, 2-mercaptoethanol 10^{-4} *M* (Fluka) was included in the culture solution. At least three dishes were used per specimen. Plates were examined for the presence of cell aggregates immediately after plating and those containing large aggregates were discarded. Cultures were incubated at 37° C with 5% CO_2 in humidified air.

Comparison of Colony Formation in Methylcellulose Versus Agar

In a series of experiments, tumor cells were plated in parallel in methylcellulose and agar. Agar cultures were established according to the technique of Hamburger and Salmon (1977). Briefly, cells were suspended at 10^5/ml in 0.3% Bacto Agar (Difco Laboratories, Detroit, Michigan) in McCoy's medium (Gibco) with 20% FCS. One milliliter of this suspension was layered over a 1-ml feeder layer of 0.5% agar in 35-mm petri dishes. No conditioned medium was added. Cultures were incubated at 37° C in humidified air with 5% CO_2.

Colony Counts

Colonies, defined as aggregates of 40 or more cells (Metcalf 1977), were counted after 7 days and at weekly intervals thereafter, using an inverted microscope at × 100 magnification.

Colony Manipulation

Individual colonies were removed from methylcellulose cultures with a sharp Pasteur pipette and transferred to fresh methylcellulose-containing medium or to liquid IMDM + 10% FCS minus methylcellulose.

Morphological and Histological Examination

Cytocentrifuge-spread slides of initial cell suspension and colonies or cells removed from the cultures were stained with Giemsa and Papanicolaou stains. Cells were also examined using a Philips EM 400 electron microscope following 1 h fixation at 4° C in 1.6 glutaraldehyde in 0.1 *M* phosphate buffer, dehydration in graded concentrations of acetone, embedding in Epon, and ultrathin sectioning using a diamond knife.
For histological examination, single colonies were fixed in 2.5% glutaraldehyde and embedded in glycolmethacrylate JB4. Sections 2 μm thick were cut on a LKB microtome (Pharmacia, Uppsala) and stained with Giemsa.

Results

The in vitro clonogenicity of cells from 124 tumors biopsy samples was examined. Of these tumor samples, 36% were mammary carcinomas, 25% lung tumors, and 15% pleural effusions of various origin. The remaining 24% were predominantly ovarian, testicular, and skin malignancies. Table 1 summarizes the results obtained when single-cell suspensions of these tumors were cultured in medium containing methylcellulose and were examined at weekly intervals. It can be seen that 55% of the samples gave rise to colonies of more than 40 cells at some stage during the first 4 weeks of culture. However, the cloning efficiencies were variable, ranging from 0.001% to 1.5%, and showed no correlation with tumor origin. Parallel cultures of 15 biopsy samples in agar-containing medium (Table 2) gave similar results to those in methylcellulose, with only minor differences in plating efficiencies. In no case was the ability of a tumor sample to form colonies dependent on the supportive agent used.
Table 1 also shows that the growth of pleural effusions was greater (72%) than that of the solid biopsies (52%). This may be due to a loss of colony-forming tumor cells during initial mechanical or enzymatic treatment of the original biopsies prior to plating, or due to an in vivo selection in effusions of cells able to survive in suspension.
Viability and numbers of cells recovered from solid tumors and pleural effusions, determined by trypan blue exclusion, is also indicated in Table 1. It can be seen that average viability varies from 45% to 60% for solid tumors and from 70% to 85% for liquid effusions.

Table 1. In vitro clonogenicity of biopsied human tumor cells (methylcellulose)

Tumor type	Number of samples	Cells × 10^6/ samples (median/ range)	Viability %/ samples (median/ range)	Number of colonies/ 10^5 cells plated			Total (%)
				1–15	16–200	200–1,500	
Breast	45	8.2/4.6–22	46/24–61	13	9	1	23 (51%)
Lung	31	11.3/2.8–18	51/42–68	8	7	1	16 (52%)
Pleural effusions	18	24.6/16–170	76/70–85	5	6	2	13 (72%)
Miscellaneous (ovary, testis, skin, kidney)	30	7.4/1.9–23	48/39–64	7	9	0	16 (53%)
Total	124						68 (55%)

Table 2. Comparison of colony formation by human tumor cells – agar vs methylcellulose-containing media

Tumor type	Samples		Plating efficiency/10^5 plated cells	
			Methylcellulose	Agar
Breast	1		15	19
	2		89	81
	3		1,672	829
	4		1	0
	5		5	6
	6		1	0
Lung	7		15	12
	8		0	0
	9		8	7
	10		0	0
	11		7	5
Testis	12		0	0
Pleural effusions	13	Breast	212	195
	14	Breast	16	21
	15	Ovary	0	1

As shown in Fig. 1, a nearly linear relationship was observed for single-cell suspensions obtained from a pleural effusion (from breast carcinoma), a solid breast carcinoma, and a lung tumor plated at different cell concentrations. This holds true at least for cell numbers in the range of 10^4 to 2×10^5. For more than 2×10^5 cells per plate, colonies were difficult to differentiate due to high numbers of cells present as background.

Figure 2 graphically illustrates the changes in numbers of colonies and clusters as a function of time following plating. The colony number increases during the 2nd and 3rd weeks and then levels off during the 4th week. After week 4, most colonies started progressively to disaggregate into separate dying cells. The pH of the culture medium changed at the same

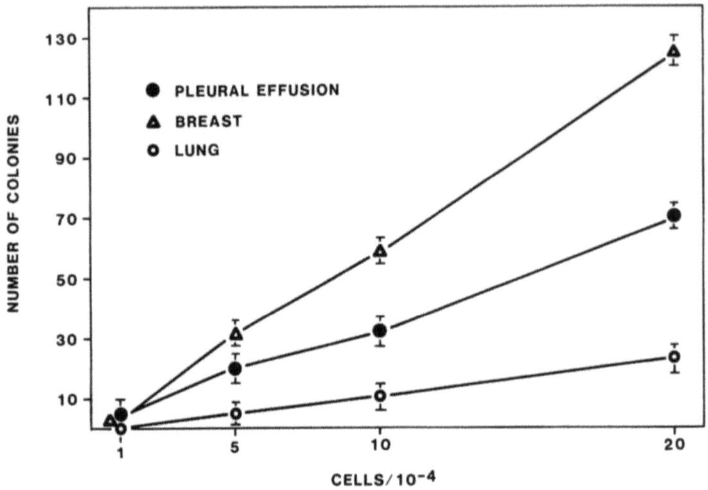

Fig. 1. Relationship between colony numbers and numbers of cells plated for three different tumor samples

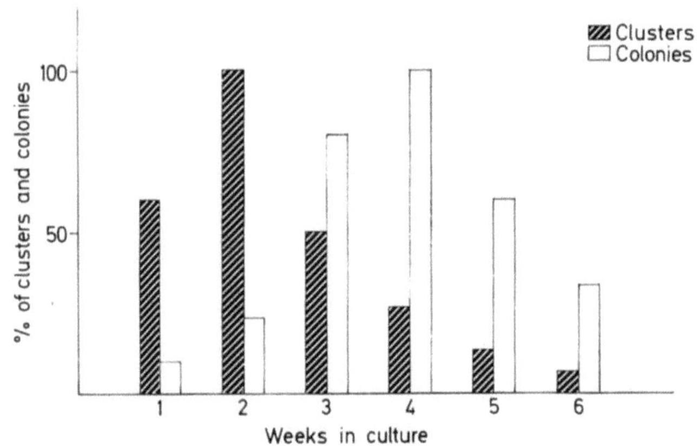

Fig. 2. Relationship between clusters or colonies formed, as a function of time after plating

time. Cell death appeared to be slightly more advanced in parallel agar culture. By weeks 6–7, all or almost all colonies had disappeared in methylcellulose cultures. During the 1st and 2nd weeks, clusters (Fig. 2) containing less than 40 cells were noted, although they decreased in number thereafter.

Fig. 3a, b. Electron micrographs of single cells (breast carcinoma). **a** a cell from the original suspensions prior to plating (× 9,000); **b** a cell from a colony after 3 weeks of culture (× 7,600). Cellular structures, particularly nucleus and nucleolus, are quite similar in both cases. Microfilaments present on the cell surface before plating **(a)** disappear after in vitro growth **(b)**. Polymerized structures of methylcellulose are also visible in **b**

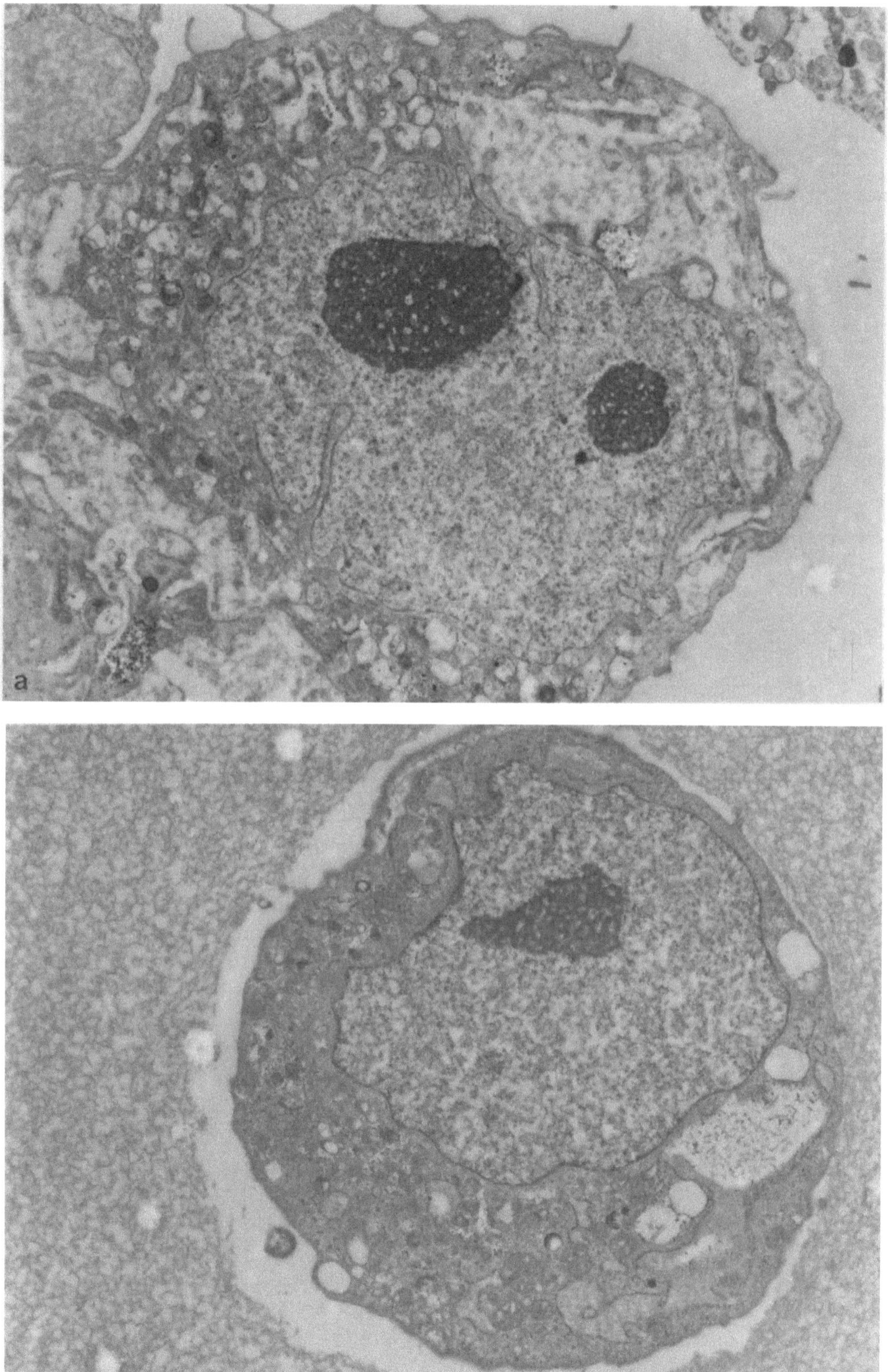
a
b

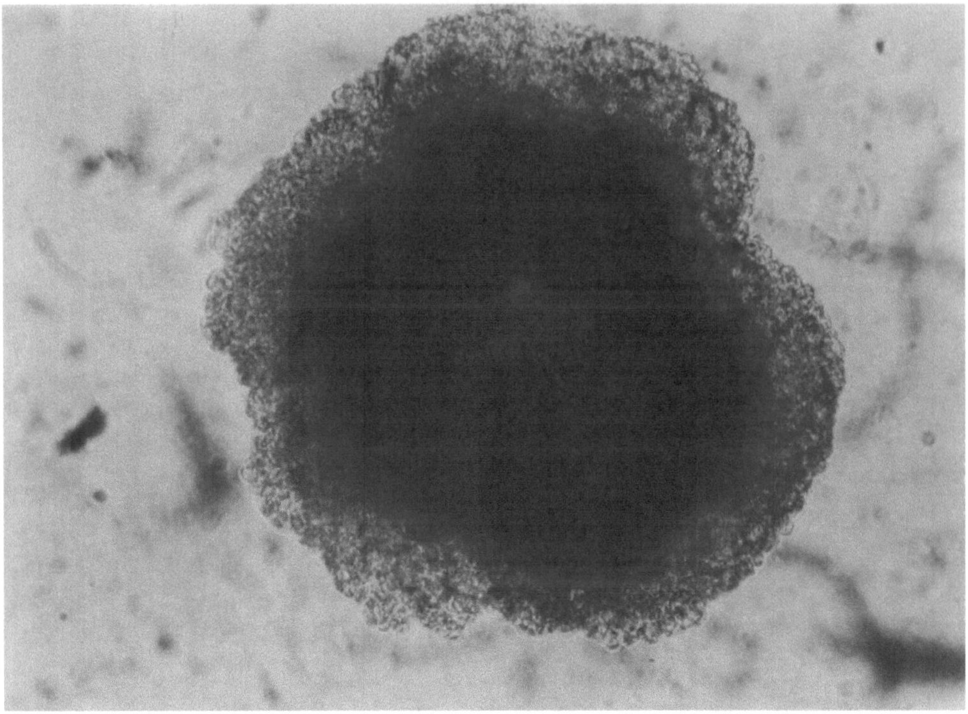

Fig. 4. Single melanoma colony after 3 weeks of growth and subsequent transfer to fresh medium for 3 more weeks (× 100)

No major morphological differences were seen between the initial tumor cell suspension and cells recovered from large single colonies or pooled smaller ones after 3 weeks of culture, as judged on cytocentrifuge smears stained with Giemsa or Papanicolaou. This preserved morphology was confirmed by electron microscopic examination of some of the samples. Figure 3 shows electron micrographs of single cells obtained from a breast carcinoma prior to (Fig. 3a) and after culture (Fig. 3b).

Some colonies from various histological types of primary tumors after 3 weeks of culture in methylcellulose were transferred to microwells containing fresh methylcellulose medium. Transferred colonies continued to grow for up to 2 months. Figure 4 is an example of such a colony, originally from a melanoma, following 3 weeks in primary culture and prior to 3 weeks in secondary culture. Some colonies were also transferred to 25-cm^3 flasks containing 5 ml fresh liquid medium and could be maintained for up to 3 months. No attempt was made to establish permanent cell lines.

The great advantage of using methylcellulose instead of agar is exemplified in Fig. 5, which shows a histological section of the same colony as Fig. 4. The overall morphology is well preserved. A central zone of cells at various stages of necrosis is in sharp contrast to the periphery, where many cells actively proliferate. Some very large cells coexist with much smaller ones, suggesting an early establishment of heterogeneity even among cells of a single colony (here of about 2,000–4,000 cells, i.e., at least 12 successive divisions, cell death not taken into account).

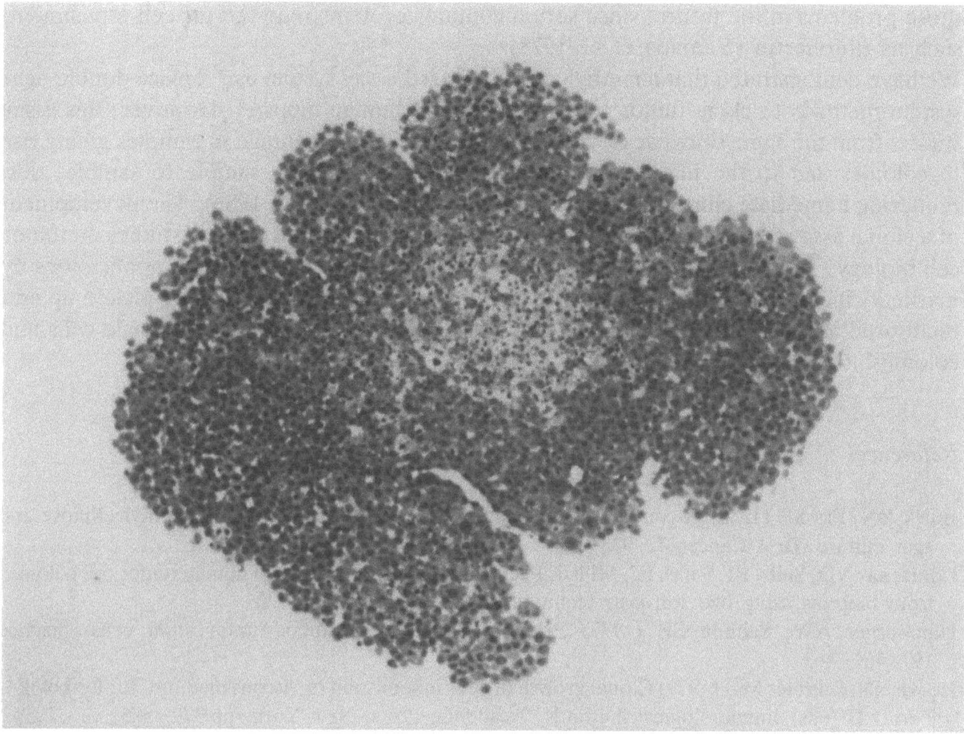

Fig. 5. Histological section of a single melanoma colony. Methacrylate embedding. Giemsa stain (× 63)

Discussion

We report the application of a methylcellulose-containing culture system originally developed for growing hemopoietic progenitor cells (Iscove et al. 1974) to the clonal growth of human tumor cells. Our results indicate that the percentage of tumor samples which form colonies in this system is similar to that reported for the double-layer agar assay (Salmon et al. 1978). In fact, when tumor cells were plated simultaneously in both assay systems, the cloning efficiencies were equivalent.

The fluidity of the methylcellulose medium allows easy handling of colonies for further studies. We have utilized this property of methylcellulose to document the fact that the colonies are composed of tumor cells. The results of light and electron microscopic examination demonstrate the morphological similarities of cells in culture to those in the original preparation. We have further shown that colonies can easily be transferred from primary cultures into new cultures and will continue to grow for at least 3 months.

On the other hand, the fluidity of the medium itself can generate some disadvantages. It has been described (Buick and Fry 1980) that neoplastic cells in methylcellulose can attach to culture dishes and give rise to adherent "pavement-type" colonies. We have avoided this problem by using petri dishes not treated for tissue culture, and batches of FCS which we had screened for their ability to support clonal growth of bone marrow progenitor cells. When some serum batches which were not prescreened were used, adherent colonies did indeed form. It is hoped that the development of serum-free media will help to eliminate those problems in the future, since serum contains proteins required for cell attachment, such as fibronectin (Yamada et al. 1978).

We have demonstrated that a methylcellulose-based assay system can replace double-agar system methods to clone tumor cells from primary human biopsies. However, this assay suffers from the same drawbacks with respect to the low percentage of samples giving rise to colonies and to the highly variable cloning efficiency from sample to sample, thus rendering immediate clinical application impractical (Selby et al. 1983). The development of a clonal assay for tumor cells, however, kemains an important tool for studies ou tumor cell biology. The use of methylcellulose media should expand its research applications by markedly increasing the number of cases in which individual clones can be picked up and recultured as well as by allowing well-founded morphological studies on single cells and colonies.

References

Buick RN, Fry SE (1980) A comparison of human tumor cell clonogenicity in methylcellulose and agar culture. Br J Cancer 42: 933–936

Courtenay VD, Selby PJ, Smith IE, Mills J, Peckham MJ (1978) Growth of human tumor cell colonies from biopsies using two soft-agar techniques. Br J Cancer 38: 77–81

Hamburger AW, Salmon SE (1977) Primary bioassay of human tumor stem cells. Science 197: 461–463

Iscove NN, Schreier MH (1979) Clonal growth of cells in semisolid or viscous medium. In: Lefkovik I, Pernis B (eds) Immunological methods. Academic Press, New York, pp 379–385

Iscove NN, Sieber F, Winterhalter KH (1974) Erythroid colony formation in cultures of mouse and human bone marrow: analysis of the requirement for erythropoietin by gel filtration and affinity chromatography on agarose-concanavalin A. J Cell Physiol 83: 309–320

Kern DH, Campbell MA, Cochran AJ, Burk MW, Morton DL (1982) Cloning of human solid tumors in soft agar. Int J Cancer 30: 725–729

Kincade PW, Ralph D, Moore MAS (1976) Growth of B-lymphocyte clones in semi-solid culture is mitogen-dependent. J Exp Med 143: 1265–1268

Metcalf D (1977) Hemopoietic colonies. In vitro cloing of normal and leukemic cells. Recent Results in Cancer Research. Springer, Berlin Heidelberg New York

Salmon SE, Hamburger AW, Soehnlen B, Durie BGM, Alberts DS, Moon TE (1978) Quantitation of differential sensitivity of human-tumor stem cells to anti-cancer drugs. N Engl J Med 298: 1322–1327

Selby P, Steel GG (1981) Clonogenic cell survival of cryopreserved human tumor cells. Br J Cancer 43: 143–148

Selby P, Buick RN, Tannock I (1983) A critical appraisal of the "human tumor stem-cell assay". New Engl J Med 308: 129–134

Trang Hoang, Iscove NN, Odartchenko N (1981) Agar extract induces release of granulocytes. Exp Hematol 9: 499–504

Von Hoff DD, Casper J, Bradley E, Sandbach J, Jones D, Makuch R (1981) Association between human-tumor colony-forming assay results and response of an individual patient's tumor to chemotherapy. Am J Med 70: 1027–1032

Yamada KM, Olden K (1978) Fibronectins-adhesive glycoproteins of cell surface and blood. Nature 275: 179–184

*The Effect of Chemotherapy on Human Bone Sarcomas: A Clinical and Experimental Study**

H.-P. Honegger, A. von Hochstetter, P. Groscurth, V. Hofmann, and M. Cserhati

Universitätsspital Zürich, Departement für Innere Medizin, Abteilung für Onkologie, Rämistrasse 100,
8091 Zürich, Switzerland

Concept of Zurich Bone Sarcoma Study

Osteosarcomas are relatively rare tumors affecting younger people, mostly men. The prognosis in these cases is uniformly bad. An analysis of the course of the disease 10 years ago revealed that roughly 10 months after removal of the primary tumor, lung metastases develop (Sweetnam et al. 1971). Before 1970, a 5-year survival of 20% or less is reported following surgical treatment of the bone tumor. Chemotherapy was essentially unsuccesful until the early 1970s, when trials with doxorubicin and high-dose methotrexate were initiated. With these two drugs, considerable response rates (mostly partial remissions) were reported in metastatic disease.

As soon as effective drugs were known, the concept of additional, that is adjuvant, chemotherapy was introduced into the management of the disease (Jaffe et al. 1974). The aim was to destroy micrometastases probably already present in the vascular bed at the time of primary resection. Several groups claimed to achieve prolonged survival with adjuvant chemotherapy after tumor resection in their study populations. These early trials were nonrandomized; their results were compared to those in historical control groups and found to be more favorable (Jaffe et al. 1974). Chemotherapy was then used preoperatively by several groups (Rosen et al. 1976). Thus it became possible to examine histologically the primary tumor after chemotherapy. It was felt that preoperative chemotherapy might allow a reduction of the extent of surgical intervention (Rosen et al. 1979).

Several groups have treated their patients preoperatively. Some claim to see response rates (assessed by the extent of necrosis) of 70%–80%, with survival rates at 4 years of 80% for responding patients. Others, including our group, were unable to reproduce these promising results. In our hands, roughly two-thirds of the patients respond in clinical terms and 50% in histological terms after preoperative chemotherapy (Honegger et al. 1983). That means, in fact, that half of our patients need other chemotherapeutic regimes, since they have a poor prognosis with respect to relapse-free and overall survival (Rosen et al. 1982; Honegger et al. 1983).

After a careful analysis of our clinical data, and taking into account our local research facilities, we embarked 3 years ago on a multidisciplinary program in order to gain more

* This study was supported in part by grants from Emdo foundation (Zürich, Switzerland) and from Cyanamid Inc. (New York). The authors are indebted to Mrs. M. Balzer, Mrs. C. Bommes, Mrs. M. Erni, and Mrs. U. Früh for excellent technical assistance and to Mrs. K. Lustig for animal care. We are grateful to Mrs. R. Fringeli for typing the manuscript

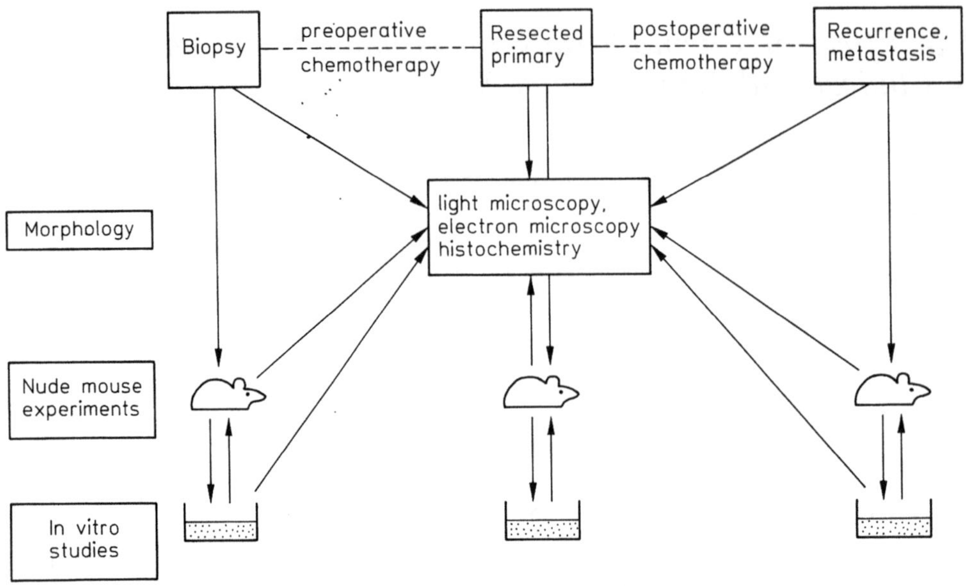

Fig. 1. Schematic presentation of Zurich Bone Sarcoma Study

insight into the biology of these rare tumors (Fig. 1). We have attempted to improve and extend the histological examination after preoperative chemotherapy by means of a semiquantitative analysis of necrotic areas. In addition, tumor specimens taken from the primary tumor before and after chemotherapy or, in several cases, from local recurrence and metastasis were transplanted into nude mice. At the same time, single-cell suspensions were prepared for culture in the clonogenic assay.

The present report will describe the potential and limitations of these methods. First, the method of semiquantitative analysis will be discussed. Biological data to support the relevance of the concept will be presented. Then the nude mouse model will be explained, take rates, growth characteristics and morphology of transplanted bone sarcomas will be shown. In a preliminary form, results of the analysis of our experiments with the mouse model as well as with the clonogenic assay with respect to clinical findings will be presented. The results will then be summarized and put into perspective.

Morphological Assessment of Necrosis in Resected Primary Tumors

Several publications have dealt with the assessment by light microscopic means of the degree of bone tumor destruction following preoperative chemotherapy (Ayala et al. 1980; Huvos et al. 1977; Rosen et al. 1976). The method described by Rosen et al. (1976) and by Huvos et al. (1977) first proposed to examine each resected specimen by numerous histological slides to determine the extent of the lesion. At the same time, the effect of treatment of the neoplastic tissue could be grouped by histological appearance into "viable" and "partially, largely, or total necrotic categories". The method received attention particularly when in 1979 it was presented as a grading system in which grades I–IV had come to replace the authors' previous descriptive categories. Follow-up of their patients suggested that a Grade III or IV response of the primary tumor to chemotherapy provided a better prognosis (Rosen et al. 1979).

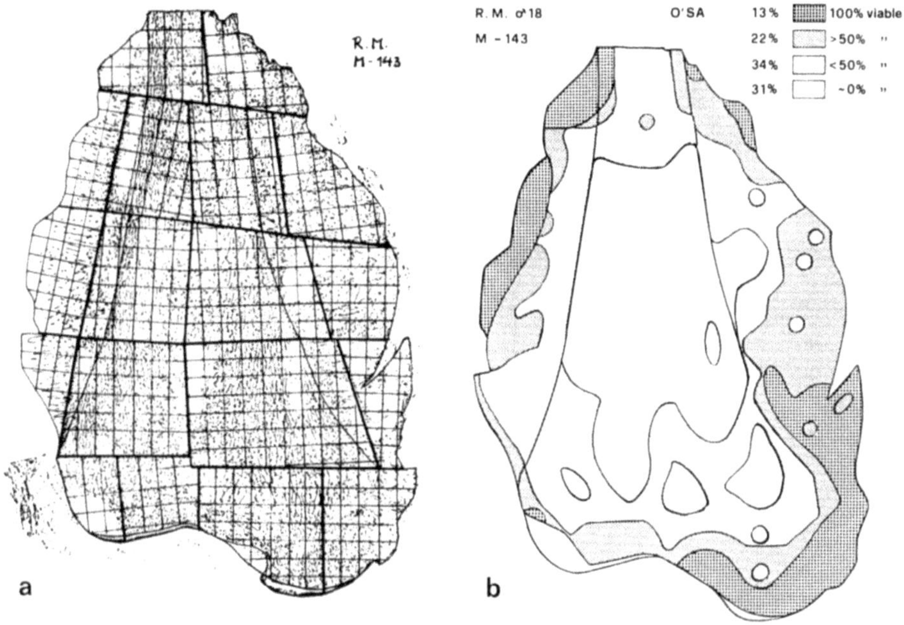

Fig. 2a, b. Cross section of an osteosarcoma of the distal femur. **a** Composite photograph reproducing the original topography; **b** "cartograph" depicting the topographic distribution from zones of different densities of tumor necrosis

About that time we began to study the effect of chemotherapy on sarcomas, particularly of bone, and found Huvos' original categories appealing in their simplicity. The attempt to quantify tissue response, however, confronted us with several difficulties. First, it is a rare neoplasm that either totally succumbs to chemotherapy or survives it without cellular loss. Tumor necrosis being patchy more often than total or absent, we obtained all categories (Grades I–IV) in nearly every specimen, depending on the size of the visual field under examination. Hence a more reliable method of quantification was needed whereby the tumor could be screened systematically by increments small enough to permit a reasonably accurate assessment of necrosis per unit area and large enough to be practicable in calculating surface area. Secondly, since in practice it is not feasible to examine a sizeable resection specimen in its entirety, the critical amount of tumor that had to be screened in such detail to yield representative data needed to be established. Consequently, we developed a method of examining our resection specimen with sufficient diligence and economy to provide reliable information in time to help direct further treatment during the patient's postoperative course.

Following preoperative chemotherapy, the resected specimen was sawn into 0.5–1 cm slices, of which at least the one showing greatest tumor extension was taken for processing. It was first photographed with a superimposed, freely adjustable grid dividing the tumor surface into maximally 5 × 6 cm areas, then cut up accordingly. The tissue blocks were fixed in Carnoy's solution and in alcohol, embedded in methacrylate, and sectioned according to the usual procedures. The histological preparations stained with Goldner and Giemsa were photographed and the prints assembled into a composite picture (Fig. 2a, heavy lines). The resultant photomosaic accurately reproduced the topography of the original slice. On a photocopy of the photomosaic we then drew in a finer grid of uniform

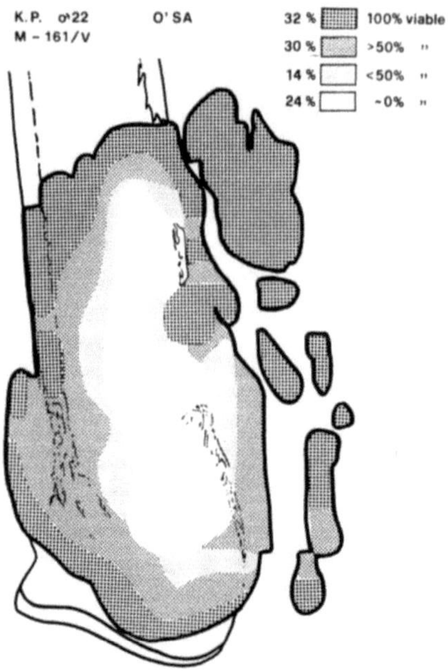

Fig. 3. "Cartograph" of an osteosarcoma of the distal femur showing major devitalization in less than 50% of the tumor's surface area

mesh size, orienting it for each block according to a given morphological reference point (Fig. 2a). The mesh size was such that one unit square corresponded to a microscopic field at low magnification. Starting with our morphological reference point we could now screen each histological preparation field by field, assess in each the amount of necrosis of the tumor cell population, and plot it on our grid.

In assessing tumor necrosis we used the original categories of "absent, partial, predominant, and total tissue necrosis," defining the middle two categories as necrosis of less and of more then 50% of the tumor cell population per visual field. We have found it impossible in practice to discern intermediate measures of necrosis with any degree of assurance (e.g., 30%, 40%, 60%). Defined as above, the categories were easy to apply and yielded reproducible results. The surface areas occupied by each of the four categories were measured with a morphometric system (MOP-AMO3), expressed in percent of total cross sectional tumor surface area, and presented graphically (Fig. 2b). In our example, 65% of the tumor's cross-sectional surface proved to be largely to totally necrotic. In another case, similarly devitalized areas covered only 38% of the examined surface area (Fig. 3).

On this basis we divided our cases arbitrarily into two groups. Group A consisted of cases where totally or predominantly viable tumor tissue made up more than 50% of the cross-sectional surface area. In Group B the zones of predominant devitalization covered more than 50% of the surface area.

An analysis of the clinical course revealed that patients in group B had significantly longer relapse-free intervals than the patients in group A (Fig. 4a). The course of group A patients did not differ appreciably from that of a historical group, treated by surgery alone. Overall survival of group B also significantly surpassed that of group A (Fig. 4b). Hence, on the basis of our quantitative analysis, we were able to conclude that the extent of necrosis in osteosarcomas following chemotherapy is of biological impact in that it relates significantly to the course of the disease.

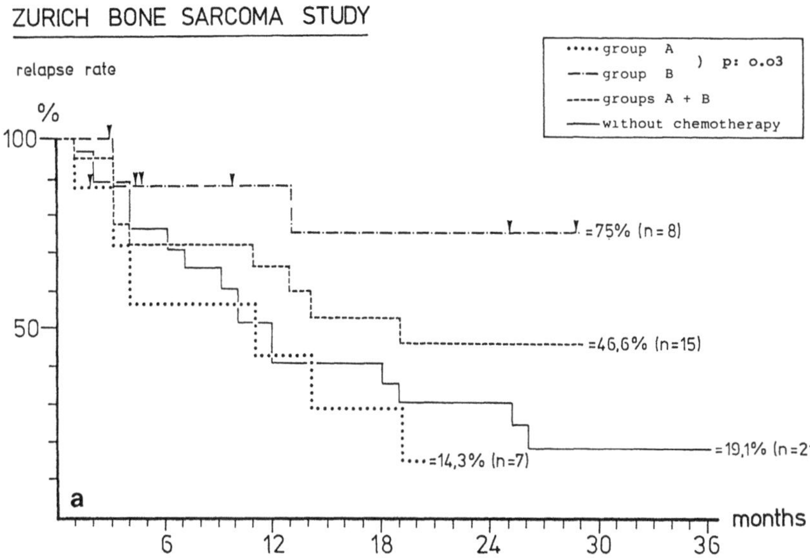

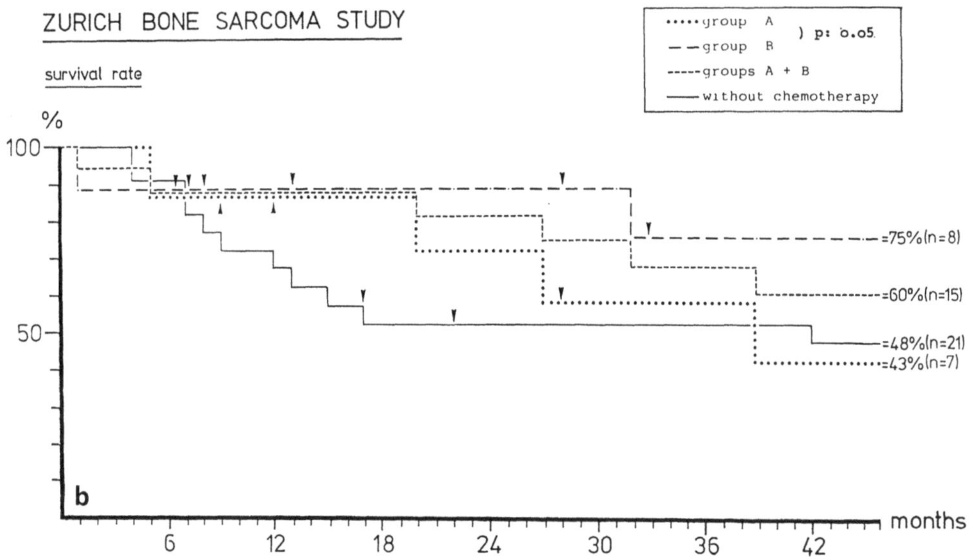

Fig. 4. a Relapse rate [Kaplan-Meier life table method (Kaplan et al. 1958)]. Statistical evaluation by log rank test (Peto et al. 1977). **b** Survival rate [Kaplan-Meier life table method (Kaplan et al. 1958)]. Statistical evaluation by log rank test (Peto et al. 1977)

Nude Mouse Experiments

As an essential part of our multidisciplinary program, we studied the biology of bone sarcomas in the nude mouse model. Here we report mainly on morphology and growth characteristics of transplanted sarcomas obtained before and/or after the patient received chemotherapy.

Table 1. Take rate of human bone sarcomas transplanted into preirradiated nude mice

Tumor type	Growth/no. of tumors implanted
Osteosarcoma	16/25 (64%)
Malignant fibrous histiocytoma	6/7 (85%)
Ewing sarcoma	1/4 (25%)
Total	23/36 (64%)

According to the literature, attempts to transplant human bone sarcomas had met with little success. The transplants either did not grow (Urist et al. 1979) or the take rate of implanted sarcomas was very low (Sharkey et al. 1978; Torhorst et al. 1976). Therefore, we tried to increase the take of xenotransplants by irradiating the recipients with 500 rad 3–5 days before implantation, with the idea of suppressing the residual immunity.

Tumor samples were obtained either prior to chemotherapy, from biopsy, or after chemotherapy, from resected primary tumor, local recurrence, or metastasis (Fig. 1). In seven patients sarcoma tissue was taken both from biopsy and resected primary, which enabled us to compare growth characteristics of transplanted bone sarcomas before and after chemotherapy. In an additional six cases tumor fragments were implanted repeatedly during the course of therapy. Transplantation techniques, plotting of growth curves, and calculation of lag phase and doubling time were performed as described (Groscurth et al. 1983). In addition fragments from the original as well as from the transplanted tumor were regularly processed for light and electron microscopy.

Using preirradiated animals the average take rate of transplants was 64% Table 1). Malignant fibrous histiocytomas grew best, whereas Ewing sarcomas displayed the lowest take. Furthermore, the take rate of transplanted bone sarcomas appeared to be related in part to the extent of necrosis in the resected primary tumor. In group A, displaying less than 50% necrosis, all six implants grew in nude mice, while in group B, with more than 50% necrosis, three out of six tumors failed to grow in the recipients.

As described previously (Groscurth et al. 1983) almost all transplanted sarcomas showed an adaptation period in early mouse passages. During the first passage the implants only maintained their size for over 50 days. In the following passages growth rate increased significantly to become constant in passage 4 or 5. Thereafter, each individual tumor tended to have characteristic growth curves. Thus comparison of growth characteristics of different bone sarcomas were meaningful only after passage 4.

The adaptation period may be accompanied by more or less distinctive morphological alterations. During mouse passages a slight shift in the cellular composition of sarcomas was occasionally noted. Such osteosarcoma transplants exhibited only one tumor cell type, whereas in the original sarcoma fragments cells of varied differentiation had been detected (Fig. 5a, b). This tendency to uniformity was associated with a substantial loss of intercellular substance.

In seven cases we could compare morphology and growth characteristics of individual bone sarcomas transplanted into nude mice prior to and following preoperative chemotherapy (Table 2). Transplants from two sarcomas failed to grow although intact tumor cells had been implanted as revealed by light and electron microscopy. In four cases sarcoma fragments obtained from both biopsy and resected primary were found to grow. Lag phase and doubling time of implants varied between 10 and 38 and 7 and 42 days respectively.

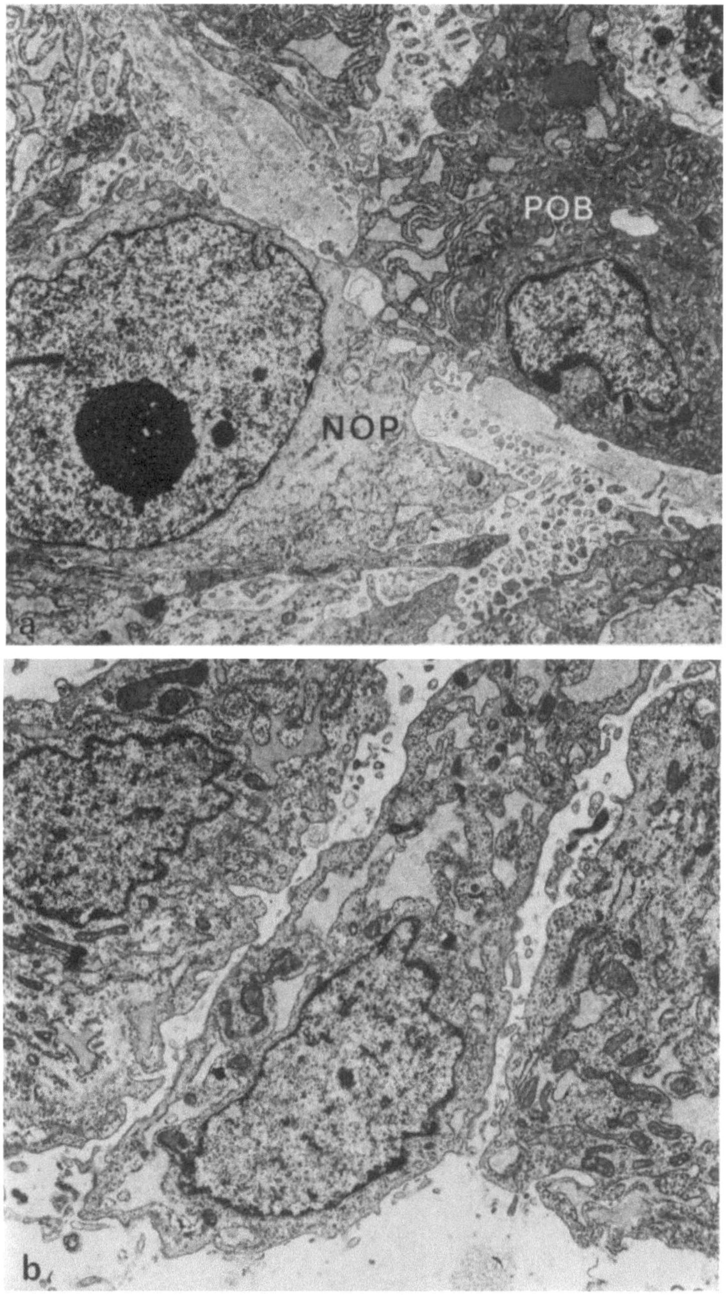

Fig. 5a, b. Ultrastructure of osteosarcoma K. P. **a** Original tumor displaying neoplastic osteoprogenitor cells *(NOP)* and preosteoblasts *(POB)*, (× 7,500). **b** Nude mouse passage 7: the transplanted tumor consists only of preosteoblastic sarcoma cells. Note the reduced amount of intercellular substance, (× 8,400)

Table 2. Growth characteristics of human bone sarcomas transplanted into nude mice before and after preoperative chemotherapy

Patient	Source of tumor samples	Nude mouse transplants (> 4 passages)	
		Lag phase (days)	Doubling time (days)
W. M. (OS)	B	No growth	
	RP	No growth	
K. D. (OS)	B	No growth	
	RP	No growth	
P. V. (OS)	B	17	20
	RP	Only necrotic tissue implanted	
K. P. (OS)	B	12	7
	RP	16	8
S. H. (MFH)	B	30	20
	RP	25	17
K. D. (OS)	B	10	20
	RP	20 (!)	42 (!)
W. R. (OS)	B	20	15
	RP	38 (!)	25 (!)

OS, osteosarcoma; *MFH,* malignant fibrous histiocytoma; *B,* biopsy (before preoperative chemotherapy); *RP,* resected primary tumor (after preoperative chemotherapy)

Table 3. Growth characteristics of human bone sarcomas transplanted repeatedly into nude mice during course of chemotherapy

Patient	Source of tumor samples	Nude mouse transplants (> 4 passages)	
		Lag phase (days)	Doubling time (days)
J. B. (OS)	B	No growth	
	RP	No growth	
	M_1	No growth	
T. C. (OS)	RP	12	8
	LR	Only necrotic tissue implanted	
Z. W. (OS)	M_1	No growth	
	M_2	30	20
	M_3	25	18
G. H. (OS)	LR_1	No growth	
	LR_2	32	20
	M_1	30	22
R. M. (OS)	RP	20	28
	LR	24	26
K. H. (OS)	RP	25	30
	M_1	30	35

OS, osteosarcoma; *B,* biopsy of primary tumor; *RP,* resected primary tumor; *LR,* local recurrence; *M,* metastasis

Two sarcomas (K. P. and S. H.) showed no differences in growth characteristics before and after chemotherapy. In two other cases (K. D. and W. R.) lag phase and doubling time increased significantly following preoperative chemotherapy. The morphological appearance of individual bone sarcomas transplanted before and after chemotherapy remained unchanged.

In addition, we studied the biology of human bone sarcomas transplanted repeatedly into nude mice during the course of therapy (Table 3). One tumor did not grow in the recipients, independent of whether fragments were obtained from biopsy, resected primary or metastasis. In four additional cases transplanted sarcoma fragments from at least two different sources were found to grow. Both morphology and growth characteristics of individual tumors remained constant.

Attempts to correlate morphology and growth characteristics of transplanted bone sarcomas with clinical data (tumor regression and relapse-free and overall survival) were so far unsuccessful. Fast growing sarcomas in the nude mouse model – characterized by short lag phase and doubling time in the animals – did not necessarily correspond to a rapid growth in the patient and vice versa. In cases K. P. and W. R. the observed changes of growth characteristics after chemotherapy were not reflected by a different clinical course.

In Vitro Studies

Only few experimental studies on the in vitro sensitivity of human bone sarcomas against anticancer drugs are presently available. One of the major reasons for this paucity of data is related to the difficulty of preparing single tumor cell suspensions. Sarcoma cells are usually surrounded by a variable amount of intercellular substance, making cell isolation by mechanical or enzymatic procedures very difficult. In addition, the individual culture conditions of the sarcoma cells may vary greatly from tumor to tumor.

To overcome some of these difficulties, we have used a stepwise approach testing several isolation procedures on sarcoma fragments obtained from different sources. The prepared single-cell suspensions were subsequently cloned in the semisolid agar system using the Hamburger-Salmon technique. The inhibitory effect of methotrexate, doxorubicin, *cis*-platinum, etc. was determined but will not be reported in detail in the present study.

Initially, we tried to isolate sarcoma cells from fresh human tumors (resected primary tumor or metastasis) by a monoenzymatic procedure (Table 4). For this purpose, minced sarcoma fragments were treated either with 0.25% trypsin or 0.2% collagenase for 30 min at 37° C. In the majority of samples plated the number of colonies formed was too low for drug sensitivity testing.

On the basis of the observation that the amount of intercellular substance is often reduced in higher nude mouse passages we then used xenogenic transplants as a source of tumor cells. By monoenzymatic cell isolation, colony formation was generally improved but only 30% of sample plated could be used for drug testing. Thereafter, we tried to produce permanent cell lines prior to the testing in the clonogenic assay. Three sarcoma cell lines were successfully established in a monolayer system. Cytogenetic examination confirmed the human origin of the cells. In the semisolid agar assay the cloning efficiency of these sarcoma cells was high, in the order of 5%, and 77% of samples plated were adequate for drug testing.

Table 4. Rate of colony formation

Source of specimens	No. of tumors	Colony formation[a]/ no. of samples plated	Adequate samples[b]/ no. of samples plated
Fresh human tumor			
Monoenzymatic cell isolation	5	2/6 (35%)	1/6 (16%)
Nude mouse transplants			
Monoenzymatic cell isolation	14	17/27 (62%)	5/17 (30%)
Established cell lines	3	11/14 (92%)	7/9 (77%)
Multienzymatic cell isolation	5	4/5 (80%)	4/5 (80%)

[a] > 5 colonies
[b] > 30 colonies in control plates

Since developing permanent cell lines is time-consuming and work-intensive, we have looked for a straightforward procedure. Thus sarcoma fragments obtained from nude mouse transplants were subjected to multienzymatic digestion modified according to Trechsel et al. (1982). Briefly, minced tumor fragments were consecutively treated with 0.1% hyaluronidase (Sigma, type I-S) for 15 min at room temperature, 0.25% trypsin (Flow) for 30 min at 37° C, 2% dispase (Boehringer) overnight at 37° C, and 0.2% collagenase (Sigma, type I) for 6 h at 37° C. After each step, single cells were harvested, seeded into plastic bottles (25 ml) and allowed to settle at 37° C using IMDM medium (Gibco) supplemented with 10% fetal calf serum and antibiotics. After 1 day, media were changed in order to remove blood cells. Adherent cells were trypsinized after 1–2 weeks and subpassaged in a ratio of 1 : 2 or 1 : 3. Follwing in vitro passage two or three sarcoma cells were processed for the clonogenic assay. By this procedure, four out of five tumors had high cloning efficiencies sufficient for chemosensitivity testing.

Our experience clearly indicates that drug testing on human sarcoma cells requires considerable preparatory procedures to obtain single-cell suspensions. The use of several enzymes in a defined order has yielded promising results on fragments explanted from nude mice. It is our purpose now to apply this approach to fresh human bone sarcomas with the intent of obtaining satisfactory single-cell preparations for subsequent cloning and drug testing.

Conclusions

First, we were able to confirm the fact that considerable necrosis in osteosarcomas following chemotherapy has impact on the course of the disease. In our study, patients with extensive necrosis manifested better overall and relapse-free survival. Secondly, we were able to achieve high take rates of sarcoma fragments transplanted into pre-irradiated nude mice. Although we have evaluated only a limited number of experiments so far, take rates in the nude mouse model and extent of necrosis seem to be inversely related. Thus our findings in the mouse model appear to confirm our pathologist's evaluations.

In the preliminary analysis, lag phase and doubling time of the transplanted tumors after four mouse passages did not correlate to clinical parameters such as overall and disease-free survival or the clinically observed growth rates of tumors in the respective patients. This lack of correlation may be due to the paucity of cases, but biological modification of the tumor during various mouse passages must also be taken into account, and the observed morphological alterations may well be an expression of this. The mouse model, on the other hand, has the advantage of providing us with sufficient material to perform clonogenic assays with sarcoma cells, especially, when multienzymatic cell separation techniques are used. It may thus open a new avenue for research into these tumors.

References

Ayala AG, Mackay B,, Jaffe N et al. (1980) Osteosarcoma: the pathological study of specimens from en bloc resection in patients receiving preoperative chemotherapy. In: Van Eys J, Sullivan MP (eds) Status of curability of childhood cancers. Raven, New York, p 127

Groscurth P, Huizink L, Cserhati M, von Hochstetter A, Hofmann V, Honegger HP (1983) Xenogeneic transplantation of human bone sarcomas before and after preoperative chemotherapy. In: Sordat B (ed) Immune-deficient animals in experimental research. Karger, Basel, p 34

Honegger HP, Cserhati M, von Hochstetter A, et al. (1983) Zürcher Erfahrungen mit präoperativem, hochdosiertem Methotrexat bei Osteosarkomen. Schweiz Med Wochenschr 113: 663

Huvos AG, Rosen G, Marcove RC (1977) Primary osteogenic sarcoma. Pathologic aspects in 20 patients after treatment with chemotherapy, en bloc resection and prosthetic bone replacement. Arch Pathol Lab Med 101: 14

Jaffe N, Frei E, Traggis D et al (1974) Adjuvant methotrexate and citrovorum factor treatment of osteogenic sarcoma. N Engl J Med 291: 994

Kaplan EL, Meier P (1958) Nonparametric estimation from incomplete observations. J Am Statist Assoc 53: 457

Peto R, Pike M, Armitage P et al. (1977) Design and analysis of randomized clinical trials requiring prolonged observation of each patient. II. Analysis and examples. Br J Cancer 35: 1

Rosen G, Murphy ML, Huvos AG et al. (1976) Chemotherapy, en bloc resection and prosthetic bone replacement in the treatment of osteogenic sarcoma. Cancer 37: 1

Rosen G, Marcove RC, Caparros B et al. (1979) Primary osteogenic sarcoma. The rationale for preoperative chemotherapy and delayed surgery. Cancer 43: 2163

Rosen G, Caparros B, Huvos AG et al. (1982) Preoperative chemotherapy for osteogenic sarcoma. Cancer 149: 1221

Sharkey F, Fogh J, Hajdu S, Fitzgerald P, Fogh J (1983) Experience in surgical pathology with human tumor growth in the nude mouse. In: Fogh J, Giovanella B (eds) The nude mouse in experimental and clinical research. Academic Press, New York, p 187

Sweetnam RA, Knowelden J, Seddon H (1971) Bone sarcoma: treatment by irradiation, amputation and combination of the two. Br Med J 2: 363

Torhorst J, Tao TW, Floersheim GL (1976) Menschliche nicht-epitheliale Tumoren bei „nude"-Mäusen: Morphologie und Wachstumscharakteristika. Schweiz Med Wochenschr 106: 757

Trechsel U, Dew G, Murphy G, Reynolds JJ (1982) Effects of products from macrophages, blood mononuclear cells and of retinol on collagenase secretion and collagen synthesis in chondrocyte culture. Biochim Biophys Acta 720: 364

Urist M, Grant T, Lindholm TS, Mirra J, Hirano H, Finerman G (1979) Induction of new-bone formation in the host bed by human bone-tumor transplants in athymic nude mice. J Bone Joint Surg 61A: 1207

Predictive Tests for Hematological Malignancies

*Experimental Approaches to Outcome Prediction in Acute Myeloblastic Leukemia**

E.A. McCulloch

The Ontario Cancer Institute, University of Toronto,
500 Sherbourne Street, Toronto, Ontario M4X 1K9, Canada

Introduction

The practice of oncology depends upon accurate staging, the process that relates the initial clinical and laboratory assessment of a patient to the subsequent history of the disease. Staging not only provides the basis for prognosis but also allows patient populations to be stratified in clinical trials. Usually, staging has statistical validity; that is, predictions of outcome are applicable to populations of patients but not to individuals. A goal of research is to develop methods sufficiently sensitive to permit such predictions.

At present, the major practical applications of staging are twofold. First, the design of clinical trials depends upon the availability of comparable groups of patients; these may then receive alternate treatments and, if differences in outcome are observed, such differences can be attributed to therapy. Comparable patient groups can be achieved by randomization; but even in randomized trials it is usual to provide evidence that differences between the groups other than treatment are not confounding the interpretation of the data. Second, staging is essential to the treatment of many forms of cancer; thus the decision to rely on local treatment alone or to use systemic therapy may be based upon the physical extent of disease.

Staging usually depends upon methods that estimate the anatomic extent of disease or the total burden of tumor cells. The former is achieved by advanced imaging techniques and exploratory surgery, the latter usually by measuring a tumor product in blood or body fluids, such as carcinoembryonic antigen (CEA) (Gold and Freedman 1965). Recent advances in tumor biology have suggested novel approaches at the cellular level; the importance of heterogeneity has become apparent and potential methods of exploiting it have been suggested. The development of monoclonal antibodies and molecular probes of the genome provide promising new avenues to explore.

In this paper two of these approaches will be considered: (a) the organization of tumor populations based upon proliferative potential, and (b) the determination of cellular phenotypes by immunologic methods. The hematologic malignancies of man, and particularly acute myeloblastic leukemia (AML), will be used as the source of most of the examples, but the problems of technology and interpretation encountered in this context are relevant to the study of many human cancers.

* The work reported in this paper was supported by the National Cancer Institute of Canada, the Medical Research Council of Canada, and the Ontario Cancer Treatment and Research Foundation

Stem Cells and Clones

At the time of diagnosis each patient with a hematologic cancer harbors a single malignant clone; that is, all of the leukemic cells are derived from the same ancestor (Fialkow 1980). This ancestor had the capacity to give rise to descendants with properties similar to, or identical with, its own. This capacity for self-renewal defines stem cells and endows each of them with the potential to initiate and maintain a clonal population, independent of recruitment from other cellular sources.

Analysis of clonal populations permits deductions about the nature of the stem cells from which they derive. In the case of the hematologic malignancies, the three lineages of myelopoiesis – granulocytopoiesis, erythropoiesis, and megakaryocytopoiesis – can often be identified as members of single malignant clones. Thus the stem cells have the property of differentiation along each of these three lineages, and may be considered to be pluripotent. The evidence is most convincing for chronic myeloblastic leukemia (CML) (Fialkow et al. 1967). Polycythemia vera (P-vera) (Adamson et al. 1976) and idiopathic myelofibrosis (IMF) (Jacobson et al. 1978), not traditionally considered to be leukemias, have been shown to have biological characteristics similar to those of CML; in these diseases, hemopoietic cells are clonal and multilineage in composition. Although hematologic progenitors not belonging to abnormal clones may be identified in patients with CML (Cunningham et al. 1979), and in patients with P-vera (Prchal et al. 1978) after radical chemotherapy or by means of culture techniques (see below), these normal progenitors do not contribute mature descendants to the populations of blood cells; the presence of the malignant clone appears to suppress their capacity for growth and differentiation, although the fibrous tissue associated with hemopoiesis may be stimulated to proliferate (myelofibrosis).

Thus apparently dissimilar diseases – CML, P-vera, and IMF – have properties in common, which may be listed as follows:

1. Clonal dominance
2. Cellular interaction leading to:
   a) Suppression of normal clones
   b) Myelofibrosis
3. Clonal progression.

These may be more significant biologically than the clinical features that serve to identify CML, P-vera, and IMF as separate entities. In particular, there is little justification for considering CML to be a leukemia but classifying P-vera and IMF as benign proliferative disorders of myelopoiesis. Rather these disorders should be grouped together; the term "clonal hemopathy", suggested in 1978, is an appropriate designation for them (McCulloch and Till 1977).

Controversy exists as to whether AML should be classified as a clonal hemopathy. A problem arises from the demonstration of heterogeneity among malignant clones in AML (Fialkow et al. 1979, 1981). In elderly patients erythropoiesis as well as granulopoiesis has been traced to a single leukemic progenitor; in contrast, in children and young adults blast populations and mature granulocytes have been shown to have a common origin, but they coexist, particularly in remission, with erythropoietic cells belonging to different and presumably normal clones. These data might be interpreted to mean that some AML clones originate in cells more differentiated than the pluripotent stem cells whose transformation yields the clonal hemopathies. However, a plausible alternative explanation exists; that either as a consequence of leukemic transformation or because of posttransformational regulatory events, leukemic pluripotent stem cells in AML may lose the capacity for erythropoiesis or may fail to express it. The dilemma illustrates a problem in clonal analysis generally. If a clonal population can be shown to have certain properties (for example, to contain cells of a recognizable differentiated phenotype), the cell of origin may properly be considered capable of yielding progeny of that phenotype. However, a failure to detect phenotypic evidence for differentiation capacity is negative evidence; it cannot be used confidently to define limitations to the potential of stem cells in malignant clones. It appears reasonable, therefore, to include at least some cases of AML among the clonal hemopathies, and it may be that all cases should be so classified.

Cellular Organization and Treatment Goals

The nature of malignant cell populations in individual patients has important consequences for clinical staging and the establishment of treatment goals. A model of cellular organization is needed before biological features can be tested for their contribution to the variance in clinical outcome.

Major features of the normal organization of myelopoiesis are maintained in the clonal hemopathies. The subject has been reviewed extensively elsewhere (Till and McCulloch 1980; McCulloch 1983); the major features are shown in Fig. 1. Normal and malignant myelopoietic clones originate in pluripotent stem cells; these may either undergo renewal or, alternatively, become committed to either erythropoiesis, granulopoiesis, or platelet formation. This second alternative, shown in the figure as determination, has a number of biological consequences. These include (a) the beginning of cellular diversity as the composition of expanding clones becomes increasingly heterogeneous; and (b) a limitation to proliferation, since following determination, progenitors committed to a specific lineage are capable of only a limited number of divisions. Thus each myelopoietic clone contains the cells of three general categories. First, a tiny minority are stem cells; but these are essential to the continuing existence of each clone. The second class consists of lineage-committed progenitors and their descendants undergoing terminal divisions. Thirdly, mature red cells, granulocytes and platelets, the majority of populations within the clones, are incapable of division.

The clinical diagnoses grouped together as clonal hemopathies are made on the basis of the cellular compositions of the clones in individual patients. In CML, P-vera, and IMF individual mature blood cells show little or no deviation from normal morphology and function. However, hemopoiesis as a whole is abnormal, often ineffective. A cellular abnormality has been identified in CML and P-vera; unlike normal pluripotent stem cells (Fig. 1), (CFU-GEMM) in CML and P-vera are found to be predominantly in cell cycle, as judged by their sensitivity to exposure to high-specific-activity tritiated thymidine (Messner and Fauser 1980). This mechanism would explain overproduction of cells in these diseases. However, the diagnostic abnormal distribution of mature elements is not explained simply

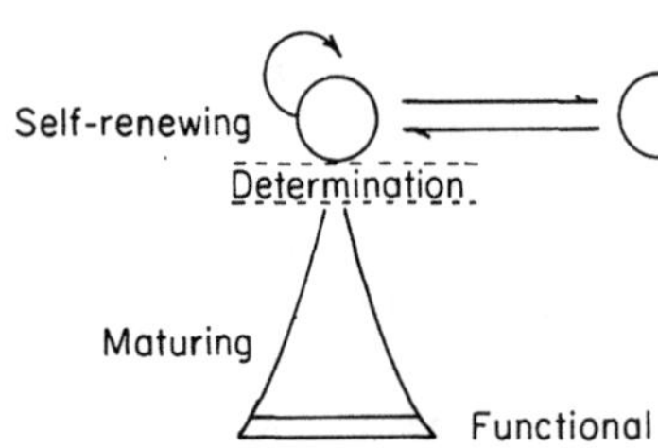

Fig. 1. A model of normal myelopoiesis. Reprinted from McCulloch (1983) by permission of the publisher

by stem cells having a low probability of entering a resting stage. In principle, the problem may be approached by considering the differentiation program that cells follow from their origins as stem cells to their terminations as mature elements. Each functional cell is the product of one such program; but, during the course of clonal expansion, each cell division may be regarded as the origin of a new program. The diversity that follows determination has already been emphasized, but even within a lineage, marked heterogeneity is found when populations are analyzed that appear to be at a similar stage in differentiation. For example, progenitors that are grouped together on the basis of their capacity to form colonies in culture are very heterogeneous by other criteria. This heterogeneity may be interpreted to mean that differentiation program within a lineage may vary with respect to the number of components they contain and the time required to pass through each maturation step. This analysis leads to the view that lineages might be considered to consist of a number of different programs, some commoner than others, but each considered normal because its end point is a functional cell (McCulloch 1983). Abnormal myelopoiesis, then, might be the consequence of a relaxation of the control that limits programs to those that are efficient. Indeed, inefficient programs might occur normally but would not affect function if they were rare. The polyclonal nature of normal hemopoiesis would serve to minimize the effect of such inefficient programs. In contrast, clonal myelopoiesis in disease would serve to emphasize inefficient programs and to allow them to be recognized as abnormal. A test of this hypothesis would require technology for following individual cells from their origin to complete maturity. At present, only a few methods are available. For example, the production in culture of colonies of mature lineage-specific descendants might reflect the expression of certain programmatic components. A few monclonal antibodies are available with lineage specificity; these may identify programmatic segments of varying lengths. However, in general, the ordering and timing of expression of immunologically defined markers is not known with precision.

Cellular organization within malignant clones bears directly upon the establishment of treatment goals (McCulloch and Buick 1982). If the goal is cure, the target for therapeutic intervention must be stem cells; if these are eliminated or sterilized their descendants will exhaust their limited capacity for growth and the malignant population will disappear. Thus the number of stem cells in a clone is an important determinant of treatment outcome. Indeed, stem cell scarcity in most malignant populations makes understandable the success of therapy using cytotoxic agents that could not be expected to destroy all the malignant cells (Bush and Hill 1975). The value of certain transplantable tumor lines for the study of chemotherapy can be appreciated for the same reason; for these, including L1210, the population is made up almost entirely of stem cells; hence a single viable tumor cell of such a line can produce a lethal tumor in a compatible host. When drugs are tested on such lines, it is their effects upon stem cells that are observed; thus information on cytotoxicity is obtained that may be relevant to the appropriate cellular target.

Treatments designed to control, rather than cure, malignant growth may also be targeted toward stem cells; reducing their number may lead to a shrinkage in total tumor mass and a reduced growth rate. However, there are cells capable of colony formation (clonogenic cells that lack renewal capacity). The committed progenitors of myelopoiesis are examples; these may be targets for treatment. CML provides a useful example of the concept; ususally this disease is managed with conservative chemotherapy regimens that have been shown not to eliminate the malignant clone (Whang et al. 1963). Nonetheless, these treatments lead to reduced production of granulopoietic elements and increased erythropoiesis. Some of the beneficial effects may arise from a reduction in the stem cell population; then, if efficient programs were commoner than inefficient ones, improved myelopoiesis would be observed during the early stages of clonal reexpansion. In addition, it may be that some of the improvements are achieved at the level of committed progenitors; for example, exposure to low concentrations of chemotherapeutic agents such as 6-mercaptopurine might convey a selective advantage upon erythropoietic as compared to granulopoietic progenitors.

Finally, end stage cells may be the appropriate targets if the goal of therapy is to achieve a rapid improvement. The most obvious example of this approach is the use of plasmaphoresis to reduce very high leukocyte concentrations in certain patients with leukemia. Small doses of radiation together with steroids as an emergency treatment of mediastinal obstruction by lymphoma may be effective by direct lytic action on lymphocytes. It is not only emergencies that can be managed by reducing the number of mature cells; repeated phlebotomy remains an effective treatment for P-vera, one that spares patients exposure to potentially toxic chemotherapy.

Clonogenic Blast Cells in Acute Myeloblastic Leukemia

Acute myeloblastic leukemia differs from other clonal hemopathies because of the presence within the malignant clones of large numbers of cells without obvious differentiation; these are called "blasts" because of their primitive morphological appearance. Examination of the nature of the blast population provides a way of testing some of the views of cellular organization described earlier.

The AML blast population contains cells with capacity to give rise to colonies in culture (Buick et al. 1977). Procedures for detecting colony-forming blast cells have features in common with many assays for clonogenic cells. Usually, the populations under examination are immobilized in viscid or semisolid media. In our laboratory, methyl-cellulose is used for this purpose and is combined with fetal calf serum and α-minimal essential medium as sources of essential nutrients. Often clonogenic assays require the presence of special stimulators; both blast progenitors and cells capable of giving rise to multilineage hemopoietic colonies depend for growth upon the presence in the cultures of media conditioned by human leukocytes in the presence of phytohemagglutinin (PHA-LCM). The method is designed to select for blast colony formation in two ways. First, peripheral blood cells from patients with AML are used; T-lymphocytes, a population capable of colony formation, are removed by exploiting their capacity to form rosettes of sheep erythrocytes; these can be separated by centrifugation through Ficoll Hypaque (Minden et al. 1979). Following this procedure few, if any, of the mononuclear cells except blasts are capable of growth. PHA-LCM is the second source of specificity; media conditioned by leukocytes without the lectin (LCM) are ineffective. Nonetheless, the blast nature of the colonies must be confirmed by morphological examination and, where feasible, by seeking metaphases with chromosomal markers characteristic of the patient from whom the cells were obtained (Izaguirre and McCulloch 1978).

Similar approaches may be used to obtain colonies of other specificities. For example, marrow, rather than blood, is the appropriate starting material for cultures designed to select for myelopoietic progenitors; effective stimulator combinations are available for the cells of origin of granulopoietic (Pluznik and Sachs 1965; Bradley and Metcalf 1966; Senn et al. 1967; Robinson et al. 1971), erythropoietic (Stephenson et al. 1971; Heath et al. 1976; Tepperman et al. 1974; Iscove et al. 1974), or mixed lineage colonies (Johnson and Metcalf 1977; Fauser and Messner 1979; Nakahata and Ogawa 1982). If the objective is to obtain clonogenic cells from solid tumors (Salmon et al. 1978; Selby et al. 1983), biopsies or surgical specimens would appear to be the best source of cells; in practice, difficulties are often encountered in obtaining adequate single-cell suspensions from solid tumors; sometimes this difficulty can be avoided by using abdominal or thoracic effusions containing malignant cells. Regardless, it is essential to distinguish between colonies of malignant origin and others developed from coexisting normal populations.

Once colony formation has been achieved and the colonies shown to be derived from single cells, several different kinds of information can be obtained. Most obvious is that derived by enumerating colony number and calculating the plating efficiency (PE). However, so many factors of little biological importance influence PE that its value is limited. For example, problems may be encountered in sampling, in preparing suitable cell suspensions, and in determinating the most appropriate culture conditions for a particular cell class. The clonal nature of the hemopoietic malignancies has been shown to complicate further the interpretation of PE values of myelopoietic progenitors. These values have been found to vary greatly from patient to patient; but the variation appears to derive from stochastic processes occurring during the expansion of individual clones rather than to reflect primary characteristics of the malignancies in each patient (Till et al. 1974). Many solid tumors are also clones, and in them tumor to tumor variation might have a similar origin.

It has often proved more valuable to use colony-forming methodology to determine the properties of clonogenic cells than to measure PE. It is also feasible to examine cellular events occurring during colony formation. Of particular importance is the determination whether or not clonogenic cells are capable of self-renewal, since it is this property that distinguishes stem cells from progenitors with limited growth potential. In the instance of clonogenic blast cells self-renewal was determined directly by replating cells from blast colonies or by pooling the cells from entire dishes (Buick et al. 1979b). By either method secondary colonies were obtained in approximately 70% of cases. However, each of the procedures yielded different kinds of information. The pooled colony method provided a quantitative estimate of the average self-renewal capacity of the blast progenitor population in each patient; the value was expressed as the PE of the pooled colony suspension (secondary plating efficiency or PE2). By separating the cells on the basis of their sedimentation velocity, subpopulations of clonogenic blast cells could be demonstrated that varied in their PE2 values (Chang et al. 1980). This finding is compatible with a hierarchical arrangement of self-renewing colongenic blast stem cells, based on the extent of their capacity for self-renewal.

Replating cells from individual blast colonies revealed a marked colony to colony variation; few clonogenic cells could be demonstrated in most colonies, while in only a few were they plentiful. The distribution of new clonogenic cells among blast colonies was similar to the distribution of new stem cells in spleen colonies (Siminovitch et al. 1963); in the latter instance, it was proposed that the asymmetrical form of distribution had a stochastic basis, where self-renewal ("birth") and determination ("death") occurred at random but with definite probabilities (Till et al. 1964). A similar interpretation might be applied to the results obtained by replating blast colonies. Such a view would extend the concept of a hierarchical arrangement in the blast population; in addition to suggesting that blast stem cells might be ranked in order of decreasing self-renewal capacity, a death process might be postulated; the existence of a death possibility is consistent with the low values of PE2 (usually $< 1\%$) found in the measurement of self-renewal using the pooled colony technique.

The data are interpreted to indicate that the blast population is organized like a lineage with stem cells, progenitors capable of limited growth, and proliferatively inert end cells. It should be emphasized, however, that the experimental distinction between the first two classes is only approximate; if, as postulated, birth and death occur at random, some stem cells might yield colonies that could not be replated successfully because, by chance, determination events occurred early during the process of colony growth.

The application of clonogenic assays to solid tumors also requires that a distinction be made between colonies derived from stem cells and others from progenitors of limited proliferative potential. Replating may not be feasible in all instances for technical reasons; other approaches to the problem are needed urgently. One such has been suggested recently by MacKillop et al. (1983); these investigators have suggested that the distinction might be made on the basis of colony size, since only progenitors undergoing renewal could be expected to give rise to large colonies. But other approaches require study; for example, the precedent of myelopoiesis would suggest that stem cells might have different culture requirements from those of lineage-restricted progeny. If similar observations could be made on cells from solid tumors, assay conditions selective for tumor stem cells might be devised.

Many other classes of progenitor properties may be determined using colony assays. These include such physical characteristics as size and density or the proportion of cells in the DNA synthetic phase of the cell cycle. Recently, much interest has been focused on the use of clonogenic assys to determine the sensitivity of progenitor cells to chemotherapeutic agents (Salmon et al. 1978). Usually, cell suspensions are exposed to various concentrations of the drug, either briefly before plating or continuously during cloning formation. The effect of the drug is evaluated by measuring reduction in colony formation

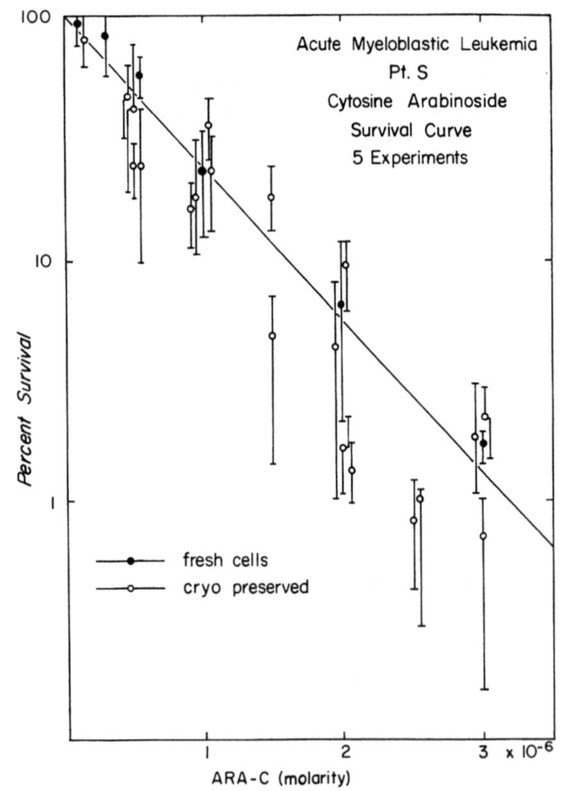

Fig. 2. The sensitivity of blast progenitors to continuous exposure to cytosine arabinoside *(ARA-C)*. *Closed symbols,* fresh specimen; *open symbols,* four repeated experiments with cryopreserved cells. *PtS.,* patient designation

as a function of dose. Figure 2 is an example of the procedure; blast cells were exposed continuously to increasing concentrations of the drug cytosine arabinoside (ara-C) and colonies were counted after 7 days of incubation. A simple negative exponential relationship was found between increasing dose of drug and colony survival. The figure shows the results of five replicate experiments, the first on fresh cells and the four repeats on cells preserved by freezing at −70° C in dimethyl sulfoxide. The data in Fig. 2 show that reproducible drug response curves could be obtained when cells from a single source were tested repeatedly. This control is important when it is considered that marked variation is found when cells from different patients are tested for drug sensitivity.

Lineage Infidelity in Acute Myeloblastic Leukemia

The concept that blast cells are organized as a hierarchy, where loss of renewal capacity is followed by terminal divisions, suggests a comparison with normal myelopoietic lineages. In these, terminal divisions yield functionally differentiated cells; proliferatively inert blasts might be considered analogous even though morphological maturation has not occurred. Indeed, we have proposed that blast cells should be considered as a lineage in which cells follow abnormal differentiation programs (McCulloch et al. 1978).

This model of AML blasts differs from those suggsted by other authors, who have postulated that the morphology of blasts in the consequence of a block in differentiation rather than the termination of an abnormal program. Models based upon blocked, truncated, or frozen differentiation find their major support from two sources. First, certain leukemic cell lines proliferate and maintain an undifferentiated morphology, but when suitable inducers are added apparently normal differentiation can be observed. Friend cells (Friend et al. 1971), the murine myeloid leukemias described in detail by Sachs (1980), and the human promyelocytic line HL60 (Rovera et al. 1979) all have these characteristics. These observations are interpreted to mean that blocks exist because they can be overcome. Second, studies of the cellular phenotypes of acute lymphoblastic leukemia (ALL) have disclosed populations in individual patients that appear to correspond with stages in normal B cell development (Greaves 1982; Seligmann et al. 1981). Thus a patient whose lymphoblasts contained cytoplasmic immunoglobulin might be considered to bear a malignant clone with differentiation "frozen" at the pre-B-cell stage.

An assumption usually included in models based upon blocked differentiation is that hemopoietic lineages are followed faithfully; that they do not reach their normal conclusion only because of an abnormal stop signal. At the molecular level, in this context, any genes that are expressed should be expressed normally. Thus the major difference between blocked differentiation models and the blast lineage hypothesis rests upon the nature of the majority of the blast cells and the differentiation programs leading up to them. The model proposed here requires that those programs be abnormal and that blast cells that failed to proliferate in culture be inert because of mechanisms similar to those associated with normal determination (Fig. 1).

This reasoning provides a motivation for describing the differentiation programs of leukemic blasts. One approach is to use markers that have been associated with either differentiation stage or lineage and to seek for order in their expression during blast cell growth. Experiments of this design were conducted by Marie et al. (1981b); these studies showed changes in antigenically defined (My-1) and histochemically detected (NASD) markers of granulopoiesis occuring during the growth of blast colonies. The data provide convincing evidence that programs are followed as blastic subclones expand. Further, differences were identified between the developmental patterns of blast colonies and normal granulopoietic colonies. These differences might be evidence for abnormal blast programs. However, the data were obtained in a small number of experiments: nor were the results in the blast colonies uniform. It remains possible that the kinetics of marker change in blast colonies described rare normal programs rather than abnormal ones.

A second approach was based on a concept of the nature of determination. It is accepted widely that normal postdetermination programs are lineage specific. That is, markers characteristic of only one lineage will be found on maturing cells. This view is the basis for the concept of lineage fidelity in leukemia (Greaves 1982). The alternate view, that blasts follow abnormal programs, would be supported if markers from different lineages could be identified on individual blast cells. A search for such cells was initiated. In between 30% and 50% of patients with acute leukemia, doubly marked blasts were found (Smith and McCulloch 1981; Smith et al. 1983). A cytoplasmic marker such as spectrin, Factor VIII, or immunoglobulin in combination with a surface marker specific for a different lineage was the commonest finding, since such combinations are technically more easily identified than associations of two surface markers. The frequency of doubly marked cells varied from less than 1% to a majority of the blasts. However, in most instances they were uncommon, although not difficult to identify unequivocally. The finding of doubly marked blast cells, termed "lineage infidelity," has been used to support an abnormal lineage model of leukemic blast cells.

The observation of lineage infidelity in leukemia raises a number of conceptual and technical issues. First, lineage specificity is only accepted widely for the postdeterministic portions of differentiation programs. Till (1976) has suggested that stem cells might express the markers of several lineages, although at low density. If this view were correct, doubly marked blast cells could be considered to be "blocked" at a predeterministic stage of normal development, rather than following abnormally assembled programs. One approach to this issue was to examine normal hemopoietic cells. Using methodologies that easily identified doubly marked leukemic blasts, thousands of cells were examined from normal marrow, regenerating marrow, and multilineage colonies (Smith et al. 1983; Marie et al. 1981b). No example of lineage infidelity was encountered. However, these negative controls might only reflect a quantitative difference in the frequency of doubly marked cells. Indeed, such a difference would be compatible with a block in differentiation with population expansion before the block.

There is, however, evidence that stem cells with B-lymphocyte differentiation potential are null in respect to immunoglobulin gene expression. At the molecular level rearrangements of Ig genes must precede synthesis of their products (Tonegawa 1983). Thus this class of lineage marker would not be found in stem cells with Ig genes in the germ line configuration. Further, to the extent that phenotypes have been obtained for early myeloid progenitors, these have usually consisted of markers without lineage specificity, such as human leukocyte antigen (HLA) or Ia antigen (Fitchen et al. 1982). Finally, in culture early cells appear to respond to stimulators with stage rather than lineage specificity, and might be expected to express receptors for such stimulators (Iscove et al. 1982).

The second issue raised by the finding of lineage infidelity relates to the specificity of immunologically defined markers. Indeed, striking examples of cross-reactivity have been seen. For example, the common acute lymphatic leukemic antigen (cALLa), as recognized by the monoclonal J5, has been found on a small proportion of kidney cells (Jani et al. 1983). Factor VIII, a lineage marker for megakaryocytes, is a component in the blood coagulation cascade and is also found on endothelial cells (Koutts et al. 1980). Specificity for erythropoiesis has been challenged by the finding of small amounts of spectrin in many somatic cells, including fibroblasts (Fitchen et al. 1982). Thus specificity may, for some markers, have a quantitative rather than a qualitative meaning.

Finally, the question may be posed: Why has infidelity not been recognized previously, especially in the numerous studies of ALL? There may be a number of technical explanations; for example, many more reagents, including monoclonals, are now available, increasing the possibility of detecting doubly marked cells. Perhaps of greater importance, a special strategy was devised to detect doubly marked cells because there was a conviction that such cells might exist. The same expectation of success led to the assignment of significance to small numbers of doubly marked cells that might have been dismissed as artifact. Recently, however, other groups have reported the finding of doubly marked cells, especially nuclear terminal deoxynucleotide transferase (TdT), in association with granulopoietic markers (Jani et al. 1983; Bettelheim et al. 1982).

After considering these issues, it still seems reasonable to conclude that the finding of lineage infidelity supports the abnormal lineage model of blasts, although an unequivocal demonstration of its validity is yet to be forthcoming.

The distinction between blocked differentiation and abnormal program assemble as mechanisms generating blast cells has more than theoretical importance. The former view makes it reasonable to search for therapeutic strategies that would improve the production of functional cells in leukemic clones, an objective that has been achieved in experimental systems. In contrast, if the lineage model is correct, increased differentiation might be beneficial by reducing the number of proliferative blasts but would not yield the increase in mature cells that defines remission. Therefore, therapeutic research should, ultimately, be directed towards primary lesions affecting the regulation of gene expression. In the short term elimination of leukemic clones followed by marrow transplantation may be the best way to improve outcome.

Contributions to Outcome Variance

The foregoing discussions of the cellular organization in AML provide a background for the development of staging procedures. At most centers patients with AML are treated with chemotherapeutic protocols developed empirically. These usually include ara-C in combination with an anthracycline; in recent protocols 6-thioguanine has often been included. The cellular morphology of blasts has not proved reliable as a predictor of outcome (van Rhenen et al. 1980). The major diagnostic problem is to ensure that patients with lymphoid leukemia receive prednisone and vincristine, a less toxic from a therapy often effective in ALL. However, the marked heterogeneity among AML patients becomes obvious from their response to standard therapy; some die rapidly; others enter remission which may be of short duration (less than 1 year) or, if long lasting, may be associated with prolonged survival (Curtis et al. 1983). Yet others experience partial responses that permit them to live for months or even years in the presence of morphologically obvious disease.

Many attempts have been made to devise ways of predicting outcome. For example, the metabolism and pharmacokinetics of ara-C vary from patient to patient; measurements of parameters of these have been proposed as predictors when ara-C was used in treatment (Rustum and Preisler 1979). Patient characteristics such as age have been shown to make important contributions to variance in some series, but not in all. As colony assays for hemopoietic cells were devised, these were used in leukemia research; in some instances, using marrow cultured in soft agar over normal leukocyte feeders, abnormal growth patterns were observed with significant predictive power (Moore et al. 1974; Spitzer et al. 1976; Vincent et al. 1977; Richman and Rowley 1983). Thus patients whose marrow failed to grow and those whose marrow cultures contained large colonies of differentiating cells were shown to have a good prognosis; in contrast, patients whose marrow grew predominantly as small clusters or large colonies of undifferentiated cells were less likely to enter remission. These studies provided important encouragement to those using cell culture as an investigative tool. They have, nonetheless, disadvantages; in particular, it was not easy to relate growth patterns to underlying cellular mechanisms. For example, some of the changes might confidently be attributable to the clonal nature of leukemia (see earlier) rather than to specific lesions in the disease. The culture conditions were not selective; they were capable of supporting the growth both of normal granulocytes and blast populations; it was difficult, therefore, to distinguish the effects of leukemia from those of normal myelopoiesis.

The assay for clonogenic blast cells has the advantage of specificity; in addition, it is sufficiently quantitative to be used in determining blast progenitor properties. Both the PE2 and sensitivities to chemotherapeutic agents (doxorubicin hydrochloride ara-C) (Buick et al. 1979a; McCulloch et al. 1982) varied widely from patient to patient but were stable in individual patients as a function fo time. These, therefore, were patient attributes that might contribute to outcome. Results of measuring PE2 and drug sensitivity in over 50 AML patients are given in Table 1. It is evident from the table that the PE2, a measure of the capacity of the blast progenitors to renew themselves, was a significant predictor of both success in remission induction and length of survival; that is, the higher the value of PE2, the worse the prognosis. In contrast, the D_{10} values (drug concentration required to reduce survival to 10%) for doxorubicin hydrochloride and ara-C were not predictive of response, although the patients were treated with these two drugs (McCulloch et al. 1982). The table also shows the significance contribution of age to outcome variation.

Other investigators, using inhibition of colony formation to measure drug sensitivity, have found correlations with clinical outcome. An important contribution is that of Marie et al. (to be published); these investigators found an excellent correlation between doxorubicin

Table 1. Univariate contributions to outcome variation of disease-related, host-related, and treatment-related factors[a]

Attribute	*n*	*p* value	
		Remission[b]	Survival[c]
PE2	53	0.002	0.003
Marrow blasts	52	0.01	0.01
Age	56	0.04	0.016
D_{10} Ara-C	49	0.42	0.46
D_{10} doxorubicin hydrochloride	45	0.68	0.58

Ara-C, cytosine arabinoside; *D_{10},* drug concentration required to reduce survival to 10%; *PE2,* secondary plating efficiency

[a] Data extracted from Table 4, McCulloch et al. (1982)

[b] Logistic regression

[c] Cox regression

hydrochloride and ara-C survival values and response when data from the two measurements of drug sensitivity were considered together. The analysis of their data depended, however, on dividing the patient population into two groups, one considered sensitive and the other resistant. Such a division may be questioned if, as seems to be the case, the distributions of the D_{10} values of the two agents are continuous. The studies of Park et al. (1980) depended on the modification of the usual method for obtaining granulopoietic colony formation; they introduced a daily feeding regimen that appeared to give increased PEs of colonies containing blast cells. They examined only nine patients; by combining results on a number of drugs they were able to suggest a correlation between outcome and drug sensitivity in this small series. Preisler (1980) also reported a correlation, but he did not regularly construct complete survival curves; further, his data analysis was complicated by the inclusion of measurements of the in vitro response of cells to tritiated thymidine.

Thus methodological and analytical differences may explain the discrepancy between the data in Table 1 and those obtained by other workers. The problem is made more complicated by the nature of the remission induction process itself; it is not sufficient to reduce the number of blasts: it is also necessary that functional blood cells be produced. A resolution of the discrepancies may be forthcoming if remission induction with ara-C alone, either at high- or low-dose schedule, is accepted as first-line therapy. Under these circumstances, the effects of a single agent could be assessed both in culture and in vivo.

Lineage infidelity provided another disease-related attribute to be tested for predictive strength. If infidelity is the consequence of a fundamental genomic lesion affecting the regulation of gene expression, its presence might be expected to be correlated with aggressive disease. Table 2 shows that, for AML, this is indeed the case; a highly significant correlation was found between failure to achieve remission and the presence of doubly marked blast cells (Smith et al. 1983; McCulloch 1983). Although infidelity was observed in ALL, a clinical correlation was not observed; perhaps this may be attributed to the availability of more effective chemotherapy for ALL than AML. Nevertheless, the striking clinical correlation in AML provides support for the biological significance of lineage infidelity. In this regard, the clinical studies contribute to the validation of the blast lineage model of cellular organization in AML.

Data from many patients are needed to enable the assessment of the relative contributions of PE2 and cell phenotype to outcome. However, it is not unreasonable to expect that these approaches may provide a valid staging procedure in AML. When patients with poor

Table 2. Effect of lineage infidelity on outcome

	Doubles (remission/total)	No doubles (remission/total)	p
AML (n = 26)	1/9	15/17	0.0002
ALL (n = 15)	6/7	5/8	0.34

AML, acute myeloblastic leukemia; *ALL*, acute lymphoblastic leukemia
Reprinted from McCulloch (1983) by permission of the publisher

prognosis are identified, they become an important target population for testing new therapeutic modalities. For them new methods may be used ethically since standard therapy has little to offer; and since a low response rate might be expected, improved efficacy of a new treatment might be obvious from the treatment results in a small number of patients. In AML, as in other tumors, accurate staging may provide for individualization of treatment; even in the absence of new methods some patients may be spared toxic therapy from which they may expect little, if any, benefit.

Conclusion

This paper has emphasized the relationship between cellular organization in cancer and treatment strategy. The view was advanced that biological attributes contribute significantly to the variance in outcome. In contrast, measurements of the sensitivity of clonogenic populations to chemotherapeutic drugs were not shown to have predictive value.

Validated risk factors are helpful in the practice of oncology if they can be used as a basis for staging. However, even very accurate staging is of limited value if it cannot be coupled with therapeutic regimens effective in identified patient subpopulations. Regrettably, drugs active against AML are limited in number and efficacy. For most solid tumors even less success has been achieved; in many instances, response can only be detected as a reduction in tumor mass. It may be that present methods of drug development will provide clinicians with more powerful tools. Alternatively, scarce research resources might be deployed effectively in basic studies of the primary molecular and cellular lesions in human cancers. Data obtained in this way might be of practical value. Recent laboratory advances provide reasoned hope that the genome itself may be manipulated advantageously. For example, 5-azacytidine, a drug that affects gene expression, has been used in the treatment of hemoglobinopathies (Ley et al. 1982). Developments in cell biology may disclose regulatory mechanisms that can be exploited to increase the probability of stem cell death and thereby limit the growth of malignant clones. In the more immediate future, advances in transplantation make this modality an attractive treatment option for some disease; but in choosing potential recipients methods that separate good from bad risk patients are needed urgently.

Finally, those disappointed by the limited value of clonogenic assays in the selection of chemotherapeutic regimens may be encouraged by an historical precedent. From their discovery until the mid-1950s, high-energy photons or particles (X-rays, gamma rays, etc.) were thought to be effective in the treatment of cancer because malignant cells were more sensitive than normal cells to their lethal effects. Clonogenic assays (Puck and Marcus 1956) and other measurements of cell proliferation (Hewitt and Wilson 1959; McCulloch and Till 1960) provided experimental systems that were used effectively to prove that all aerated mammalian cells have very similar radiation dose-response curves. This knowledge did not inhibit radiation therapy; rather it gave it vigorous new life. The contribution of cellular heterogeneity in tumors to the response to treatment became known; tumors outside the radiation field or the presence of anoxic cells, rather than cellular radioresistance, were identified as significant causes of local failure. With these and other leads, research proceeded rapidly and on rational grounds. A similar surge forward might be anticipated as medical oncologists begin to appreciate the new insights coming from research into the cellular and molecular biology of neoplasia.

References

Adamson JW, Fialkow PF, Murphy S, Prchal JE, Steinmann L (1976) Polycythemia vera: stem cell and probable clonal origin of the disease. N Eng J Med 295:913–916

Bettelheim P, Paiatta E, Majdic O, Gadner H, Schwarzneir J, Knap W (1982) Expression of a myeloid marker on TdT-positive acute lymphocytic leukemic cells: evidence by double-fluorescence staining. Blood 60:1392–1396

Bradley TR, Metcalf D (1966) The growth of mouse bone marrow cells in vitro. Aust J Exp Biol Med Sci 44:287–299

Buick RN, Till JE, McCulloch EA (1977) Colony assay for proliferative blast cells circulating in myeloblastic leukaemia. Lancet 1:862–863

Buick RN, Messner HA, Till JE, McCulloch EA (1979a) Cytotoxicity of Adriamycin and daunorubicin for normal and leukemic progenitor cells of man. J Natl Cancer Inst 62:249–255

Buick RN, Minden MD, McCulloch EA (1979b) Self renewal in culture of proliferative blast cells circulating in myeloblastic leukaemia. Lancet 1:862–863

Bush RS, Hill RP (1975) Biologic discussions augmenting radiation effects and model systems. Laryngoscope 85:1119–1133

Chang LJA, Till JE, McCulloch EA (1980) The cellular basis of self renewal in culture by human acute myeloblastic leukemia blast cell progenitors. J Cell Physiol 102:217–222

Cunningham I, Gee T, Dowling M, Chaganti R, Bailey R, Hopfan S, Bowden L, Turnbull A, Knapper W, Clarkson B (1979) Results of treatment of Ph[1]+ chronic myelogenous leukemia with an intensive treatment regimen (L-5 protocol). Blood 53:375–395

Curtis JE, Smith L, Messner HA, Senn JS, McCulloch EA (1983) Contribution of lineage-specific markers to outcome of acute leukemia. Abstract, Annual Meeting of the American Association for Cancer Research, San Diego

Fauser AA, Messner HA (1979) Identification of megakaryocytes, macrophages and eosinophils in colonies of human bone marrow containing neutrophilic granulocytes and erythroblasts. Blood 53:1023–1027

Fialkow PF (1980) Clonal and stem cell origin of blood cell neoplasms. In: LoBue J (ed) Contemporary hematology/oncology, vol 1. Plenum, New York, pp 1–46

Fialkow PJ, Gartler SM, Yoshida A (1967) Clonal origin of chronic myelogenous leukemia in man. Proc Natl Acad Sci USA 58:1468–1471

Fialkow PJ, Singer JW, Adamson JW, Berkow RL, Friedman JM, Jacobson RJ, Moohr JW (1979) Acute nonlymphocytic leukemia: expression in cells restricted to granulocytic and monocytic differentiation. N Eng J Med 301:1–5

Fialkow PJ, Singer JW, Adamson JW, Vaidya K, Dow LW, Ochs J, Moohr JW (1981) Acute nonlymphocytic leukemia: heterogeneity of stem cell origin. Blood 57:1068–1073

Fitchen JH, LeFevre C, Ferrone C, Cline MJ (1982) Expression of Ia-like and HLA-A, B antigens on human multipotential hematopoietic progenitor cells. Blood 59:188–197

Friend C, Scher W, Holland JG, Sato T (1971) Hgb synthesis in murine virus-induced leukemic cells in vitro: stimulation of erythroid differentiation by DMSO. Proc Natl Acad Sci USA 68:378–382

Gold P, Freedman SO (1965) Demonstration of tumor specific antigens in human colonic carcinomata by immunological tolerance and absorption techniques. J Exp Med 121:439–462

Goodman SR, Zagon IS, Kulikowski RR (1981) Identification of a spectrin-like protein in non-erythroid cells. Proc Natl Acad Sci USA 78:7570–7574

Greaves MF (1982) "Target" cells, cellular phenotypes, and lineage fidelity in human leukaemia. J Cell Physiol [Suppl 1] 111:113–125

Heath DS, Axelrad AA, McLeod DL, Shreeve M (1976) Separation of the erythropoietin-responsive progenitors BFU-E and CFU-E in mouse bone marrow by unit gravity sedimentation. Blood 47:777–792

Hewitt HB, Wilson CW (1959) A survival curve for mammalian leukaemia cells irradiated in vivo (implications for the treatment of mouse leukaemia by whole-body irradiation). Br J Cancer 13:69–75

Iscove NN, Sieber F, Winterhalter K (1974) Erythroid colony formation in cultures of mouse and human bone marrow: analysis of the requirement for erythropoietin by gel filtration and affinity chromatography on agarose-concanavalin A. J Cell Physiol 83: 309–320

Iscove NN, Roitsch CA, Williams N, Guilbert LJ (1982) Molecules stimulating early red cell, granulocyte, macrophage, and megakaryocyte precursors in culture: similarity in size, hydrophobicity, and charge. J Cell Physiol [Suppl 1] 111: 65–78

Izaguirre CA, McCulloch EA (1978) Cytogenetic analysis of leukemic clones. Blood [Suppl 1] 52: 287 (Abstract)

Jacobson RJ, Salo A, Fialkow PJ (1978) Agnogeneic myeloid metaplasia: a clonal proliferation of hematopoietic stem cells with secondary myelofibrosis. Blood 51: 189–194

Jani P, Verbi W, Greaves MF, Bevan D, Bollum F (1983) Terminal deoxynucleotidyl transferase in acute myeloid leukaemia. Leuk Res 7: 17–29

Johnson GR, Metcalf D (1977) Pure and mixed erythroid colony formation in vitro stimulated by spleen conditioned medium with no detectable erythropoietin. Proc Natl Acad Sci USA 74: 3879–3882

Koutts J, Howard MA, Firkin BG (1980) Factor VIII physiology and pathology in man. Progr Hematol 11: 115–145

Ley TJ, DeSimone J, Anagnou P, Keller GH, Humphries RK, Turner PH, Young NS, Heller P, Nienhuis A (1982) 5-Azacytidine selectively increases γ-globin synthesis in a patient with $\beta+$ thalassemia. N Engl J Med 307: 1469–1475

MacKillop WJ, Ciampi A, Till JE, Buick RN (1983) A stem cell model of human tumor growth: implications for tumor cell clonogenic assays. J Natl Cancer Inst 70: 9–16

Marie JP, Izaguirre CA, Civin CI, Mirro J, McCulloch EA (1981a) Granulopoietic differentiation in AML blasts in culture. Blood 58: 670–674

Marie JP, Izaguirre CA, Civin CI, Mirro J, McCulloch EA (1981b) The presence within single K-562 cells of erythropoietic and granulopoietic differentiation markers. Blood 58: 708–711

Marie JP, Zittoun R, Thevenin D, Mathieu M, Viguie F (to be published) In vitro culture of clonogenic leukaemic cells in acute myeloid leukaemia. Growth pattern and drug sensitivity. Br J Haematol

McCulloch EA (1983) Stem cells in normal and leukemic hemopoiesis. Blood 62: 1–13

McCulloch EA, Buick RN (1982) The cellular basis of treatment goals in human neoplasia. Med North Am 26: 2557–2562

McCulloch EA, Till JE (1960) The radiation sensitivity of normal mouse bone marrow cells, determined by quantitative marrow transplantation into irradiated mice. Radiat Res 13: 115–124

McCulloch EA, Till JE (1977) Stem cells in normal early haemopoiesis and certain clonal haemopathies. In: Hoffbrand AV, Brain MC, Hirsch J (eds) Recent advances in haematology, vol 2. Churchill Livingstone, Edinburgh, pp 85–110

McCulloch EA, Buick RN, Minden MD, Izaguirre CA (1978) Differentiation programmes underlying cellular heterogeneity in the myeloblastic leukemias of man. In: Golde DW, Cline MJ, Metcalf D, Fox CF (eds) Hematopoietic cell differentiation, ICN-UCLA Symposium on hemopoietic cell differentiation. Academic Press, New York, pp 317–333

McCulloch EA, Curtis JE, Messner HA, Senn JS, Germanson TP (1982) The contribution of blast cell properties to outcome variation in acute myeloblastic leukemia (AML). Blood 59: 601–608

Messner HA, Fauser AA (1980) Culture studies of human pluripotent hemopoietic progenitors. Blut 41: 327–333

Minden MD, Buick RN, McCulloch EA (1979) Separation of blast cell and T-lymphocyte progenitors in the blood of patients with acute myeloblastic leukemia. Blood 54: 186–195

Moore MAS, Spitzer G, Williams N, Metcalf D, Buckley J (1974) Agar culture studies in 127 cases of untreated acute leukemia: the prognostic value of reclassification of leukemia according to in vitro growth characteristics. Blood 44: 1–18

Nakahata T, Ogawa M (1982) Identification in culture of a class of hemopoietic colony-forming units with extensive capability to self-renew and generate multipotential hemopoietic colonies. Proc Natl Acad Sci USA 79: 3843–3847

Park CH, Amare M, Savin MA, Goodwin JW, Newcomb MM, Hoogstraten B (1980) Prediction of chemotherapy response in human leukemia using an in vitro chemotherapy sensitivity test on the leukemic colony-forming cells. Blood 55: 595–601

Pluznik DH, Sachs L (1965) The cloning of normal "mast" cells in tissue culture. J Cell Comp Physiol 66: 319–324

Prchal JF, Adamson JW, Murphy S, Steinmann L, Fialkow PJ (1978) Polycythemia vera: the in vitro response of normal and abnormal stem cell lines to eryhtropoietin. J Clin Invest 61: 1044–1047

Preisler HD (1980) Prediction of response to chemotherapy in acute myelocytic leukemia. Blood 56: 361–367

Puck TT, Marcus PI (1956) Action of x-rays on mammalian cells. J Exp Med 103: 653–666

Richman CM, Rowley JD (1983) Correlation of in vitro culture pattern and Q-banded karyotype in acute nonlymphocytic leukemia. Am J Hematol 14: 37–47

Robinson WA, Kurnick JE, Pike BL (1971) Colony growth of human leukemic peripheral blood cells in vitro. Blood 38: 500–508

Rovera G, Santoli D, Damsky C (1979) Human promyelocytic leukemia cells in culture differentiate into macrophage-like cells when treated with phorbol diester. Proc Natl Acad Sci USA 76: 2779–2783

Rustum Y, Preisler HD (1979) Correlation between leukemic cell retention of 1-B-D-arabinofuranozyl-cytosine-5′-triphosphate and response to therapy. Cancer Res 39: 42–49

Sachs L (1980) Constitutive uncoupling of pathways of gene expression that control growth and differentiation in myeloid leukemia: a model for the origin and progression and malignancy. Proc Natl Acad Sci USA 77: 6152–6156

Salmon SE, Hamburger AW, Soehnlen B, Durie BG, Alberts DS, Moon TE (1978) Quantitation of differential sensitivity of human tumor stem cells to anticancer drugs. N Engl J Med 298: 1321–1327

Selby P, Buick RN, Tannock I (1983) A critical appraisal of the "human tumor stem-cell assay." N Engl J Med 308: 129–134

Seligmann M, Vogler LB, Preud'Homme JL, Guglielmi P, Brouet JC (1981) Immunological phenotypes of human leukemias of the B-cell lineage. Blood Cells 7: 237–246

Senn JS, McCulloch EA, Till JE (1967) Comparison of the colony forming ability of normal and leukemic human marrow in cell culture. Lancet 2: 597–598

Siminovitch L, McCulloch EA, Till JE (1963) The distribution of colony forming cells among spleen colonies. J Cell Comp Physiol 62: 327–336

Smith LJ, McCulloch EA (1981) Phenotypic classification of acute myeloblastic leukemia (AML) blasts: the presence in single cells of erythropoietic and granulocytic/monocytic markers. Blood [Suppl 1] 58: 153a (Abstract)

Smith LJ, Curtis JE, Messner HA, Senn JS, Furthmayr H, McCulloch EA (1983) Lineage infidelity in acute leukemia. Blood 61: 1138–1145

Spitzer G, Dicke KA, Gehan EA, Smith T, McCredie KB, Barlogie B, Freireich EJ (1976) A simplified in vitro classification for prognosis in adult acute leukemia: the application of in vitro results in remission-predictive models. Blood 48: 795–807

Stephenson JR, Axelrad AA, McLeod DC, Shreeve MM (1971) Induction of colonies of hemoglobin-synthesizing cells by erythropoietin in vitro. Proc Natl Acad Sci USA 68: 1542–1546

Tepperman AD, Curtis JE, McCulloch EA (1974) Erythropoietin colonies in cultures of human marrow. Blood 44: 659–669

Till JE (1976) Regulation of hemopoietic stem cells. In: Cairnie AB, Lala PK, Osmond DG (eds) Stem cells of renewing cell populations. Academic Press, New York, pp 143–145

Till JE, McCulloch EA (1980) Hemopoietic stem cell differentiation. Biochim Biophys Acta 605: 431–459

Till JE, McCulloch EA, Siminovitch L (1964) A stochastic model of stem cell proliferation, based on the growth of spleen colony forming cells. Proc Natl Acad Sci USA 51: 29–36

Till JE, Messner HA, Price GB, Aye MT, McCulloch EA (1974) Factors affecting normal and leukemic hemopoietic cells in culture. In: Clarkson B, Baserga R (eds) Control of proliferation in animal cells. Cold Spring Harbor Laboratory, New York, pp 907–913

Tonegawa S (1983) Somatic generation of antibody diversity. Nature 302: 575–581

van Rhenen DJ, Verhulst JC, Huijgens PC, Langenhuijsen MMAC (1980) Maturation index: a contribution to quantification in the FAB classification of acute leukaemia. Br J Haematol 46: 581–586

Vincent PC, Sutherland R, Bradley M, Lind D, Gunz FW (1977) Marrow culture studies in adult acute leukemia at presentation and during remission. Blood 49: 903–912

Whang J, Frei E, Tjio JH, Carbone PP, Brecher G (1963) The distribution of the Philadelphia chromosome in patients with chronic myelogenous leukemia. Blood 22: 664–673

*Experimental Approaches to Drug Testing and Clonogenic Growth: Results in Multiple Myeloma and Acute Myelogenous Leukemia**

B.G.M. Durie**

The University of Arizona, Health Sciences Center, Section of Hematology and Oncology, Tucson, AZ 85724, USA

Introduction

With the introduction of clonogenic assay methods for human multiple myeloma and acute myelogenous leukemia, it has become possible to carry out in vitro drug testing in these diseases. Major goals of developing such techniques have included both the selection of the best treatment for individual patients and preclinical drug screening and new drug development. In this analysis, results of in vitro testing using both clonogenic assay methods and thymidine incorporation techniques will be summarized. Correlations between the in vitro methods and the subsequent clinical response of patients will be critically evaluated and the potential for these types of approaches discussed.

Materials and Methods

Patients with plasma cell myeloma as defined by the Chronic Leukemia-Myeloma Task Force of the National Cancer Institute were studied (Chronic Leukemia-Myeloma Task Force, National Cancer Institute 1973). All patients had baseline staging and follow-up as previously reported (Durie and Salmon 1975a; Durie et al. 1980). Bone marrow aspiration was performed at least 3 weeks following pulse chemotherapy and from sites not currently or previously irradiated. Bone marrow samples were divided into aliquots for comparative studies of the tritiated thymidine labeling index, soft agar culture (including thymidine suicide, and drug sensitivity studies). Previously untreated patients with multiple myeloma received bifunctional oral alkylating agents singly or in combination. Our standard protocol for patients in relapse from oral alkylating agents was with carmustine (BCNU) 30 mg/m^2 combined with doxorubicin hydrochloride (Adriamycin) 30 mg/m^2. Southwest Oncology Group criteria of response were used as previously reported (Durie et al. 1980).

* Supported in part by Grants CA 17094, CA 21839, and CA 14102 from the National Cancer Institute, National Institutes of Health, Bethesda, MD 20205, USA

** I wish to acknowledge that many of the presented studies have been carried out in close collaboration with Sydney E. Salmon and with the technical assistance of Barbara Soehnlen, Linda Norris-Kloos, Marti McFadden, Yvette Frutiger, and Judy Christian

Tritiated Thymidine Labeling Index Studies

Labeling index studies were performed on aliquots of the single-cell suspensions prepared for soft agar culture. Heparinized bone marrow cell suspensions ($0.5 \times 10^6 - 1.0 \times 10^6$ cells per milliliter cell suspensions) were incubated for 1 h at 37° C with high-specific-activity (40–60 Ci/mmol) tritiated thymidine (dose, 5.0 μCi/ml cell suspension). Cytocentrifuge smears were then made on gelatin-coated slides and fixed with methanol. The tritiated thymidine labeling index was then determined using our previously published high-speed scintillation autoradiography method (Durie and Salmon 1975).

Soft Agar Culture

The culture system has previously been described in detail (Hamburger and Salmon 1977a, b; Salmon et al. 1978; Durie et al. 1983). In brief, cells to be tested were suspended in 0.3% agar in enriched CMRL 1066 medium, supplemented with 15% horse serum to yield a final concentration in the range of $2-5 \times 10^5$ cells per milliliter. Freshly prepared 2-mercaptoethanol was added at a concentration of $5 \times 10^{-5}\,M$ immediately before triplicate plating of the cells. One milliliter of the mixture was pipetted into a 35-mm plastic petri dish containing conditioned medium in a 1-ml agar feeder layer. Because the plating efficiency in myeloma was 0.01%–0.2%, a plating concentration of 5×10^5 cells was normally used. Cultures were incubated at 37° C in 5% carbon dioxide in humidified air. Colonies (collections of more than 30 cells) appeared in 10–21 days, and plates were counted for drug effects after 14–21 days, with use of an inverted phase microscope.

At least 30 tumor colonies per plate were required in the control plates to assure an adequate range for measurement of drug effect. Drug effects were scored only for reduction in the number of colonies.

In Vitro Incorporation of ^{3}H-Thymidine After Short-Term Culture

The method for the tritiated thymidine labeling index (LI%) in myeloma specimens has been described previously (Durie et al. 1980; Durie and Salmon 1975b). The use of measuring thymidine uptake as a marker of drug efficacy has also been described (Sanfilippo et al. 1979; Daidone et al. 1981). A modification of this method was used in the experiments described in this paper. Briefly, bone marrow cells from patients with multiple myeloma were incubated for 3 h at 37° C, with and without interferons, bisantrene, or other agents, using concentrations similar to those used for the human tumor clonogenic assay. Five microcuries ^{3}H-TdR per milliliter was then added and the cell suspension incubated for 1 further hour at 37° C. After washing the cells free of unincorporated ^{3}H-TdR, slides for autoradiography and microscopic examination were prepared with a cytocentrifuge. After methanol fixation, autoradiographs were prepared as detailed above. All marrow cells that were morphologically in the lymphoid-plasma cell series were defined as myeloma cells. Plasma cells containing five grains over the nucleus were considered labeled. One thousand cells were counted to determine the LI%, which was then expressed as a percentage.

In Vitro Exposure of Myeloma Cells to Drugs

The drug assay procedure has been reported previously (Hamburger and Salmon 1977a, b; Salmon et al. 1978; Durie et al. 1983). Stock solutions of intravenous formulations of melphalan, carmustine, doxorubicin, vinblastine, methotrexate, vincristine, and other agents were prepared in sterile buffered saline or water and stored at −70° C in aliquots sufficient for individual assays. Subsequent dilutions were made in saline for cell incubation. Myeloma cell suspensions were transferred to tubes and adjusted to a final concentration of 10^6 cells per milliliter in the presence of the appropriate drug dilution or control medium. Each drug was tested at a minimum of three dose levels, including low concentrations as previously published. Cells were incubated with drugs for 1 h at 37° C in Hank's balanced salt solution. The cells were then centrifuged at 150 *g* for 10 min, washed twice in the balanced salt solution, and prepared for culture.

Leukemia Cell Culture

Leukemic cells were obtained from samples of bone marrow or peripheral blood and cultivated in methylcellulose as previously described (Buick et al. 1977; Trent et al. 1983). In brief, mononuclear cells obtained by Ficoll-Hypaque density centrifugation were rosetted with neuraminidase-treated sheep red blood cells (ERF-C) to remove rosette-forming T-lymphocyte precursors from precursors of leukemic blast cells to be cultured. Two hundred thousand cells per plate were set up in triplicate for each patient sample. Following a 1-h exposure to control media, cells were washed thoroughly and plated in 0.8% (v/v) methylcellulose in Alpha Medium supplemented with 10% fetal bovine serum and 20% (v/v) PHA (phytohemagglutinin) stimulated leukocyte-conditioned medium (PHA-LCM) (Buick et al. 1977). Cultures were then incubated at 37° C in a humidified atmosphere of 5% CO_2 with air for 7 days. Plates were examined for growth and categorized as showing no growth, minimal growth (e.g., cell doublings), clusters (< 30 cells), or colonies (≥ 30 cells). Routine morphological and histochemical analysis with Wright-Giemsa, Sudan Black B, and chloracetate esterase staining was also used to confirm the leukemic nature of colonies. Drug studies were carried out as for the myeloma clonogenic assay.

Method for the Rapid ^3H-Thymidine Assay for Acute Myelogenous Leukemia

For the ^3H-thymidine assay, the leukemia blast cells were set up as for the methylcellulose blast cell assay, except that the cells were placed in Linbro 12-well flat-bottomed tissue culture plates. At the end of 48−72 h incubation, [methyl-^3H]-thymidine of specific activity 60−80 Ci/mmol, 1 μCi/ml in a 0.1-ml volume was added to each well. After an additional 24 h incubation at 37° C, 2 ml phosphate-buffered saline (PBS) was added to each well and incubated at 37° C for 1−2 h. This diluted the methylcellulose to allow for subsequent removal. At the end of the 1−2 h incubation period, the plates were inserted into a microtiter plate carrier for centrifugation at 1,250 rpm for 10 min to pellet the cells. The supernatant methylcellulose/PBS mixture was removed by suction. Subsequently, 2.5 ml PBS was added to each well, which was again centrifuged at 1,250 rpm for 10 min. The supernatant was again removed and the remaining cells resuspended in PBS. These cells were then harvested onto filter paper for evaluation using the multiple automatic sample harvester (MASH) harvester unit. The recovered filter paper disks were inserted into single vials for scintillation counting.

Results

Myeloma Clonogenic Growth – In Vitro Drug Testing

The results of in vitro drug testing with myeloma using the clonogenic assay at the University of Arizona have recently been published (Durie et al. 1983). In the study, samples from 97 patients with multiple myeloma were serially evaluated. Of those 33 patients had data sufficient for in vitro drug sensitivity analysis. It was of interest to note that the likelihood of colony growth was highly correlated with the tritiated thymidine labeling index of the plated myeloma cells. The higher the labeling index, the greater the likelihood of colony growth. The overall in vitro/in vivo correlations are shown in Table 1. It can be seen that there was excellent prediction of both sensitivity and resistance. Nonetheless, there were several false positive and false negative predictions. For the previously untreated patients, the correlations were particularly good, as is shown in Table 2. All eight of the patients sensitive in vitro to the primary alkylating agent, melphalan, had significant regression with induction treatment and survived for at least a year, with several of the patients surviving for over 5 years. Conversely, no patient resistant in vitro to melphalan has yet had long disease-free survival.
Patients were tested for in vitro sensitivity to melphalan both at the time of initial diagnosis and when early relapse supervened. A plot of the in vitro chemosensitivity survival/concentration curves for melphalan is shown in Fig. 1, illustrating the marked heterogeneity in sensitivity observed from patient to patient. The marked sensitivity in many patients does, however, correlate with the known activity of this agent in myeloma. Other drugs tested have included BCNU, doxorubicin, vincristine, vinblastine, vindesine, *cis*-platinum, bleomycin, imidazole carboxamide-dimethyl triazeno (NSC-45388) (DTIC), leukocyte interferon, and clone A interferon. Significant sensitivity has been noted with BCNU, doxorubicin, vinblastine, or vindesine and the interferons. Occasional patients have also been sensitive to the other agents. In vitro testing with a variety of other new agents is

Table 1. Overall in vitro/in vivo correlations[a]

	SS	SR	RS	RR
No. of patients	11 (73%)	4 (27%)	3 (17%)	15 (83%)

With the Fisher exact test, the association between in vitro and in vivo results was significant ($p < 0.01$)

[a] In vitro sensitive includes intermediate also. In vivo response defined as > 50% tumor regression. Only clinical trials immediately following the in vitro test are included. Early deaths (< 2 courses of treatment) are excluded

SS, sensitive in vitro and in vivo (true positive); *SR*, sensitive in vitro but resistant in vivo (false positive); *RS*, resistant in vitro but sensitive in vivo (false negative); *RR*, resistant in vitro and in vivo (true negative)

With permission from BLOOD

Table 2. Previously untreated patients (16): interrelationships between stage, type, tritiated thymidine labeling index (LI%), in vitro sensitivity, response to induction, and survival duration

Stage	Type	LI (%)	In vitro sensitivity[a]		Regression (%)	Survival duration (months)
			Sensitive or intermediate	Resistant		
A. Sensitive						
III A	IgA κ	2	M, A, B, V	–	90	74 +
III A	IgA λ	0.5	M, A, B, V	–	86	68
II A	IgG λ	2	M, A, B, V	–	60	64
II B	IgA κ	3	M	–	80	57
III A	IgG κ	1	M, A	–	80	40
III A	IgG κ	0	M, A	–	75	24 +
III A	IgG λ	3	M, IL	A, B, Cis-p	90	18 +
III A	IgG κ	2	M, A, B, Mitox, IL, Cis-p	Bis	30	12 +
B. Resistant						
III A	IgG λ	2	–	M, A	54	48 +
III B	IgG κ	1.6	–	M	56	49 +
III A	IgG λ	1.4	–	M, A, B, V	30	17
III A	IgG λ	2.4	–	M, A, B, V	25	12
III A	IgG λ	3.2	–	M	20	6
III B	IgG λ	18	–	M	90	5
III B	IgA κ	5	–	M, A, B, V	20	2
III A	IgG κ	1	–	M, A, B, V	15	2

[a] Sensitivity defined in Materials and Methods
M, melphalan; *A*, doxorubicin; *B*, BCNU; *V*, vincristine; *IL*, interferon (leukocyte); *Bis*, bisantrene; *Mitox*, mitoxantrone; *Cis-p*, *cis*-platinum
With permission from BLOOD

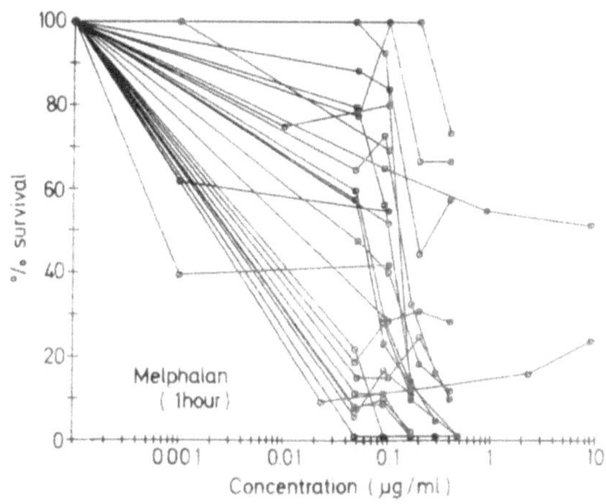

Fig. 1. In vitro survival/concentration curves for myeloma colony forming units (MCFU) from a series of 26 patients tested with various doses of melphalan for 1 h prior to culture. Marked heterogeneity in sensitivity is apparent from patient to patient. Those curves with reduction in survival of MCFU to 30% or less at the 0.1 µg/ml dose level would be classed as sensitive

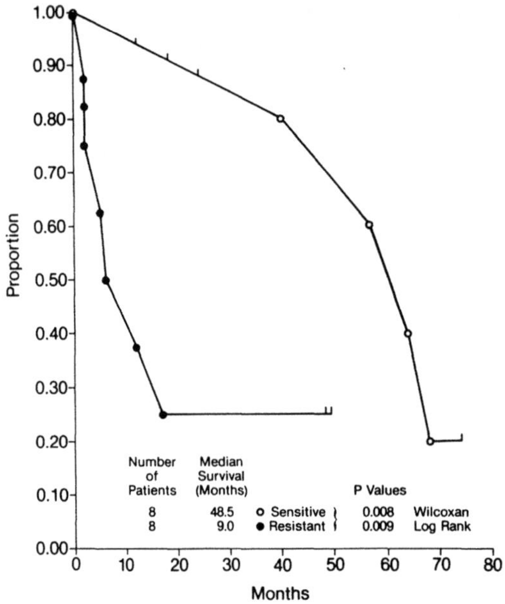

Fig. 2. Displayed are the Kaplan-Meier survival curves for 16 previously untreated patients with multiple myeloma categorized by in vitro sensitivity (○) or resistance (●) to melphalan. The median survival values are 48.5 (○) and 9.0 (●) months respectively. Survival is significantly different between the two groups by both the Wilcoxan and log-rank methods ($p < 0.01$)

continuing in an effort to select therapy for individual patients as well as part of a more general drug-screening evaluation.

The significance of the in vitro information is illustrated by the survival plot as shown in Fig. 2. Displayed are the Kaplan-Meier survival curves for 16 previously untreated patients with multiple myeloma categorized by in vitro sensitivity or resistance to melphalan. As can be seen, in vitro sensitivity to melphalan correlated with a significantly superior survival duration. This type of survival difference was also found with testing of patients at the time of relapse.

In Vitro Drug Evaluation by Thymidine Suppression in Multiple Myeloma

Because in vitro clonogenic growth is currently feasible in only a percentage of patients with myeloma and also because of the time delay in obtaining results (approximately 2 weeks), an alternate method of drug testing using thymidine suppression has been evaluated in patients with multiple myeloma (see Materials and Methods). The overall results with initial screening of this technique are summarized in Table 3. As can be seen, there has been excellent correlation between in vitro testing by the clonogenic assay method and thymidine suppression techniques. Thus far, all four patients treated with interferons who have been shown to be sensitive by both techniques in vitro have responded to treatment. Two of three patients sensitive in vitro to bisantrene, who have been treated with bisantrene, have also shown responses to treatment. Thus the preliminary information suggests that the thymidine suppression technique, which can give information in approximately 3 or 4 days from the time of the bone marrow aspiration, is a clinically useful tool. Further testing with interferon and bisantrene as well as other agents will be necessary to determine the overall utility of this method.

Table 3. Summary of results

	Myeloma stem cell assay	Thymidine suppression
Bisantrene		
Total	10	22
Sensitive	4 (40%)	12 (55%)
Stimulation	0	1 (4%)
Leukocyte interferon		
Total	13	24
Sensitive	5 (38%)	10 (42%)
Stimulation	0	2 (8%)
Clone A interferon		
Total	11	21
Sensitive	5 (45%)	7 (33%)
Stimulation	2 (20%)	4 (20%)

Number represent numbers of studies

Studies in Acute Myelogenous Leukemia

A large number of studies have now been carried out using peripheral blood and bone marrow blast cells from patients with acute myelogenous leukemia (Durie and Brooks 1982). The overall growth observed with the first 159 samples evaluated is summarized in Table 4. It can be seen that some type of growth was observed in 79% of cases with sufficient growth for in vitro drug evaluation in 59% of patients. Thus the majority of patient information can be used for clinical correlations. Thus far, in vitro drug testing has been carried out with the standard agents such as cytosine arabinoside and daunomycin as well as the experimental agents 4'(9-acridinylamino) methane sulfon-M-anisidide (m-AMSA), mitoxantrone, bisantrene, 4-deoxydoxorubicin, and various types of interferons. Because of the many variables which can influence outcome in patients with acute myelogenous leukemia, it has proved difficult to draw final conclusions as to the utility of the in vitro cell kill information. However, preliminary analysis has suggested a high degree of correlation between combined sensitivity to both daunomycin (1 μg/ml, 1 h exposure) and cytosine arabinoside (≥ 1 μg/ml, continuous exposure) and the likelihood of obtaining a hypocellular marrow cleared of blasts with induction chemotherapy. However, the correlation with subsequent remission duration and survival is less clear cut and multivariant regression analyses including all of the known prognostic factors in accute myelogenous leukemia are currently being carried out to delineate more clearly the role of the in vitro information. If some correlation with likelihood of long-term disease-free survival can be obtained, this should prove particularly useful in the selection of bone marrow transplantation for patients in first remission. If good correlation with clinical response is obtained, this will also be encouragement to use the assay more extensively for new drug screening.

As for multiple myeloma, it is possible to use a thymidine incorporation assay to obtain the results of in vitro testing more rapidly (see Materials and Methods). In this group of patients we have evaluated the effect of adding the tritiated thymidine to the clonogenic

Table 4. Overall leukemia (AML) growth[a]

Tested	159 sample		%
Colonies	88	Drug	
Clusters	6	studies	59
(sufficient for drug study)			
Small clusters	32	Growth	79
No growth	33		21

[a] Includes patients with preleukemia
AML, acute myelogenous leukemia

assay plates after the initial 48–72 h of growth. We have, therefore, been able to determine the effect of drugs upon this early clonogenic growth phase. After 48–72 h there is a striking incorporation of tritiated thymidine into the rapidly proliferating blast cell clusters and colonies. This incorporation is readily inhibited by active agents such as daunomycin and cytosine arabinoside. In the first 19 patients from whom samples were evaluated there has been excellent cross-correlation for sensitivity to both daunomycin and cytosine arabinoside between the clonogenic assay method and the method evaluating thymidine incorporation. The major advantage of the thymidine incorporation method has been the fact that the results can be available more rapidly. In addition, a significant technical advantage is that instead of having to count the colonies and/or clusters in the methylcellulose culture by eye (since the methylcellulose cultures cannot be counted using the automatic Fas II Colony Counter), the in vitro drug efficacy can be evaluated by scintillation counting and the automatic counter. This is obviously a tremendous saving in technician time. The thymidine incorporation assay is also considerably more sensitive than the visual assessment of clusters and colonies. Effects upon thymidine incorporation can be assessed in patients with minimal growth-routine evaluation is impossible in this subpopulation.

Although final evaluation of in vitro testing in acute myelogenous leukemia is not yet available, the thymidine incorporation method can potentially provide comparable clinically useful information.

Conclusions

In this paper I have attempted to highlight some of the applications of in vitro drug sensitivity testing in both multiple myeloma and acute myelogenous leukemia. A better understanding of growth factors required to enhance in vitro clonogenicity will be important for further progress in this field. The current cloning efficiencies greatly limit the number of patients whose cells can be studied effectively. However, the thymidine incorporation methods which appear to give comparable information can be used to provide rapid chemosensitivity information. Further developments in evaluation of clonogenicity will be extremely important to a more basic understanding of the biology of both leukemia and myelogenous leukemia. Better understanding of the mechanisms of chemosensitivity in resistance should also provide leads for the development of useful new therapeutic agents.

References

Buick RN, Till JE, McCulloch EA (1977) A colony assay for proliferative blast cells circulating in myeloblastic leukemia. Lancet 1: 862

Chronic Leukemia-Myeloma Task Force, National Cancer Institute (1973) Proposed guidelines for protocol studies-II. Plasma cell myeloma. Cancer Chemother Rep 4: 145–158

Daidone MG, Silvestrini R, Sanfilippo O (1981) Clinical Relevance of an in vitro antimetabolic assay for monitoring human tumor chemosensitivity. In: Salmon SE, Jones SE (eds) Adjuvant therapy of cancer III. Grune & Stratton, New York, pp 25–32

Durie BGM, Brooks RJ (1982) In vitro drug sensitivity testing of leukemia clonogenic cells; clinical correlations. Third Conf Human Tumor Cloning, Jan 22–30

Durie BGM, Salmon SE (1975a) A clinical staging system for multiple myeloma: correlation of measured myeloma cell mass with presenting clinical features, response to treatment and survival. Cancer 36: 842–854

Durie BGM, Salmon SE (1975b) High speed scintillation autoradiography. Science 190: 1093–1095

Durie BGM, Salmon SE, Moon TE (1980) Pretreatment tumor mass, cell kinetics, and prognosis in multiple myeloma. Blood 55: 364–372

Durie BGM, Young YA, Salmon SE (1983) Human myeloma in vitro colony growth: interrelationships between drug sensitivity, cell kinetics, and patient survival duration. Blood 61: 929–934

Hamburger AW, Salmon SE (1977a) Primary bioassay of human tumor stem cells. Science 197: 461–463

Hamburger AW, Salmon SE (1977b) Primary bioassay of human myeloma stem cells. J Clin Invest 60: 846–854

Salmon SE, Hamburger AW, Soehnlen BJ, Durie BGM, Alberts DS, Moon TE (1978) Quantitation of differential sensitivity to human tumor stem cells to anticancer drugs. N Engl J Med 298: 1321–1327

Sanfilippo O, Daidone MG, Silvestrini R (1979) Antimetabolic effect of drugs in short-term culture as a potential tool for monitoring tumor chemosensitivity. Chemother Oncol 4: 261–265

Trent JM, Davis JR, Durie BGM (1983) Cytogenetic analysis of leukemic colonies from acute and chronic myelogenous leukemia. Br J Cancer 47: 103–109

*In Vitro Assessment of Drug Sensitivity in Acute Nonlymphocytic Leukemia**

H.D. Preisler and A. Raza

Roswell Park Memorial Institute, New York State Department of Health, Leukemia Service, 666 Elm Street, Buffalo, NY 14263, USA

Introduction

Acute nonlymphocytic leukemia (ANLL) is an excellent model system for developing in vitro assays for the following reasons: the neoplastic cell population is readily accessible to repeated sampling, the cells do not have to be disaggregated prior to study, a variety of "effective" chemotherapeutic agents are available, and response to therapy is relatively unambiguous. All of these characteristics make ANLL much easier to study than solid tumors, and since there are considerable biological similarities among neoplastic cells it is likely that the biological principles defined by the study of ANLL will be applicable to solid tumors.

Response to therapy in ANLL is usually defined as complete remission (CR) or treatment failure. For the purposes of understanding the nature of response and especially for evaluating drug sensitivity assays, this method of analysis is inadequate since the "treatment failure" category is heterogeneous and includes patients in whom the chemotherapy fails to affect the leukemic cells at all, patients in whom the therapy reduces the number of leukemic cells but in whom leukemic cells repopulate the marrow, patients who die soon after the initiation of chemotherapy, and patients who die during an aplastic phase without evidence of persistent leukemia (Preisler 1978). Clearly one cannot know whether the patients who died early in treatment or who died while aplastic would have entered CR, had they lived long enough, or whether leukemic cells would have repopulated the marrow. Since one cannot expect a drug sensitivity assay to predict for death due to bleeding or infection, the outcome of therapy for these patients cannot be used to assess the efficacy of a drug sensitivity assay (Preisler 1980a).

In the discussion which follows, a CR outcome is used to define drug-sensitive disease. Patients who survive long enough for a determination to be made that treatment failure is due to persistent leukemia are used to define resistant disease (RD), and all other patients are referred to as "other" failures (Preisler 1978, 1980a).

General Considerations in the Design of in Vitro Drug Sensitivity Tests

There are several basic questions which must be addressed when in vitro drug sensitivity assays are being designed: which drug effects should be measured, which cells should be

* Supported by Grants CA 5834 and CA 28734-01

Recent Results in Cancer Research. Vol. 94

studied, what are the appropriate conditions of in vitro exposure, and how does one assess the efficacy of a putative assay?

The interaction of a chemotherapeutic agent with a cell involves a series of steps including the entry of the drug into the cell, in some cases activation of the drug, and ultimately interference by the drug with a critical metabolic pathway resulting in the death of the cell. Ideally, for the purpose of assessing the drug sensitivity the assay should measure a drug effect or interaction which is as close as possible to the locus of action of the drug, or one should directly quantitate cell killing.

The question as to which cells should be studied is difficult to answer (Preisler and Rustum 1978). Theoretically, one should restrict studies to the tumor stem cell and its immediate proliferating progeny and one need not (and perhaps should not) study end stage tumor cells which have ceased to proliferate or whose remaining replications are self-limited. Unfortunately, since the ability to clone in vitro is not the sine qua non of stem cellness, there is no way of knowing whether any of the putative stem cell assays are in fact stem cell assays. Even if stem cells do proliferate in vitro, there is no way of knowing whether all stem cells proliferate or whether proliferation in vitro is restricted to a subset of stem cells whose properties are not representative of the population as a whole.

Another significant problem relates to designing in vitro conditions of drug exposure which approximate those which occur in vivo. To accomplish this optimally, one would need to have precise information regarding the actual concentration X time values for drug levels. One would also have to have information regarding the possible presence of normal metabolites which may compete with the agent(s) being administered, as well as information regarding the presence or absence of factors which may enhance or reverse the potentially lethal effects of chemotherapeutic agents.

Clearly, the information necessary to address the questions posed above is not currently available. The information provided in the next section summarizes our studies of several different approaches to assessment of the drug sensitivity of ANLL cells. These studies demonstrate that different approaches are necessary when different chemotherapeutic agents are used to treat the same disease. The penultimate section provides information regarding some of the limitations inherent in the clonogenic assays, assays which are in fashion today.

In Vitro Assays Predictive of Response to Therapy

Nonspecific in Vitro Studies Predictive of Response to Therapy

While in vitro drug sensitivity assays usually refer to the sensitivity of tumor cells to specific agents, two biological assays not involving chemotherapeutic agents assess neoplastic characteristics which are reflective of biological properties associated with poor responsiveness to therapy.

The ability of ANLL cells to clone in vitro (Moore et al. 1978; Spitzer et al. 1976; Goldberg et al. 1979; Preisler et al. to be published a) and the cytogenetic constitution of the leukemic cells (Sakukrai and Sandberg 1973; Golomb et al. 1978; Preisler et al. 1983a) are general biological characteristics which serve to identify patients who are likely to do well or poorly with cytosine arabinoside/anthracycline (araC/anthracycline) therapy (Preisler et al. 1979a, 1979b). Patients whose leukemic cells grow well in vitro are less likely to enter remission than patients whose cells do not clone in vitro (CR rate = 41% and 68% respectively) (Table 1). Furthermore, the median duration of remission for patients whose

Table 1. Relationship between leukemic cell growth characteristics and cytogenetic type and outcome of remission induction therapy

	n	CR	RD	Other	p-difference in CR
AraC/anthracycline remission induction therapy					
Growth characteristics					
No growth	22	15 (68%)	3 (14%)	4 (18%)	0.056
20 clones, 20 cells[a]	27	3 (41%)	7 (26%)	9 (33%)	
Growth[b]	54	29 (54%)	13 (24%)	12 (22%)	
Cytogenetic type					
Normal metaphases (NN)	28	20 (71%)	4 (14%)	4 (14%)	0.02
Abnormal metaphases (NA/AA)	19	6 (32%)	7 (36%)	6 (32%)	
Combined					
No growth					
NN	9	7 (78%)	0	2 (22%)	0.02
NA/AA	6	4 (66%)	1 (17%)	1 (11%)	
20 clones, 20 cells[a]					
NN	9	5 (56%)	3 (33%)		
NA	9	1 (11%)	3 (33%)	5 (56%)	
HDaraC remission induction therapy					
Growth characteristics					
No growth	5	2 (40%)	2 (40%)	1 (20%)	
Growth[b]	20	9 (45%)	7 (35%)	4 (20%)	
Cytogenetic type					
Normal metaphases					
Abnormal metaphases					

[a] 20 clones, 20 cells = specimen which produced 20 cluster/colonies with at least one cluster/colony consisting of 20 cells

[b] Growth = production of at least one cluster consisting of > 20 cells

n, number of patients; *CR*, complete remission; *RD*, resistant disease; *AA*, only abnormal metaphases detected; *NA*, both normal and abnormal metaphases detected

cells fail to clone in vitro was 93 weeks, while it was 27 weeks for patients whose cells produced at least one clone consisting of > 20 cells ($p = 0.02$) (Preisler et al. to be published a). Similarly, patients in whom conventional bone marrow cytogenetic studies detect only normal metaphases (NN patients) are more likely to enter remission than patients whose marrow contains abnormal metaphases, with or without normal clones with a CR rate of 71% and 32% respectively (Table 1) (Preisler et al. 1983a). If both adverse prognostic features are simultaneously present the prognosis is still worse (CR = 11%), indicating that these two features reflect different biological characteristics (Table 1) Preisler et al., to be published a).

The ability or inability of leukemic cells to clone in vitro is probably a reflection of their ability to adapt to growth under adverse conditions, conditions which we attempt to create in vivo by administering chemotherapeutic agents. Hence the tendency of patients whose leukemic cells clone in vitro to have drug-resistant disease should not be unexpected. With respect to the cytogenetic studies, we have found that patients whose marrow contains only abnormal metaphases are likely to die during remission induction therapy (Preisler et al.

1983a) and we have failed to demonstrate that aneuploid leukemias are more likely to be associated with drug resistance than euploid leukemias (Preisler et al., to be published a). Hence a summation of effects of these two adverse prognostic signs might be expected. As illustrated in Table 1, adverse growth characteristics in vitro are also associated with an increase likelihood of other (non-drug-resistant) failure and the presence of aneuploidy is also associated with drug-resistant failure. It is unclear as to why growth pattern should also be associated with early death and why the presence of aneuploidy should also be associated with drug resistance in vivo.

These nonspecific prognostic assays provide useful information regarding the probability that a patient will or will not enter CR. There are nevertheless patients within the poor prognostic categories who do quite well with araC/Anthracycline therapy. Given that growth and cytogenetic assays are weak predictors of response for individual patients, neither can be used by itself to identify patients who should or should not receive araC/anthracycline therapy, and much larger numbers of patients must be studied to dermine whether the simultaneous assay of growth characteristics and cytogenetic type can be used for this function. Neither appears to be an adverse prognostic factor for remission induction therapy with high-dose cytosine arabinoside (HDaraC) (Karanes et al. 1979; Early et al. 1982) (Table 1). Hence HDaraC may be one such alternate induction regimen for these "poor prognosis" patients (Preisler et al. 1983b)

Uptake and Activation of Drugs

Cytosine arabinoside must be taken up by cells and activated to the triphosphate form (ara-CTP) for it to be effective. Several groups of investigators have attempted to measure the araCTP content within cells and use this information to predict response to therapy. These attempts were unsuccessful (Smyth et al. 1976; Chou et al. 1977). In 1976 we initiated studies looking at this same issue and we found that while high in vitro cellular retention of araCTP was associated with long remissions, cellular metabolism of araC in vitro was not predictive of response to remission induction therapy with araC/anthracycline (Rustum and Preisler 1979 (Tables 2 and 3). Furthermore, recent studies suggest that it also is not predictive of response to single-agent high-dose araC remission induction therapy (H. D. Preisler and Y. Rustum, unpublished observations) (Table 4). While cells cannot be killed by araC if the drug is not taken up and phosphorylated, the presence of intracellular araCTP does not by itself guarantee that cell death will occur because there may be competing normal cellular metabolites, DNA polymerase may be insensitive to araC, or cells may have the ability to remove or be unaffected by araC which has been incorporated into DNA. The prognostic significance of araCTP retention with respect to remission duration may be due to pharmacokinetic-cellular interactions rather than solely to the properties of leukemic cells (Preisler et al., to be published b).

It is likely that the anthracycline antibiotics must also be taken up if they are to produce toxic effects in cells. We have taken advantage of their fluorescent properties to quantitate cellular uptake of these agents (Fig. 1) (Preisler and Raza, to be published). Comparison of Fig. 1b and 1c demonstrates that there can be marked inter patient differences in the doxorubicin hydrochloride uptake by their leukemic cells. Additionally, one can discern two different cell populations in Fig. 1b, a population which fails to take up doxorubicin and which comprises approximately one-quarter of the cells present and a second population which clearly takes up the drug, though to a small extent. Inspection of Fig. 1c demonstrates the presence of three populations: one which takes up substantial amounts of

Table 2. Relationship between outcome of remission induction treatment with cytosine arabinoside (*araC*)/anthracycline in patients with acute nonlymphocytic leukemia and in vitro phosphorylation of araC and retention of araC triphosphate (*araCTP*) by leukemic marrow cells

	araCTP (pm/10^7 cells)					
	n	$t = 0$	n	$t = 4$ h	n	% Retention
CR	39	61[a]	33	9	33	17
RD	7	43	6	12	6	14
Other	16	64	11	19	11	19

[a] Median value

Table 3. Relationship between duration of remission in patients with acute nonlymphocytic leukemia treated with cytosine arabinoside (*araC*) and in vitro retention of araC triphosphate (*araCTP*) by leukemic marrow cells

% araCTP retention	P950501 ( )		P970701 ( )	
20	30[a]	$p = 0.02$	50	$p = 0.04$
20	159		66[a]	

[a] Median duration of remission (weeks)

Table 4. Outcome of remission induction treatment with high-dose cytosine arabinoside triphosphate (*araCTP*) in patients with acute nonlymphocytic leukemia and its relationship to in vitro araCTP retention by leukemic marrow cells

	araCTP (pm/10^7 cells)					
	n	$t = 0$	n	$t = 4$ h	n	% Retention
CR	8	26	6	18	6	38
RD	16	88	14	13	12	20
Other	10	56	9	25	9	32

drug and two lesser populations, one of which fails to take up the drug and one of which takes up only small amounts of the drug. Preliminary analysis of the data suggest that low uptake of doxorubicin may be associated with an increased likelihood of failure to enter remission with araC/doxorubicin therapy.

These studies will require a large number of patient entries, since a CR in a patient whose cells fail to take up doxorubicin in vitro could, of course, have been the result of the effects of araC. Cytofluorographic studies have an inherent advantage in that the properties of individual cells are discerned rather than the usual mean population value which conventional drug sensitivity studies provide.

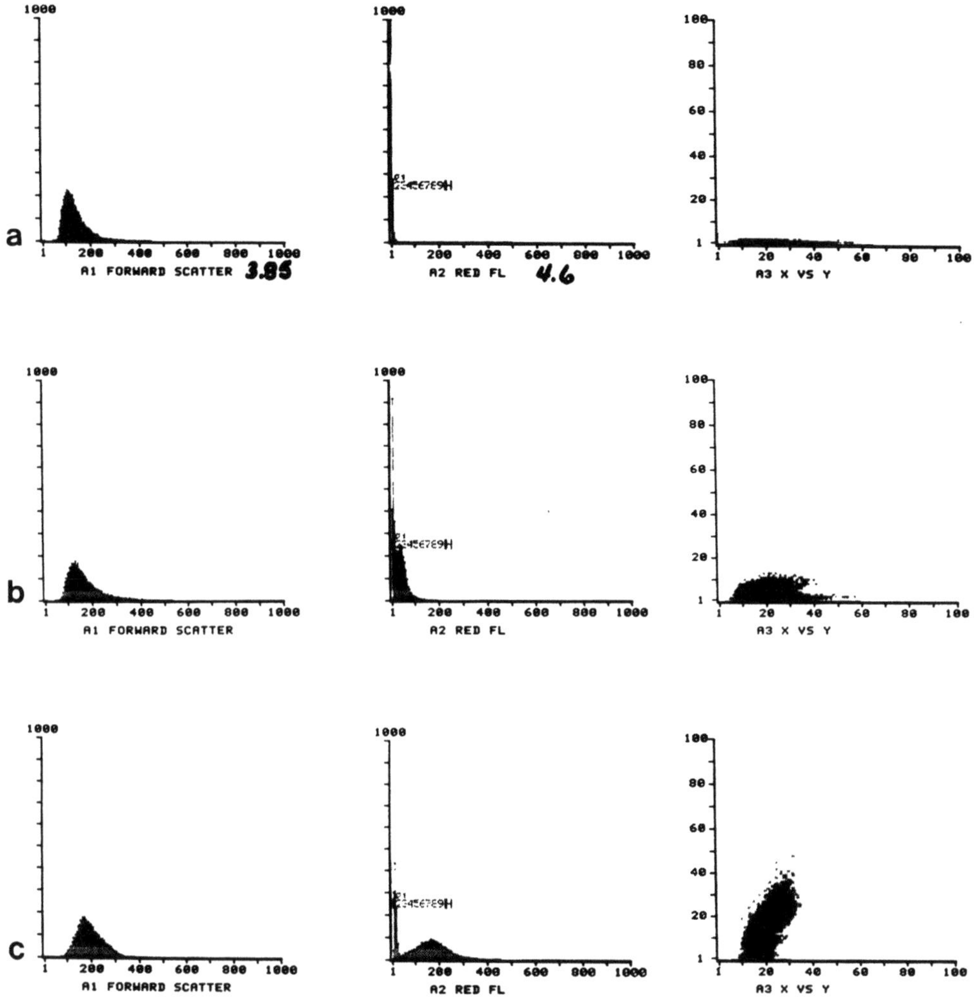

Fig. 1a–c. Uptake of doxorubicin hydrochloride by leukemic marrow cells incubated for 4 h with doxorubicin hydrochloride at 10 μg/ml. **a** cells not incubated with doxorubicin; **b** cells from patient 1; **c** cells from patient 2. In each part of the figure, the *upper left-hand panel* indicates forward light scatter *(abscissa)*, the *upper right-hand panel* indicates red fluorescence *(abscissa)*, and the *lower left panel* indicates light scatter *(abscissa)* vs fluorescence intensity *(ordinate)*. The *ordinates of the upper panels* indicate cell number

Effects of Chemotherapeutic Agents on Cellular Metabolism or on DNA

The selection of the assay for measuring cell sensitivity to an agent must be based upon an understanding of the mechanism of action of the agent. For example, araC produces lethal effects when it is incorporated into DNA (Kufe et al. 1980) and araC incorporation can be recognized by its effects on DNA synthesis. With this in mind we evaluated response to HDaraC therapy and the effects of araC on DNA synthesis in vitro (Preisler et al., to be published c).

The relationship between the effects of araC on DNA synthesis by leukemic cells in vitro and the outcome of remission induction therapy for patients being treated with HDaraC therapy is provided in Table 5. These data demonstrate that cellular resistance to the effects of araC on DNA synthesis was associated with treatment failure due to persistent leukemia. It should be noted that significant in vitro sensitivity to araC did not guarantee that a CR would be induced, since pretherapy leukemic cell mass and the percentage of cells in S phase also play a role in determining response to therapy, with a high tumor cell mass and/or a low ^{3}HTdR labeling index being associated with treatment failure even in the face of high sensitivity of DNA synthesis to araC (Preisler, to be published a; Preisler et al., to be published c; Preisler and Azarnia, to be published b). We have demonstrated that a course of HDaraC therapy produces the same absolute reduction in leukemic cell mass whether the pretherapy mass was high or low (Preisler, to be published b). Hence, for patients with a high pretherapy cell mass, a course of HDaraC therapy does not reduce the mass sufficiently to permit a resumption of normal hematopoiesis. With respect of the adverse prognosis of low pretherapy ^{3}HTdR labeling index, even if the cells are very sensitive to araC, if only few cells are in S phase the absolute number of leukemic cells killed by araC will be low. Similar studies of the effects of 5-azacytidine on both DNA and RNA synthesis of leukemic cells should be performed because of the similarity between this agent and araC.

Doxorubicin intercalates into DNA, causing distortion of the double helix and DNA breaks. We performed a pilot study of the relationship between the in vitro sensitivity of ANLL cells to doxorubicin as indicated by DNA strand breaks and response to remission

Table 5. Relationship between agents which interact with DNA and the outcome of remission induction therapy

Cytosine arabinoside (araC) inhibition of DNA synthesis (patients treated with high-dose araC)	
CR	n = 10 (88%)
RD	n = 19 (78%)
Other failure	n = 14 (85%)
AraC inhibition of DNA synthesis (patients treated with araC/anthracycline therapy)	
CR	n = 23 (90%)
RD	n = 22 (92%)
Other failure	n = 16 (60%)

CR, complete remission; *RD*, resistant disease; *n*, no. of patients
Percentages represent the median % inhibition of DNA synthesis

Table 6. Doxorubicin-induced DNA breaks in patients treated with cytosine arabinoside/doxorubicin

N	*n*	CR	RD
< 0.15	5	0	5
> 0.15	6	5	1

N, no. of breaks and alkali labile regions per alkaline unwinding unit of single-stranded DNA; *n*, no. of patients; *CR*, complete remission; *RD*, resistant disease induction therapy with araC/doxorubicin (Schwartz et al. 1981). These studies suggested that patients in whom strand breaks were detected had a greater likelihood of entering CR than patients in whom such breaks did not occur (Table 6). This study, however, involved few patients and requires repeating on a larger scale. This methodology should also be explored for its ability to predict the outcome of therapy with other agents which produce DNA damage, such as AMSA and mitoxantrone.

Recognition of the Lethal Effects of Chemotherapeutic Agents

To the present time the routine laboratory methods for recognizing "dead" cells are all based on a loss of membrane integrity. In such cases the cell admits otherwise excluded substances such as trypan blue, eosin, or fluorescein diacetate. Unfortunately, the clinically relevant effects of chemotherapeutic agents are too subtle to be detected in this manner, since the relevant event, loss of reproductive ability, may occur in the absence of a loss of membrane integrity. The only method capable at the present time of recognizing the effects of chemotherapeutic agents on cellular reproductive ability is the clonogenic assay.

The data presented in Table 7 demonstrate that the percentage of leukemic clonogenic cells (LCFUc) killed by araC/anthracycline is significantly greater for patients who enter CR than for individuals in whom therapy fails because of persistent leukemia (Preisler 1980b; Preisler and Azarnia, to be published a). This appears to be the case for patients studied at the time of initial diagnosis or at first relapse. When presented in this manner these data are somewhat misleading, since they suggest that this in vitro assay system is more useful than it really is. Several problems exist with this assay. Firstly, the marrow cells of 35% of patients failed to produce > 10 clusters/10^5 cells plated in vitro while the marrow cells of 55% of patients failed to produce > 20 clusters/10^5 cells. Hence this assay does not provide drug sensitivity data for approximately one-half of the patients. Additionally four of 16 patients whose cells were completely unaffected by exposure to araC/anthracycline in vitro nevertheless entered CR. With respect to the outcome of HDaraC therapy, the lower the percentage of LCFUc killed by araC, the greater the likelihood of response (Table 7). This paradoxical observation is without explanation at the present time, since the data are unchanged even when the ^{3}HTdR suicide indices of the LCFUc are included in the calculations (Preisler 1980b). These data serve to emphasize our ignorance of the in vivo conditions which determine response to therapy.

When one considers the high CR rates for ANLL at initial diagnosis (Preisler et al. 1979a, b) and at first relapse (Goldberg et al., to be published), together with the fact that only a small percentage of the treatment failures are ascribable to "resistant" leukemia (Table 8), on can see that the clonogenic assay is of little practical utility in this setting. On the other hand, its role in helping to select appropriate treatment for multiply relapsed patients or in projecting likely remission durations remains undefined.

Comments on Drug Sensitivity Assay Based upon Alterations in In Vitro Clonogenicity

In the past it has generally been assumed that the reported correlations between in vitro drug sensitivity as measured by clonogenicity assays and clinical response to therapy existed because the in vitro assay was assessing the effects of chemotherapeutic agents upon the relevant cell, the tumor stem cell. The other putative in vitro predictive assays were believed to be ineffective because they assessed the properties of the tumor cell population

Table 7. Relationship between the killing of leukemic clonogenic cells (LCFUc) and the outcome of remission induction therapy

	n CR	*n* RD	
araC/anthracycline[a]	20 (48%)[b]	11 (11%)	$p = 0.04$
HDaraC[b, c]			
0.3 µg/ml	5 (0.3%)	9 (30%)	$p = 0.003$
3.0 µg/ml	7 (0.3%)	7 (40%)	$p = 0.01$
30 µg/ml	7 (15%)	7 (43%)	NS

n, number of patients; *CR*, complete remission; *RD*, resistant disease; *araC*, cytosine arabinoside
Percentages represent median percentage decrease in LCFUc
[a] Patients treated with araC/anthracycline remission induction therapy
[b] Patients treated with high-dose araC remission induction therapy
[c] Leukemic marrow cells exposed to indicated concentration of araC for 1 h prior to plating

Table 8. Remission induction outcome for previously untreated and first-relapse patients with acute nonlymphocytic leukemia

	No. of patients	CR	RD	Other failure
No prior therapy	112	63 (56%)	15 (13%)	34 (31%)
First relapse	40	20 (50%)	7 (18%)	13 (33%)

CR, complete remission; *RD*, resistant disease

Table 9. Effects of pulse exposure (1 h) to doxorubicin on P388 murine leukemic cells

	% Clonogenic cells killed (mean)
1. Cells were exposed to doxorubicin for 1 h	82
2. Cells were exposed to doxorubicin for 1 h, washed, and cloned for 10 days; then 20 individual colonies were picked and divided in half and half were reexposed to doxorubicin for 1 h and cloned	92
3. Cells were cloned in the presence of doxorubicin	98.4
4. Colonies which grew in the presence of doxorubicin (3 above) were picked and dispersed and half the cells were recloned in the presence of doxorubicin	81
5. Colonies which grew under 4 above were picked and recloned in the presence of doxorubicin	30

Experiments 1 and 2 demonstrate that the progeny of the clonogenic cells which survived 1 h exposure to doxorubicin were just as sensitive as the cell population prior to initial exposure
Experiments 3–5 demonstrate the stepwise increase in doxorubicin resistance when cells were continuously exposed to the drug. It should be noted that for experiment 4, the cells of five of the 20 colonies which were picked after experiment 3 and recloned in the presence of doxorubicin failed to grow, despite the fact that they grew during the initial exposure, indicating that transient resistance had occurred

as a whole, which differed from those of the stem cells (Preisler and Rustum 1978). The data presented above demonstrate that this assumption was incorrect. Depending on the agent under consideration and the regimen employed, study of the population as a whole may provide clinically useful information. In fact, at least with respect to HDaraC, the fact that araC inhibition of DNA synthesis of the population as a whole correlates with the presence of drug-sensitive disease makes much more intuitive sense than the fact that the lower the percentage LCFUc killed by araC, the greater is the likelihood that a patient will enter CR.

Detailed study of the clonogenicity drug sensitivity assay has demonstrated several unexpected characteristics of this system. It might be logically assumed that cells which survived exposure to a chemotherapeutic agent would be drug resistant. In fact, the nature of the surviving cells depends upon the conditions of drug exposure. When P388 murine leukemic cells sensitive to doxorubicin were exposed to the drug for 1 h, the surviving cells were just as sensitive as the cells which were killed (Table 9). In contrast, if cells were exposed continuously to doxorubicin, most of the cells which formed colonies in the presence of doxorubicin were doxorubin resistant (Table 9). Serial cloning in the presence of doxorubicin resulted in a stepwise increase in doxorubicin resistance. Interestingly, some of the colonies formed during the first period of growth in the presence of doxorubicin consisted of cells which were incapable of growth when recloned in the presence of the drug. This "reversible" or "transient" drug resistance and the stepwise incremental "permanent" resistance suggests the presence of at least two mechanisms of resistance with differing therapeutic implications, since patients whose cells manifest the former type of resistance may benefit from retreatment with doxorubicin, while patients in whom an initial course of therapy fails because of the latter type of resistance would not.

Further consideration of these data suggest that clonogenicity assays which employ exposures to drugs for transient periods of time provide what is in effect an "average" sensitivity value for the population as a whole and those assays are not capable of detecting the rare drug-resistant cell. The cells which survive a pulse drug exposure consist of cells whose drug sensitivities are indistinguishable from the cells which were killed together with any drug-resistant cells which may be present. The former phenomenon, the differential killing and survival of equally sensitive cells, reflects the stochastic nature of the killing of sensitive cells, which in turn is reflective of phenomena such as the uptake of drugs, the interaction of drugs with intracellular targets, and other critical events which are affected by random phenomena.

On the other hand, the inclusion of the drug in the agar permits, in general, the growth only of drug-resistant cells. A potential disadvantage of this approach is that some drug effects which impair cell proliferation may be reversible upon removal of the drug. If this were the case, clonal growth would be inhibited, but the reduction in the number of colonies produced would not necessarily be indicative of cell killing.

One of the purposes of the clonogenic assay is to detect differences in drug resistance among different subpopulations which may be present. Unfortunately, the assay is incapable at present of performing this task. In the experiment reported in Table 10, cell mixtures consisting of different proportions of doxorubicin-sensitive and -resistant cells were exposed to doxorubicin and then cloned in vitro. In every instance, the assay underestimated the proportion of resistant cells which were present because the cloning efficiency of the doxorubicin-resistant cells was lower than that of the sensitive cells (WeiDong et al., to be published). Since this is an inherent characteristic of the system, clonogenic drug sensitivity assays can provide information about the proportion of

Table 10. Detection of doxorubicin-resistant cell subpopulation by clonogenic drug sensitivity assay

Actual % resistant cells	% Resistant cells estimated by clonogenic assay
100	100
75	58
50	30
25	15
0	0

For these studies, two subclones of P388 cells were used, one sensitive and one resistant to doxorubicin. They were mixed in varying proportion, exposed to doxorubicin for 1 h and then cloned in vitro, and the percentage of resistant cells was estimated by the percentage reduction in colony growth

different subpopulations of cells which are present only if the cloning efficiencies of each subpopulation are known before the assay is performed, clearly an impossibility when clinical specimens are studied.

Taken together, these data demonstrate several potential problems inherent in clonogenic drug sensitivity assays. The potential seriousness of these problems becomes clear when one considers the fact that just as some tumor specimens may not clone in vitro some tumor subpopulations may not be clonogenic in vitro. If this is the case, then the clinical relevance of the drug sensitivity assay will depend upon the clinical relevance of the populations which do or do not clone in vitro. This is simply an extreme example of the problems encountered when tumor subpopulations having different cloning efficiencies are cloned in vitro.

Concluding Remarks

Since the purpose of developing in vitro drug sensitivity assays is to develop the means of designing rational therapeutic regimens, it must be remembered that the drug sensitivity of the neoplastic cells is but one factor which plays a role in determining response to therapy. Figure 2 illustrates the three major determinants of response; the interaction of each with the other two determines the outcome of therapy. The biological status of the patient determines the ability of the patient to survive the leukemia per se and also determines which drugs should be used and the maximum intensity of therapy which is compatible with patient survival. The drug sensitivity of the leukemic cells determines the potential sensitivity of the leukemia to chemotherapeutic agents, but not the actual effects of drug administration on leukemic cells in vivo, since the killing of sensitive cells depends upon delivery of the drug to the target. Intergroup leukemia studies of plasma doxorubicin levels have demonstrated wide variability in plasma drug levels among patients treated with identical doses of doxorubicin (Preisler et al., to be published d). In some patients very high levels were associated with death, while in other patients the plasma levels were so low that only patients with extremely sensitive leukemia were likely to benefit from therapy. Hence, knowledge of all three parameters is necessary if truly rational chemotherapeutic regimens are to be designed for individual patients.

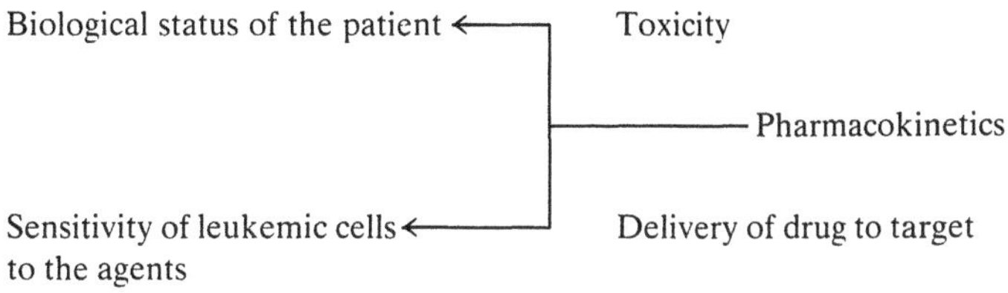

Fig. 2. Determinants of outcome of chemotherapy

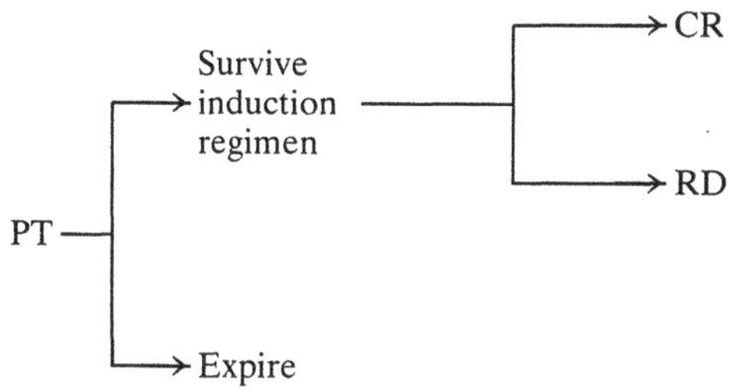

Fig. 3. The decision-making sequence for treatment of patients with acute nonlymphocytic leukemia. *PT,* patient; *CR,* complete remission; *RD,* resistant disease

The decision-making sequence for treating ANLL patients is given in Fig. 3. The first question which should be asked is whether or not the patient is likely to survive the regimen(s) under consideration. If the answer is yes, than the second question becomes significant: can the regimen(s) under consideration produce an adequate reduction of the leukemic cell numbers to permit a resumption of normal hematopoiesis? The questions are formulated in this sequence rather than in the reverse, since as the data in Table 8 demonstrate, death during remission induction therapy is a more common cause of treatment failure than is persistent leukemia. Hence, while intensive efforts are directed toward developing in vitro assays, equally intensive efforts should be directed toward identifying patients who are at high risk of dying during induction therapy, as well as the likely reasons for death, so that measures can be taken to prevent a fatal outcome.

References

Chou TC, Arlin Z, Clarkson BD et al. (1977) Metabolism of 1-B-d-arabinofuranosylocytosine in human leukemic cells. Cancer Res 37: 3561–3570

Early AP, Preisler HD, Slocum S, Rustum YM (1982) A pilot study of high dose 1-B-arabinofuranosylcytosine for acute leukemia and refractory lymphoma: clinical response and pharmacology. Cancer Res 45: 1587–1594

Goldberg J, Tice D, Nelson DA, Gottlieb AJ (1979) Predictive value of in vitro colony and cluster formation in acute nonlymphocytic leukemia. Am J Med Sci 277: 81–84

Goldberg J, Grunwald H, Vogler R, Browman G, Miller K, Gottlieb A, Brennan J, Chervenick P, Azarnia N, Priore RL, Preisler HD (to be published) Treatment of patients with acute nonlymphocytic leukemia in relapse: a leukemia intergroup study. Amer J Hematol

Golomb HM, Vardiman JW, Rowley JD et al. (1978) Correlation of clinical findings with quinacrine-banded chromosomes in 90 adults with acute nonlymphocytic leukemia. N Eng J Med 299: 613–619

Karanes C, Wolff SM, Herzig GP, Phillips GL, Lazarus HM, Herzig RL (1979) High dose cytosine arabinoside in the treatment of patients with refractory acute nonlymphocytic leukemia. Blood 54: 191a

Kufe DW, Major PP, Egan EM, Beardsley GP (1980) Correlation of cytotoxicity with incorporation of cytosine arabinoside into DNA. J Biol Chem 255: 8997

Moore MAS, Spitzer G, Williams N, Metcalf D, Buckley J (1978) Agar culture studies in 127 cases of untreated acute leukemia: the prognostic value of reclassification of leukemia according to in vitro growth characteristics. Blood 44: 1–18

Preisler HD (1978) Failure of remission induction in acute myelocytic leukemia. Med Pediatr Oncol 4: 275–276

Preisler HD (1980a) Evaluation of in vitro predicitve assays for acute myelocytic leukemic. Blut 41: 393–396

Preisler HD (1980b) Prediction of response to chemotherapy in acute myelocytic leukemia. Blood 56(3): 361–367

Preisler HD (to be published a) Intergroup leukemia study: inhibition of DNA synthesis by Cytosine Arabinoside: relation to response of acute nonlymphocytic leukemia to remission induction therapy and to stage of the disease

Preisler HD (to be published b) Intergroup leukemia study: changes in the characteristics of the bone marrow during therapy for acute nonlymphocytic leukemia: relationship to response to remission induction therapy

Preisler HD, Azarnia N (to be published a) Assessment of the drug sensitivity of acute nonlymphocytic leukemia using the in vitro clonogenic assay Br J Haematol in press

Preisler HD, Azarnia N (to be published b) Mechanisms of resistance to six day "high dose" cytosine arabinoside (araC) therapy in patients with acute nonlymphocytic leukemia (ANLL). Am Assoc Cancer Res

Preisler HD, Raza A (to be published) Uptake of Adriamycin by human leukemic cells as measured by flow cytometry. Med Oncol Pharmacol

Preisler HD, Rustum Y (1978) Prediction of therapeutic response in acute myelogenous Leukemia. In: Neth R, Gallo RC, Hofschneider P-H, Mannweiler K (eds) Modern trends in leukemia III. Springer, Berlin Heidelberg New York, pp 93–98

Preisler HD, Henderson ES, Bjornsson S et al. (1979a) Treatment of acute nonlymphocytic leukemia: use of anthracycline-cytosine arabinoside induction therapy as a comparison of two maintenance regimens. Blood 53: 455–464

Preisler HD, Bjornsson S, Henderson ES, Hyrniuk W, Higby D (1979b) Remission induction in acute nonlymphocytic leukemia: comparison of a 7-day and 10-day infusion of cytosine arabinoside in combination with Adriamycin. Med Pediatr Oncol 7: 269–278

Preisler HD, Reese PA, Marinello M, Pothier L (1983a) Adverse effects of aneuploidy on the outcome of remission induction therapy for acute nonlymphocytic leukaemia: analysis of types of treatment failure. Br J Haematol 53: 459–466

Preisler H-D, Early AP, Raza A, Vlahides G, Marinello MJ, Stein AM, Browman G (1983b) Therapy for secondary acute nonlymphocytic leukemia with cytosine arabinoside. N Eng J Med 308: 21–22

Preisler HD, Azarnia N, Marinello MJ (to be published a) Relationship of growth of leukemic cells in vitro to the outcome of therapy for acute nonlymphocytic leukemia. Cancer Res

Preisler HD, Rustum Y, Priore R (to be published b) Relationship between leukemic cell metabolism of cytosine arabinoside and the outcome of chemotherapy for AML. Eur J Cancer and Clin Oncol

Preisler HD, Epstein J, Barcos M, Priore R, Browman G, Vogler R, Grunwald H, Brennan J, Goldberg J, Chervenick P, Tricot G (to be published c) Prediction of response of patients with acute nonlymphocytic leukemia to remission induction therapy: use of clinical measurements. Eur J Cancer and Clin Oncol

Preisler HD, Guessner T, Azarnia N, Bolanska W, Epstein J, Early AP, D'Arrigo P, Browman G, Miller K, Vogler R, Grunwald H, Chervenick P, Joyce R, Larson R, Lee H, Steele R, Brennan J, Goldberg J (to be published d) Relationship between Adriamycin plasma levels and the outcome of remission induction therapy for ANLL. Cancer Chemother Pharmacol

Raza A, Preisler HD, Kuliczkowski K, Zhao S (1983) Studies of the sensitivity of leukemic cells to cytosine arabinoside in vitro (Abstract). International Conference on Predictive Drug Testing on Human Tumor Cells, Zurich, July 20–22

Rustum YM, Preisler HD (1979) Correlation between leukemic cell retention of 1-B-D-arabinosylocytosine-5'-triphosphate and response to therapy. Cancer Res 39: 42–49

Sakukrai M, Sandberg AA (1973) Prognosis of acute myeloblastic leukemia: chromosomal correlation. Blood 41: 93–104

Schwartz HS, Preisler HD, Kanter PM (1981) DNA damage in AML cells exposed to Adriamycin: correlations with clinical response to therapy. Leuk Res 5: 363–366

Smyth JF, Robins AD, Leese CL (1976) The metabolism of cytosine arabinoside as a predictive test for clinical response to the drug in acute myeloid leukemia. Eur J Cancer 12: 567–573

Spitzer G, Dicke KA, Gehan EA, Smith T et al. (1976) A simplified in vitro classification for prognosis in adult acute leukemia: the application of in vitro results in remission – predictive models. Blood 48: 795–807

WeiDong G, Preisler HD, Priore T (to be published) Potential limitations of in vitro clonogenic drug sensitivity assays. Cancer Chemother Pharmacol

*Short-Term In Vitro Sensitivity Testing in Acute Leukemia**

J.D. Schwarzmeier, R. Pirker, and E. Paietta

I. Medizinische Universitätsklinik, Lazarettgasse 14, 1090 Wien, Austria

Introduction

The treatment of acute leukemia with highly cytotoxic drug regimens has made well-planned protocols necessary in order to avoid undue exposure of the patients. It has therefore become more and more desirable to find determinant initial factors related to possibly good or poor responses of the patients to cytostatic therapy. Numerous investigators have attempted to establish such factors and have stressed the role of pretreatment cytokinetics, leukemic cell mass, cytologic criteria, chromosomal aberrations, etc. Previous studies in our laboratory have emphasized the predictive implications of cyclic adenosine monophosphate in leukemic cells and have reconfirmed the prognostic value of proliferation kinetics (Paietta et al. 1980). A more direct approach to detect sensitivity of leukemic blast cells, however, is their in vitro exposure to specific chemotherapeutic agents and the determination of reliable parameters of cell viability (Dow 1980; Preisler 1980; Spiro et al. 1981; Von Hoff and Weisenthal 1980; Weisenthal 1981). Since the course of the disease in patients with acute leukemia very often requires prompt therapeutic decisions, the application of short-term tests seems to be more feasible than the use of time-consuming but perhaps more sensitive assays. We therefore investigated the in vitro effect of various cytostatic agents on the incorporation of nucleoside precursors into leukemic blast cells and correlated the in vitro results with the clinical response of the patients to chemotherapy.

Materials and Methods

Cell Lines

Human cell lines established from a promyelocytic leukemia (HL-60) and from lymphomas (Raji, Daudi) were used in order to standardize the assay procedure. The cells were kindly provided by Dr. W. Knapp (Institute of Immunology, University of Vienna).

* Supported by the Fonds zur Förderung der wissenschaftlichen Forschung in Austria, Project No. 4782

Recent Results in Cancer Research. Vol. 94

Freshly Isolated Leukemic Cells

Human leukemic blast cells were obtained from patients with acute myelocytic (AML), myelomonocytic (AMMOL), monocytic (AMOL), lymphocytic (ALL), and undifferentiated (AUL) leukemia. The blast cells were isolated from heparinized bone marrow aspirates by Ficoll-Hypaque gradient centrifugation according to Bøyum (1968) and resuspended in TC Medium RPMI (Difco). If bone marrow samples yielded insufficient numbers of blast cells, heparinized peripheral blood was used.

Radioactive Substances

5-^3H-Uridine (sp. act. 5 Ci/mmol, 2.5 μCi/ml cell suspension), methyl-^3H-thymidine (sp. act. 25 Ci/mmol, 2.5 μCi/ml), and deoxy-6-^3H-uridine (sp. act. 17.5 Ci/mmol, 2.5 μCi/ml) were purchased from the Radiochemical Centre, Amersham, England.

Cytostatic Agents

The following cytostatic agents were used:
Doxorubicin (adriamycin) (Adriblastin, AESCA, Vienna; max.conc. 5.26 μg/ml cell suspension)
Cytosine arabinoside, ara-C (Alexan, Mack, Vienna; max.conc. 52.6 μg/ml)
VP-16213, etoposide (Vepesid, Bristol-Laevosan, Linz; max.conc. 10 μg/ml)
4-Hydroperoxycyclophosphamide (ASTA-Werke AG, Bielefeld, FRG, kindly provided by Dr. H. Burkert; max.conc. 5 mg/ml)
6-Mercaptopurine (Puri-Nethol, Wellcome Fdn, Vienna; max.conc. 65.8 μg/ml)
Thioguanine (Thioguanin, Wellcome Fdn, Vienna; max.conc. 52.6 μg/ml)
Methotrexate (Methotrexat, Lederle, Wolfratshausen, FRG; max.conc. 12.5 μg/ml)
Prednisone (Prednisolon, Chemie-Linz, Linz; max.conc. 65.8 μg/ml)
Betamethasone (Solu-Celestan, AESCA, Vienna; max.conc. 20 μg/ml).

Patients

During a period of 3 years a total of 64 adults were evaluated, of whom 12 had to be excluded from the study because of early death during the course of therapy or because of inadequate test results. Remission induction chemotherapy in patients with acute nonlymphocytic leukemia consisted of doxorubicin and cytosin arabinoside, either in 5-day or 7-day cycles, and in patients with ALL of doxorubicin, vincristine, and prednisone as outlined elsewhere (Paietta et al. 1980). The clinical response to therapy was divided into complete remission (normal cellular M_1 bone marrow, peripheral granulocytes more than 1,000/μl, platelets more than 100,000/μl, hemoglobin concentration more than 10 g/dl for a minimum of 30 days), clinical response (clearance of bone marrow from blast cells, but death during the period of aplasia), and failure to respond.

Test Procedure

The effect of chemotherapeutic agents on leukemic cells (cell lines or freshly isolated blast cells) was evaluated by their ability to inhibit the incorporation of nucleoside precursors. A method similar to the one originally described by Volm et al. (1970) for solid tumors was adapted for leukemic cells (Schwarzmeier et al. 1984). Briefly, after a 30 min preincubation period 1×10^6 cells were incubated for 2 h at 37° C in the absence (controls) or presence of cytostatic agents at concentrations of 1/10, 1/100, and 1/000 of their maximum concentrations indicated above. After this period one of the radioactive precursors (^{3}H-uridine, ^{3}H-deoxyuridine, ^{3}H-thymidine) was added for another hour. Aliquots (100 µl) of the cell suspensions were pipetted onto glass fiber filters (Whatman GF/C), the nonincorporated radioactivity was extracted (5% icecold TCA), and the filters counted in a liquid scintillation counter. The inhibition of precursor incorporation by cytostatic drugs was expressed as percentage values by reference to 0% inhibition in cells not exposed to the drugs (controls). All samples were run in triplicate.

Results

Nucleoside Precursor Incorporation in Leukemic Cell Lines

Several established cell lines (HL-60, Raji, Daudi) were tested to find the optimal assay conditions for leukemic cells and to determine the radiolabeled nucleoside precursors which reflect the effect of a given drug in a specific and dose-related fashion. Figure 1

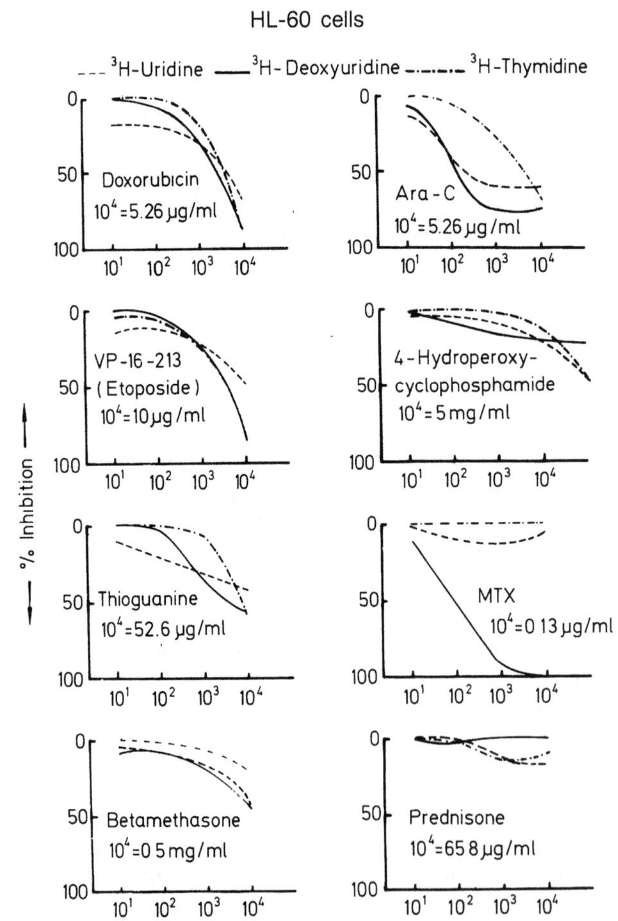

Fig. 1. Effect of chemotherapeutic agents on the incorporation of ^{3}*H*-uridine (– – –), ^{3}*H*-deoxyuridine (———), and ^{3}*H*-thymidine (-· · · -) into HL-60 cells. Results are expressed as percentage inhibition compared to cells not exposed to the drugs (0%). Mean values of three experiments. *Ara-C*, cytosine arabinoside; *MTX*, methotrexate

shows the results on HL-60 cells obtained with a battery of cytostatic agents frequently used in the treatment of acute leukemia. Typical dose-response curves were observed for doxorubicin, ara-C, 4-*OH*-cyclophosphamide, and thioguanine on the incorporation of ^3H-uridine as well as ^3H-deoxyuridine and ^3H-thymidine. Of the two corticosteroids tested, betamethasone but not prednisone yielded clear dose-response curves. The effect of methotrexate, as expected, was best reflected by ^3H-deoxyuridine. Since the experiments with Raji and Daudi cells yielded results very similar to that obtained with HL-60 cells, in subsequent tests with freshly isolated leukemic cells ^3H-deoxyuridine was used to monitor the effect of methotrexate, and ^3H-uridine the effects of the other drugs mentioned above. In addition, the rate of cellular DNA synthesis in the blast cells was evaluated, measuring the uptake of ^3H-thymidine in the absence of cytostatic agents.

In a number of experiments we examined the effect of drug combinations to find out whether the in vitro test also detects synergistic as well as antagonistic drug effects. Figure 2 summarizes some of the results obtained with HL-60 and Raji cells. A constant dosis of VP-16 leading to 15% and 25%inhibition of ^3H-uridine incorporation was combined with various concentrations of ara-C, doxorubicin, 4-*OH*-cyclophosphamide, or thioguanine. As can be seen, in no case was the combination of a pair of drugs more effective than the single agent. In contrast, combinations of methotrexate with ara-C, doxorubicin, VP-16, or 4-*OH*-cyclophosphamide, when tested on Raji cells, showed a greater inhibition of ^3H-deoxyuridine incorporation than either drug alone.

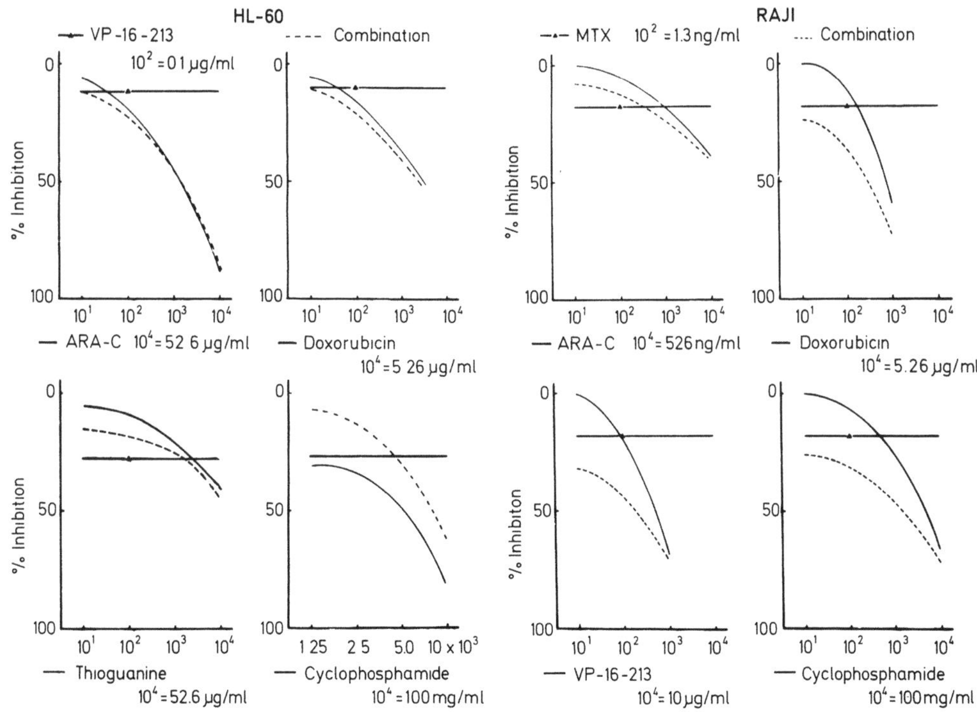

Fig. 2. Comparison of the effect of single chemotherapeutic agents (———) versus the effect of drug combinations [(pairs of drugs simultaneously added to the cell suspension (– – –)] on the incorporation of ^3H-deoxyuridine into Raji cells and of ^3H-uridine into HL-60 cells. Combinations were: methotrexate (MTX) + ara-C, MTX + doxorubicin (adriamycin), MTX + VP-16, MTX + 4-*OH*-cyclophosphamide; VP-16 + ara-C, VP-16 + doxorubicin, VP-16 + thioguanine, VP-16 + 4-*OH*-cyclophosphamide; *Ara-C,* cytosine arabinoside

Pretherapeutic In Vitro Testing: Correlation with In Vivo Response

Figure 3 retrospectively correlates the response of 48 acute leukemia patients to chemotherapy with the in vitro inhibition of ^{3}H-uridine incorporation into their blast cells by doxorubicin. In 23 out of 24 patients who showed response to therapy, the pretherapeutic in vitro inhibition by doxorubicin was above 30%. False positive in vitro results, i.e., in vitro doxorubicin effect above 30% but lack of therapeutic response, applied to four out of 25 failures. Thus, in the great majority of patients, in vitro sensitivity as well as resistance to doxorubicin reflected the in vivo situation. The level of in vitro inhibition that could differentiate between doxorubicin-sensitive and -resistant disease appeared to be 30%.

With respect to ara-C, 36 patients receiving this drug were evaluated. Four cases of ALL were included into this group, since they were given ara-C in combination with other drugs (doxorubicin, vincristine, prednisone) during reinduction chemotherapy. In 17 of the cases which showed clinical response to polychemotherapy including ara-C, the in vitro inhibition by ara-C had been at least 20%. In four out of 19 failures to respond to therapy, the ara-C inhibition had also exceeded this threshold level; thus they represented false positive in vitro results. In contrast, no false negative results, i.e., in vitro resistance but in vivo responsiveness, were observed.

Further substances included in the test scheme were 6-mercaptopurine, thioguanine, and 4-*OH*-cyclophosphamide. Since these drugs were not a constituent part of our induction therapy regimens, their in vitro effects could not be regularly correlated with the in vivo responses. As outlined below, however, in individual patients, the clinical relevance of in vitro test results with these substances could be demonstrated.

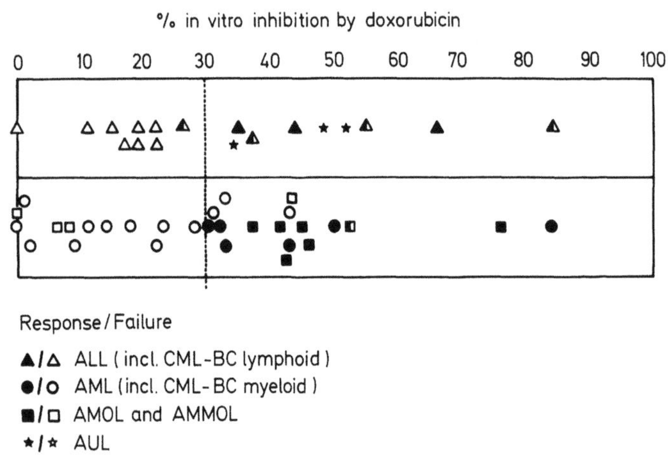

Fig. 3. Correlation of in vitro inhibition of ^{3}H-uridine incorporation by doxorubicin (adriamycin) (5.5 µg/ml) with the clinical outcome in acute leukemia patients following polychemotherapy including doxorubicin. The various types of leukemias tested are as follows: acute lymphocytic leukemia *(ALL),* acute myelocytic leukemia *(AML),* acute monocytic leukemia and myelomonocytic leukemia *(AMOL, AMMOL),* and acute undifferentiated leukemia *(AUL). Closed symbols* represent patients who achieved complete remission; *half-closed symbols* represent patients who died during a period of severe marrow hypoplasia without evidence of leukemia; *open symbols* represent failure to respond. *CML-BC,* chronic myelocytic leukemia → blast crisis

Correlation of Cell Proliferation with In Vitro Sensitivity

The dependence of in vitro sensitivity on the proliferative state of a leukemic cell population was evaluated by comparing the in vitro inhibition of ^3H-uridine incorporation by doxorubicin with the cellular uptake of ^3H-thymidine as a measure of DNA synthesis. As shown in Fig. 4, these two parameters were not related positively to each other.

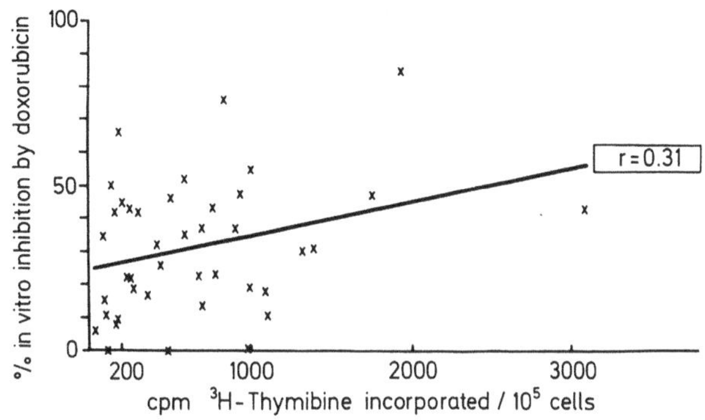

Fig. 4. Correlation of ^3H-thymidine incorporation with the inhibitory effect of doxorubicin (adriamycin) on the ^3H-uridine uptake of acute leukemic cells in vitro

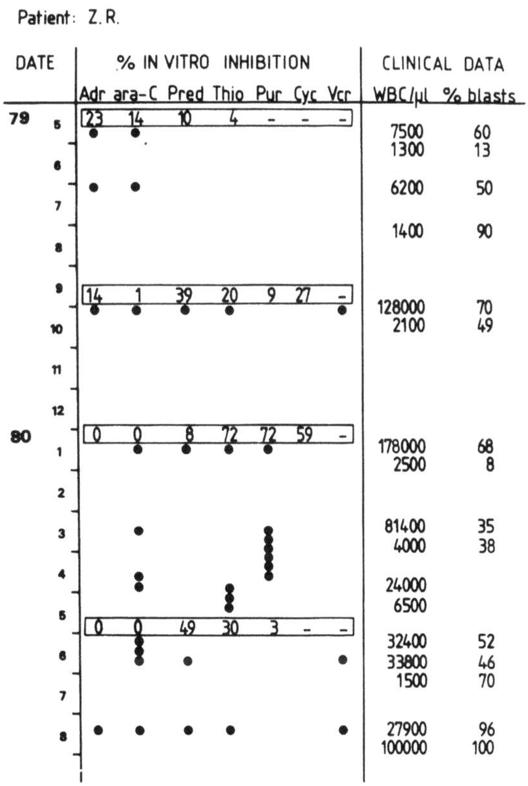

Fig. 5. Clinical course in a patient with acute myelocytic leukemia. The results of in vitro sensitivity testing expressed as percent inhibition of ^3H-uridine incorporation (arabic numbers) were compared with the clinical response of the patient to chemotherapy (●) as indicated by the white blood cell count *(WBC/μl)* and the percentage of blast cells in the peripheral blood. The abbreviations used for the cytostatics are: *Adr,* adriamycin; *ara-C,* cytosine arabinoside; *Pred,* prednisone; *Thio,* thioguanine; *Pur,* 6-mercaptopurine; *Cyc,* 4-*OH*-cyclophosphamide; *Vcr,* vincristine. Maximum concentrations of cytostatics for in vitro testing were as indicated in Methods

Serial In Vitro Testing During the Course of Acute Leukemia

In particular cases who did not achieve complete remission or who developed drug resistance during the course of their disease, in vitro test results obtained with a series of cytostatic agents proved to be especially helpful. Figure 5 gives an example: in a patient with AML, the maximal inhibitory effect of doxorubicin in vitro initially was only 23% and that of ara-C only 14%. These values were below the threshold levels established for these drugs (30% and 20% respectively), thereby pointing toward a probable in vivo resistance of the blast cells. In fact, the clinical response to chemotherapy was poor. After two induction therapy courses, the peripheral blast cell count was still 90%. When the leukocyte count rose to 128,000/µl, doxorubicin and ara-C had become less effective in vitro than in the initial testing. Addition of thioguanine to the cytostatic regimen in accordance with its strong in vitro effect yielded a partial in vivo response. At relapse, a high in vitro sensitivity to thioguanine- 6-mercaptopurine, and cyclophosphamide was again observed, whereas doxorubicin and ara-C had lost their in vitro effectiveness completely. Thioguanine and 6-mercaptopurine were included in the therapeutic regimen during the following months, resulting in a significant reduction of the blast cell count and a considerable improvement of the performance status of the patient. Further progression of the disease, however, revealed an increasing in vitro resistance to all the drugs tested, paralleled by a gradual loss of responsiveness in vivo.

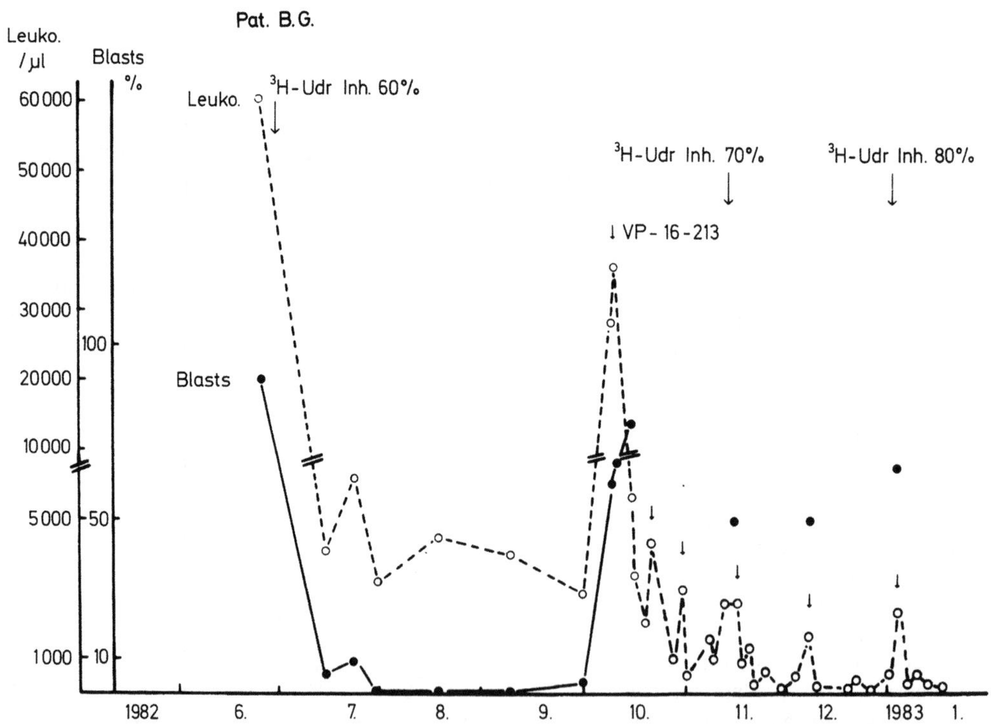

Fig. 6. Leukocyte count (O– – –O) and percentage of blast cells (●——●) in a patient with acute monocytic leukemia during the course of chemotherapy. The in vitro sensitivity test was performed three times within a period of 6 months, the inhibition of ³*H*-uridine (³H-Udr) incorporation by VP-16 is indicated as percentage values. Courses of therapy with VP-16 are symbolized by ↓

Another example is shown in Fig. 6. A patient with AMOL, having responded to induction chemotherapy with complete hematological remission, relapsed 3 months later. Since the blast cells had initially expressed in vitro sensitivity to VP-16 (even with a thousandfold dilution of the drug and its potentially cytotoxic solvent), the substance was added to the following treatment cycles. As can be seen, the patient responded with a marked reduction of blast cells. The effect was transient, however, and even after the induction of an extended period of bone marrow aplasia by repeated courses of VP-16, normal hematopoiesis could not be restored. Although the test results accurately predicted the in vivo responsiveness to the drug, an obvious lack of normal stem cells permitted the reappearance and the uncontrolled growth of the leukemic cell population.

Discussion

Aggressive, high-dose chemotherapy and adequate supportive care have markedly increased the remission rate of patients with acute leukemia. However, the overall disease-free survival has remained disappointingly low. This necessitates the exploration of other treatment modalities. Tailoring the treatment by drug sensitivity tests might be one of the tools to extend remission durations and to improve substantially the therapeutic results.

Short-term in vitro tests based upon the incorporation of nucleic acid precursors into tumor cells have been used by several groups to study chemosensitivity of solid human tumors, and reasonably good correlations have been reported between in vitro and in vivo results (Group for Sensitivity testing of Tumors 1981; Sanfilippo et al. 1982, Volm et al. 1975). This prompted us to apply the method to leukemic cells. Cell damage, a serious problem in studies with solid tumors, is minimal and artifacts due to the isolation procedure can be neglected in leukemic cells.

Our results obtained with leukemic cell lines indicate that the effect of most of the drugs employed in this study is reflected in a dose-dependent fashion by the short-term test. As expected, ^{3}H-deoxyuridine is the appropriate nucleoside precursor for methotrexate, and ^{3}H-uridine as well as ^{3}H-thymidine can be used to monitor the effects of doxorubicin, ara-C, cyclophosphamide, VP-16213, thioguanine, and betamethasone. We decided to use ^{3}H-uridine and ^{3}H-deoxyurine to monitor the drug effects and ^{3}H-thymidine to measure the DNA synthesis in the cells prior to the drug exposure. To avoid inhomogeneity of the test results through the influence of various factors (e.g., differences in nucleotide pool size) we expressed the results as inhibition in percent of the incorporation rate of control cells not exposed to the cytostatics.

Since no data were available on the capability of the test system to monitor the effect of drug combinations, we investigated the influence of a number of cytostatic agents on the inhibitory effect of a constant dosis of a given drug. From the many experiments performed, two examples are given. They show either a low degree of synergism (combinations of methotrexate with doxorubicin, 4-*OH*-cyclophosphamide, ara-C, VP-16) or no potentiation of the single drug effect (combinations of VP-16 with doxorubicin, 4-*OH*-cyclophosphamide, ara-C, thioguanine). Although these results are preliminary and much work has still to be done, we believe that the short-term test in its present form is not suited to simulate the in vivo effect of drug combinations.

In vitro sensitivity tests cannot be expected to predict the clinical outcome of remission induction therapy in acute leukemia. The success of any treatment protocol ultimately depends on a number of clinical factors, such as the biological characteristics of the patient, the behavior during drug-induced marrow hypoplasia, and the capacity of the bone marrow to recover adequately. The results can, however, predict the ability of drugs to reduce the leukemic cells in vivo. This was fully confirmed by our study. With doxorubicin we found that in vitro inhibition of ^3H-uridine uptake by more than 30% corresponded significantly to an excellent in vivo response in terms of massive reduction of leukemic blast cells. Similarly, with ara-C, an inhibition rate of more than 20% indicated high drug sensitivity in vivo. Clear dose-related inhibition was observed in all cases whose blast cells showed responsiveness to doxorubicin and ara-C in vivo. Retrospectively, 83% of the patients with more than 30% doxorubicin-induced inhibition of precursor incorporation into leukemic cells in vitro came into complete hematological remission or showed clinical response under a treatment protocol including this drug. We were not able, however, to draw any conclusions with respect to remission durations. Since only single drugs were tested in vitro, no evidence could be derived for additive, synergistic or antagonistic effects within a polychemotherapy regimen in vivo.

Cline and Rosenbaum (1968), studying the influence of ara-C, vincristine, and cortisol on uridine incorporation into leukemic blast cells, found a positive correlation between in vitro and in vivo cytotoxicity but failed in predicting the likelihood of hematological remission. This might have been due to clinical as well as methodological factors. Because of the relatively long incubation periods, the in vivo situation might not have been reflected adequately. Studies from Brody (1979) comparing 72- and 4-h in vitro drug exposure of peripheral lymphocytes support the idea that short-term incubations are more appropriate for testing drug sensitivity of blood cells. It was shown that interference with intracellular DNA, RNA, and protein synthesis occurs promptly and is related quantitatively to drug concentration rather than to duration of contact.

Even though other substances besides doxorubicin and ara-C had also been included into our test system, we were not able statistically to compare the in vitro effects of these substances with the in vivo results, since they were not a constituent part of our induction therapy regimen. In individual cases, however, the effects of various drugs were monitored at different stages of the disease. A good correlation was seen between the ability of drugs to affect the leukemic blast cells in vitro and in vivo. Increased drug resistance observed with progression of the disease was always paralleled by a diminished in vitro response. This indicates that the test can actually help to avoid unnecessary exposure of the patients to ineffective drugs.

Since depression of DNA synthesis is a common manifestation of damage of proliferating cells, several investigators favor the measure of inhibition of DNA synthesis over that of RNA synthesis. Zittoun et al. (1975), studying the in vitro effect of cytostatic agents on blast cells from acute nonlymphocytic leukemias, did not find a parallel between the inhibition of incorporation of labeled thymidine and that of uridine. They concluded that prediction primarily concerns the sensitivity of the blast cells in the proliferative compartment. To test whether our in vitro results merely reflected the proliferative behavior of the cells or if other, proliferation-independent parameters were also recorded, we measured the incorporation of ^3H-thymidine into the cells and compared it with the doxorubicin-induced inhibition of ^3H-uridine uptake. When we related the rate of DNA synthesis to the inhibitory effect of doxorubicin in vitro, no correlation was detected. This indicated that the effect of doxorubicin was not solely dependent on the proliferative state of the cells. The concentrations of doxorubicin effective in vitro were several fold higher than the therapeutic levels normally achieved in man. Similarly, Dosik et al. (1981), determining dose-dependent suppression of DNA synthesis in leukemic cells by ara-C and doxorubicin, found that the anthracycline doses necessary in vitro by far exceeded the concentrations attained in vivo. Longer drug exposures and ^{3}H-nucleoside labeling times may perhaps increase the chance of detecting suppression of a greater portion of cells from the generally slowly cycling pool of leukemic blast cells as suggested by Raich (1978). The pitfalls associated with long incubations periods, however, favor the use of short-term tests as outlined above.

Summary

To detect sensitivity or resistance of leukemic cells to chemotherapy prior to treatment, a short-term incubation method was employed. Blast cells from the peripheral blood or bone marrow of adult patients with different forms of acute leukemia were analyzed for in vitro responsiveness to cytostatic agents in terms of suppression of nucleoside precursor (^{3}H-uridine, ^{3}H-deoxyuridine) incorporation into the cells. Retrosepctively, the in vitro data were compared to the clinical response of the patients to polychemotherapy. In the majority of patients, in vitro cytotoxic effectiveness of doxorubicin and ara-C reflected the in vivo situation. The levels of in vitro inhibition that could distinguish between drug-sensitive and drug-resistant diseases appeared to be 30% for doxorubicin and 20% for ara-C. No correlation was found between the doxorubicin effect in vitro and the proliferative state of the leukemic cell populations. Serial in vitro testing during the course of the disease of various patients proved the ability of the test system to detect acquired resistance to chemotherapeutic agents. Studies with established cell lines (HL-60, Raji) indicated that the short-term test in its present form is probably not suited to monitor the effect of drug combinations in vitro.

References

Bøyum A (1968) Isolation of mononuclear cells and granulocytes from human blood. Scand J Clin Lab Invest [Suppl 97] 21: 77–89

Brody JI (1979) In vitro effects of single alkylating agents on normal peripheral blood lymphocytes. J Lab Clin Med 94: 114–122

Cline MJ, Rosenbaum E (1968) Prediction of in vivo cytotoxicity of chemotherapeutic agents by their in vitro effects on leukocytes from patients with acute leukemia. Cancer Res 28: 2516–2521

Dosik GM, Barlogie B, Johnston D, Mellard D, Freireich EJ (1981) Dose-dependent suppression of DNA snythesis in vitro as a predictor of clinical response in adult myeloblastic leukemia. Eur J Cancer 17: 549–555

Dow LW (1980) Sensitivity of normal and neoplastic cells to chemotherapeutic agents in vitro. Adv Intern Med 25: 427–452

Group for Sensitivity Testing of Tumors (KSST) (1981) In vitro short term test to determine the resistance of human tumors to chemotherapy. Cancer 48: 2127–2135

Paietta E, Mittermayer K, Schwarzmeier JD (1980) Proliferation kinetics and cyclic AMP as prognostic factors in adult acute leukemia. Cancer 46: 102–108

Preisler HD (1980) Prediction of response to chemotherapy in acute myelocytic leukemia. Blood 56: 361–367

Raich PC (1978) Prediction of therapeutic response in acute leukaemia. Lancet 1: 74–76

Sanfilippo O, Silvestrini RS, Daidone MG (1982) In vitro anitmetabolic assay for the prediction of human tumor chemosensitivity. Proc ASCO 1: 24

Schwarzmeier JD, Paietta E, Mittermayer K, Pirker R (1984) Prediction of the response to chemotherapy in acute leukemia by a short term test in vitro. Cancer 53: 390–395

Spiro TE, Mattelaer MA, Efira A, Stryckmans P (1981) Sensitivity of myeloid progenitor cells in healthy subjects and patients with chronic myeloid leukemia to chemotherapeutic agents. J Natl Cancer Inst 66: 1053–1059
Volm M, Kaufmann M, Hinderer H, Goerttler K (1970) Schnellmethode zur Sensibilitätstestung maligner Tumoren gegenüber Cytostatika. Klin Wochenschr 48: 374–376
Volm M, Kaufmann M, Mattern J, Wayss K (1975) Möglichkeiten und Grenzen der prätherapeutischen Sensibilitätstestung von Tumoren gegen Zytostatika im Kurzzeittest. Schweiz Med Wochenschr 105: 74–82
Von Hoff DD, Weisenthal L (1980) In vitro methods to predict for patient response to chemotherapy. Adv Pharmacol Chemother 17: 133–156
Weisenthal LM (1981) In vitro assays in preclinical antineoplastic drug screening. Semin Oncol 8: 362–376
Zittoun R, Bouchard M, Facquet-Danis J, Percie-du-Sert M, Bousser J (1975) Prediction of the response to chemotherapy in acute leukemia. Cancer 35: 507–513

Non-Clonogenic Assays for Drug Testing

*Development of a Nucleotide Precursor Incorporation Assay for Testing Drug Sensitivity of Human Tumors**

O. Sanfilippo, M.G. Daidone, N. Zaffaroni, and R. Silvestrini

Istituto Nazionale per lo Studio e la Cura dei Tumori, Oncologia Sperimentale C,
Via Venezian 1, 20133 Milano, Italy

Introduction

Experimental research carried out in the past few years has led to different kinds of in vitro assays for drug screening on human tumors at the preclinical level. At present, many different systems, newly developed or derived from the improvement of previous ones (Daidone et al. 1981; Group for sensitivity testing of tumors 1981; Salmon 1980; Sanfilippo et al. 1981; Von Hoff et al. 1981; Weisenthal et al. 1983), have proved to be specific enough and potentially useful in different fields of application. Some of these systems appear more suitable for basic studies of biochemical and genetic aspects of cellular sensitivity and of the relationship between activity and pharmacokinetics at the cellular level; others can be used for studies more directly related to clinical application, such as the screening of new drugs and of the more active drugs on different tumor types or on individual patients for a tailored therapy.

Our main purpose was to set up an assay which would be mainly useful for immediate clinical use. To achieve this goal, we developed an in vitro short-term assay in which the effect of the drugs was evaluated through the interference of a central metabolic event, i.e., nucleic acid metabolism. The choice of this system arose mainly from its following characteristics: the relative technical simplicity, the potential applicability to a large percentage of tumors, and the possibility of maintaining the representativeness of the original tumor during the very short time of in vitro treatment.

In this chapter we summarize our efforts to set up an assay with all these characteristics and to verify its reliability, specificity, and potential clinical applicability through the analyses of in vitro results. All the analyses of clinical predictivity based on clinical correlations on individual patients, in part already previously reported (Daidone et al., to be published; Sanfilippo et al. 1983a, b), are detailed and discussed elsewhere in this volume.

Basic Methodology

Our assay was set up starting from the previous experience of other groups (Sky-Peck 1971). The choice of a very short assay in which the effect of the drugs could be evidenced as interference with central cellular metabolic events, such as nucleic acid metabolism,

* Supported in part by Grant PFCCN n° 83.00946.96

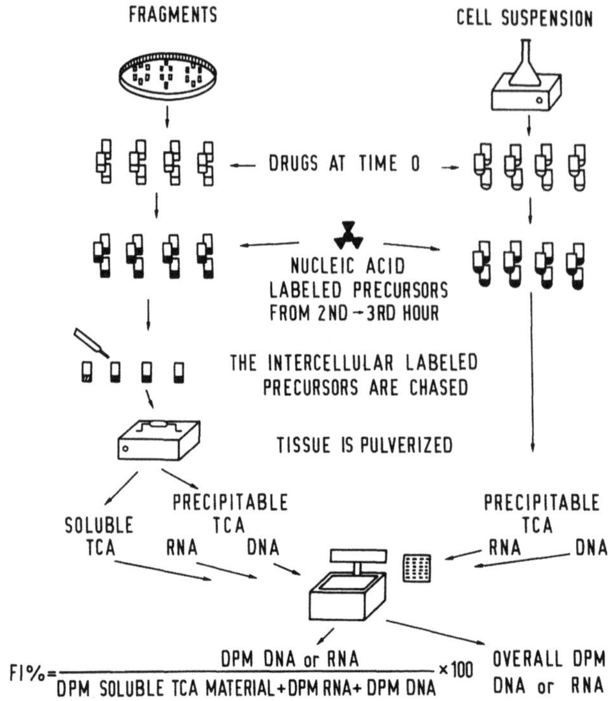

Fig. 1. Flow chart of the antimetabolic assay. The effect of the drugs is calculated as the percentage variation of the incorporation of treated in relation to untreated samples. *FI%*, fractional incorporation; *TCA*, trichloroacetic acid; *DPM*, disintegrations per minute

instead of cytocidal effect, which can be evidenced with longer assays, appeared to satisfy the basic characteristics we were looking for. Figure 1 summarizes the basic steps of the assay. Each step was checked to avoid as far as possible the interference of non-drug-related variations (such as the consequences of tumor material management, culture, or labeling conditions) with the evaluation of drug-induced effects.

Tumor material was derived from the sampling of different areas of a representative surgical biopsy in order to maintain closely the representativeness of the overall tumor cell population. For each tumor assayed, fragments or cell suspensions derived from all of the aforementioned samples were pooled and routinely submitted to histological verfication of correct tumor sampling and autoradiographic analysis of ^{3}H-thymidine labeling of tumor and non-tumor cells. The use of fragments of solid tumors avoids selections and modifications that may be induced in the overall tumor cell population by disaggregation procedures, which are necessary to obtain single-cell suspensions. The incubation time of 3 h was chosen as the time in which no significant modification of nucleic acid metabolism was observed as the consequence of culture conditions (Costa et al. 1977; Silvestrini et al. 1983).

The choice of the concentration of drugs to be used is one of the most controversial points for all in vitro assays. A general problem with all these assays is the best criterion for this choice, since there is no way to mimic with an in vitro assay the in vivo pharmacokinetic pattern of each drug. Moreover, the need for a method as simple and standardizable as possible for routine clinical use makes it impossible to conceive different schedules for in vitro testing of the different drugs. In our assay, the concentrations of drugs to be tested in vitro was derived from those used in clinical protocols, according to the formula:

$$\mu g/ml = \frac{mg \times \text{median body surface}}{\text{median weight}} \times \frac{100}{60}$$

which is based on the theoretical assumption of an equal distribution of the drug throughout the patient's body fluids (Tisman et al. 1973). Although absolutely theoretical, the use of this criterion has some advantages, such as the derivation of concentration values that are always in the ranges of plasma peak levels (Alberts et al. 1983) and the possibility of being applied to all drugs, both conventional and relatively new, such as those derived from phase I and II clinical studies. In any case, the reliability of such criteria, as well as those chosen for other assays, can be demonstrated only by the analysis of the specificity and clinical predictivity of in vitro variations induced by the different drugs.
The concentration of exogenous precursors was defined for each tumor type to guarantee the maintenance of steady state conditions for their incorporation into nucleic acids during the hour of in vitro labeling. These conditions prevent small variations in the labeled precursor uptake consequent to the modification of the specific activity of the intracellular pool that may mimic drug-induced variations in DNA and RNA synthesis. The expression of the incorporation of labeled precursors as fractional incorporation in tumor fragments, i.e., as the ratio between the incorporated and overall labeled precursors taken up by the cells, makes it possible to compare the unitary incorporation by cells able to incorporate because of physiological or technical conditions and overcomes heterogeneity among the different samples due to tumor cell content, penetration of labeled precursors within the fragments, presence of microareas of necrosis and stromal tissue.

Incorporation of Labeled Precursors

The incorporation of labeled precursors in control, untreated samples was generally sufficiently high in solid and nonsolid tumors; more than 90% of tumors had more than 100 cpm in each scintillation sample and therefore reliable incorporation values. The variation coefficients among triplicate samples (Table 1) ranged from very low to very high values and were similar for all the tumor types tested. Variation coefficients of drug-treated samples were occasionally analyzed and were similar to those of control samples. The assays in which the variation coefficients of control samples were less than 40% (about 80% of the cases) and those with markedly different or overlapping values of incorporation in treated vs control samples (about 10%) were considered reliable.

Table 1. Coefficients of variation (*CV*) among triplicate control samples

Tumor type	CV (%)	
	Median	Range
Non-Hodgkin's lymphoma	20	6–52
Breast cancer	19	2–67
Testicular tumor	27	3–67
Ovarian cancer	28	7–66
Melanoma	22	5–68
Overall	22	2–68

Table 2. Agreement between drug interference on ^3H-thymidine and ^3H-uridine incorporation

Tumor type	Drug	No. of tumors tested	Agreement of the effect on nucleic acids	
			%	*p*
Testicular	Doxorubicin hydrochloride	45	89	< 0.00003
	4-H-cyclophosphamide	44	75	$= 0.00014$
	Bleomycin	89	71	$= 0.0006$
	cis-Platinum	111	70	$= 0.0016$
	Actinomycin D	33	64	NS
	Vinblastine	109	60	NS
	Etoposide (Vp16-213)	41	89	NS
Breast	Doxorubicin hydrochloride	118	81	< 0.00003
	Vincristine	32	81	$= 0.002$
	4-H-cyclophosphamide	62	79	$= 0.00006$
Ovarian	Doxorubicin hydrochloride	28	86	$= 0.0002$
	cis-Platinum	32	78	$= 0.002$
	4-H-cyclophosphamide	29	66	NS

4-H-cyclophosphamide, 4-hydroperoxycyclophosphamide

Relationship Between the Variations Induced on RNA and DNA Metabolism

The variations induced on the incorporation of ^3H-thymidine and ^3H-uridine into DNA and RNA respectively were associated for most of the drugs tested (Table 2) regardless of their mechanism of action. The lack of significant agreement was mainly ascribed to a more frequent effect on RNA metabolism. The variations induced by the drugs on the two precursors were separately considered in all the analyses but, since they usually showed quite similar results, only those related to ^3H-thymidine incorporation are reported in this chapter.

Distribution of Drug-Induced Variations and Definition of In Vitro Sensitivity

As shown from reports for other in vitro assays (Moon 1980), also in our assay the variations induced by the drugs on nucleotide precursor incorporation are continuous rather than all-or-nothing effects and can be described by normal (bell-shaped) distribution. Parameters of the distribution of ^3H-thymidine, expressed as percentage of the controls, are reported in Table 3. Mean values were often different from one drug to another, and for the same drug sometimes differed among the different tumor types. Among potential factors that may affect these differences, one is the intrinsic sensitivity of the tumor type to the drug, as demonstrated by the lower mean values in more sensitive tumor types; moreover, phase-nonspecific drugs showed lower mean values, suggesting a relationship between the mechanism of action of the drug and the possibility of evidencing its interaction with the tumor. The drug concentration tested, for those drugs in which a strict dose-related effect can be evidenced in vitro, may also influence the drug-tumor interaction. On the basis of these observations, the parameters of the distributions of each

Table 3. Parameters (mean ± SD) of the distributions[a] of the variations in ^3H-thymidine incorporation induced by some conventional agents in different tumor types

Tumor type	Doxorubicin hydrochloride[b]	4-Hydroperoxy-cyclophosphamide[c]	Etoposide (Vp16-213)[d]	Bleomycin
Breast	99 ± 34	99 ± 33	88 ± 33	
Testicular	101 ± 39	77 ± 33	72 ± 25	90 ± 37
NHL	66 ± 32	31 ± 16	45 ± 25	102 ± 36
Ovarian	99 ± 46	91 ± 50		

NHL, non-Hodgkin's lymphoma
[a] Expressed as percentage of control
[b] Ovarian vs NHL, $p < 0.001$; breast vs NHL, $p < 0.001$; testicular vs NHL, $p < 0.001$
[c] Ovarian vs NHL, $p < 0.001$; breast vs NHL, $p < 0.001$; testicular vs NHL, $p < 0.001$
[d] Testicular vs NHL, $p < 0.001$; breast vs NHL, $p < 0.001$; testicular vs breast, $p < 0.025$

drug on each tumor type seem have a definite biological significance for the analysis of the interaction between the drug and the tumor.

As a consequence of this continuity of the drug effects, one of the problems in analyzing in vitro drug sensitivity is how to define the cutoff value of the variations induced on nucleotide precursors above which a tumor can be defined as sensitive and below which it has to be defined as resistant. In our previous studies we defined in vitro sensitivity on the basis of comparative analyses between in vitro and clinical sensitivity of individual patients according to the retrospective criterion generally used by other authors (Moon 1980; Daidone et al. 1981; Sanfilippo et al. 1981). However, this retrospective criterion requires large in vitro-clinical correlations, possibly employing patients treated with monochemotherapeutic clinical regimens; moreover, since the interaction with the tumor seems to be peculiar to each drug, this kind of study should be carried out for each individual drug. The difficulties of application of such studies to conventional drugs and the impossibility of using new drugs are evident.

The mean value of the variations induced by each drug on ^3H-thymidine incorporation for each tumor type appeared to be a suitable indicator of the biological and experimental factors that influence drug-tumor interactions, and was therefore regarded as the cutoff value of in vitro sensitivity; when intertumor variability expressed by the distribution was lower than the intratumor variability expressed by the variation coefficient, the latter was taken into account for the definition of sensitivity. The biological correctness of such a prospective criterion was evaluated through the analysis of in vitro results and the relationship between in vitro and clinical chemosensitivity of individual patients, as reported in detail elsewhere in this volume.

Dose-Effect Plots

Mean plots of the induced variations in ^3H-thymidine incorporation by sensitive and resistant tumors, defined according to the prospective criterion, are given in Figs. 2–4. Plots of 4-hydroperoxycyclophosphamide in non-Hodgkin's lymphomas (Fig. 2) are reported as representative of the behaviors we observed for typical phase-nonspecific drugs. A dose-related effect was generally observed without overlapping of plots at the

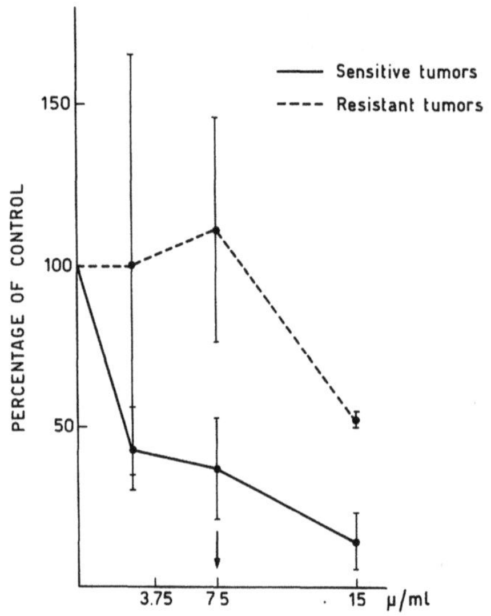

Fig. 2. Mean plots of the variations induced by 4-hydroperoxycyclophosphamide on ^{3}H-thymidine incorporation into DNA in non-Hodgkin's lymphomas. *Bars,* standard deviation of the mean; *arrow,* calculated clinical concentration

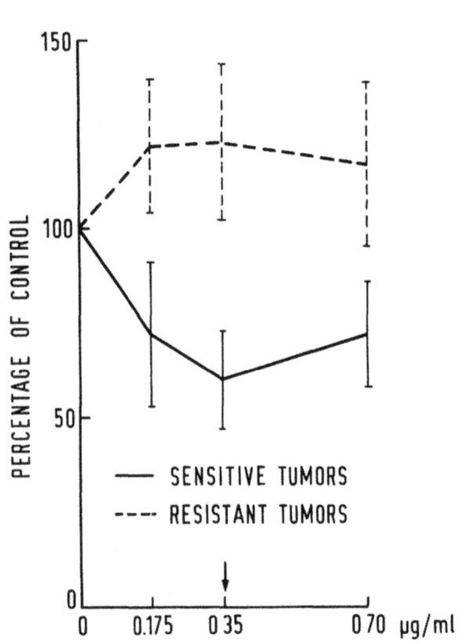

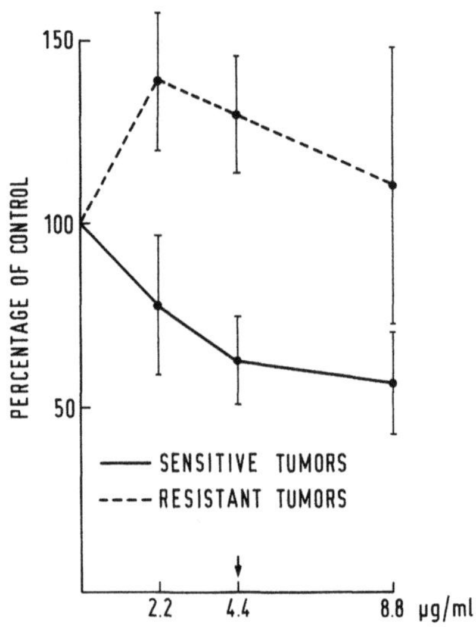

Fig. 3 (Left). Mean plots of the variations induced by vinblastine on ^{3}H-thymidine incorporation into DNA in testicular tumors. *Bars,* standard deviation of the mean; *arrow,* calculated clinical concentration

Fig. 4 (Right). Mean plots of the variations induced by *cis*-platinum on ^{3}H-thymidine incorporation into DNA in ovarian cancers. *Bars,* standard deviation of the mean; *arrow,* calculated clinical concentration calculated clinical and higher concentrations. Vinblastine in testicular tumors (Fig. 3) showed the patterns of sensitivity of the typical phase-specific drugs, for which an inhibition was observed in sensitive tumors at the equivalent clinical concentration, and no increasing effects were observed with increasing drug concentration. *cis*-Platinum in ovarian cancers (Fig. 4) showed a behavior intermediate between the aforementioned ones. The wide standard deviations clearly indicate the large variability in response among tumors within the same tumor type. Even if no conclusions about the mechanism of action of the drugs can be drawn in such short-term, unstable systems, the behaviors of the plots seem to fit the known mechanism of action of these drugs in the cell cycle, and the variations induced of ^{3}H-thymidine incorporation are a good direct or indirect marker of the interference of drugs in the cell cycle.

Comparison Between Drug Sensitivity Evaluated as Interference with Thymidine Incorporation and as Cytocidal Effect

To evaluate whether the effect on nucleic acid precursors indicates a cytocidal effect, in vitro sensitivity evaluated according to our assay was compared to that evaluated with the tumor stem cell assay (TSCA) according to Hamburger and Salmon (1977). The assays were performed on the same tumor material, which consisted of cell suspensions obtained mechanically or enzymatically from non-Hodgkin's lymphoma, ovarian carcinoma, and melanoma. Eighteen comparisons were possible (Table 4) of different drugs with different mechanism of action [doxorubicin hydrochloride, *cis*-platinum, vincristine, bleomycin, 4-hydroperoxycyclophosphamide, VP-16-213 (etoposide)]. The optimal conditions for each of the assays were used: drug concentrations used for the antimetabolic assay were those calculated from the clinical ones as previously detailed; concentrations used for the clonogenic assay were those reported by Alberts and Chen (1980) and usually corresponded to about one-tenth of our calculated clinical concentration. Whenever possible, lower and higher log concentrations were tested. Sensitivity was defined according to the prospective criterion for our assay and as 70% inhibition of colony growth at the basic concentration as reported by Von Hoff et al. (1981). Among the overall 18 trials, an agreement between drug responses evaluated with the two assays was observed in 15 cases (83%, $p = 0.008$). Drug sensitivity and resistance evaluated with the antimetabolic assay correlated in 100% and 80% of the cases, respectively, with the results of the TSCA. In contrast, the correlative accuracy of sensitivity was only 50%, which indicates a lower frequency of "sensitivity" in our assay than in the clonogenic assay and could in part explain the differences in clinical predictivity of the two assays: i.e., the higher incidence of false negative tumors in our assay and the higher incidence of false positives in the TSCA.

Table 4. Relationship between chemosensitivity evaluated with the antimetabolic assay (AA) and the tumor stem cell clonogenic assay (TSCA)

Sensitivity (S) or resistance (R) evaluated with the AA/S or R evaluated with the TSCA				*p* value
S/S	S/R	R/R	R/S	
3	0	12	3	= 0.008

Table 5. Potential of the antimetabolic assay in the experience of the Istituto Nazionale Tumori

Tumor type	No. of tumors received by the laboratory	% of tumors in which the assay was possible	Median no. of drugs tested	% of tumors in which the assay was evaluable
Breast				
Operable	1,285	15	3	93
Advanced	145	71	3	93
NHL	423	60	7	95
Testis	259	78	4	87
Colorectal	79	92	6	93
Ovarian	109	48	9	81

NHL, non-Hodgkin's lymphoma

Feasibility of the Assay

The feasibility of the assay was evaluated by a retrospective study of tumors received by our laboratory between January 1974 and May 1983; results of this study are reported in Table 5. The possibility of performing an in vitro assay on at least one drug was strictly dependent on the tumor material available for the assay: other simultaneous studies, such as hormone receptors or immunologic studies, limited the possibility of performing the assay in operable breast cancer, ovarian tumors, and non-Hodgkin's lymphomas, whereas for testicular and gastrointestinal tumors the main limitation was the clinical extension of the disease. The median number of drugs tested was also generally sufficiently high. The percentage of unevaluable results was generally low; the main factors of unevaluability were high values of variation coefficients of labeling incorporation in control samples, low incorporation values, and lack of representativeness of samples used for the assay of overall tumor histology.

Relationship Between In Vitro Effects of Different Drugs

The relationship between in vitro sensitivity of different drugs with similar or different mechanism of action was analyzed for a large number of drugs; some of the results are reported in Table 6. Although the regression analysis of in vitro responses of pairs of drugs showed a linear correlation for most of the drugs tested, the poor regression coefficients prompted us to consider the results of the kappa test more reliable. The relationship between the activity of drugs seemed to be unrelated to their mechanism of action and was unpredictable on the basis of any evident biological characteristic of the tumor.

Tumor-Type Specificity of In Vitro Responses

To be clinically useful, in vitro assays must first reproduce the variability of clinical patterns of response to drugs of the different tumor types. In our assay, in vitro sensitivity seemed to fit very well to the variability of clinical response to conventional agents (Table 7), and reproduced both the general chemoresponsiveness of the different tumor types and the

Table 6. Linear regression analysis (LR) and agreement rate (AR) between the in vitro effects of different drugs

Drugs	Tumor type							
	Breast		Ovarian		Testicular		Non-Hodgkin's lymphoma	
	LR	AR (%)	LR	AR (%)	LR	AR (%)	LR	AR (%)
A vs 4HC	$r = 0.268$ (< 0.02)	64 (NS)	$r = 0.239$ (NS)	59 (NS)	$r = 0.105$ (NS)	66 ($= 0.04$)	$r = 0.45$ (NS)	59 (NS)
A vs CDDP			$r = 0.5$ (< 0.02)	67 (NS)	$r = 0.581$ (< 0.001)	63 (NS)		
A vs Vcr	$r = 0.718$ (< 0.001)	68 ($= 0.04$)					$r = 0.390$ (< 0.005)	54 (NS)
A vs EpiA	$r = 0.355$ (NS)	74 (NS)						
4HC vs M	$r = 0.424$ (NS)	70 (NS)						
4HC vs CDDP			$r = 0.507$ (< 0.02)	68 (NS)				
B vs Vb					$r = 0.581$ (< 0.001)	71 (< 0.005)		
B vs Vcr							$r = 0.629$ (< 0.001)	71 ($= 0.05$)

4HC, 4-hydroperoxycyclophosphamide; *CDDP*, *cis*-platinum; *A*, doxorubicin hydrochloride (Adriamycin); *Vcr*, vincristine; *EpiA*, 4-epi-doxorubicin; *M*, melphalan; *B*, bleomycin; *Vb*, vinblastine

The statistical evaluation of the linear regression coefficients and of the agreement rates were assessed by means of Student's *t*-test and the kappa test respectively

Table 7. In vitro response rates of different tumor types to conventional agents

Drug	Breast cancer (%)	Testicular tumor (%)	Ovarian cancer (%)	Non-Hodgkin's lymphoma (%)
Doxorubicin hydrochloride	28	26	39	52
4-Hydroperoxycyclophosphamide	30	50	43	60
cis-Platinum	16	43	50	–
Etoposide (Vp16-213)	48	54	–	54
CCNU	8	–	–	38
Bleomycin	–	36	–	25
Vincristine	19	–	–	26
Vinblastine	;	40	–	–
Mitomycin C	26	–	–	–
Actinomycin D	–	27	–	–
Prednisolone	–	–	–	39
Melphalan	19	–	–	–
Procarbazine	–	–	–	16

CCNU, 1-(2-chloroethyl)-3-cyclohexyl-1-nitrosourea

absolute response rates of the individual drugs. In fact, tumor types known to be clinically responsive to chemotherapy, such as non-Hodgkin's lymphomas or testicular tumors, showed very high response rates to most of the drugs tested, whereas more chemoresistant tumors, such as breast cancers, were responsive to only a few drugs. Moreover, for all the drugs a good concordance was shown between in vitro and clinical response rates reported in the literature for monochemotherapy regimens, which accounts for the specificity of the assay to reproduce the clinical patterns of sensitivity to individual drugs. Only VP16-213 (etoposide) showed in breast cancers an in vitro activity higher than that expected on the basis of clinical results of monochemotherapeutic regimens (Schmoll 1982): pharmacologic problems, such as drug concentration and time of exposure, may have influenced the different in vitro activity of the drug.

Interlesion Heterogeneity in Drug Sensitivity

Tumor heterogeneity appears to be a fundamental limitation for an accurate prediction of individual tumor sensitivity. It was not possible to study intralesion heterogeneity because of technical problems related to the requisite of sufficient tumor material for each assay. Interlesion heterogeneity, i.e., differences in sensitivity that may occur among different tumor sites of the same patient, were studied in breast, testicular, and gastrointestinal cancers (Table 8). The sensitivity of the primary tumor was analyzed in relation to that of its lymph node, lung, or liver metastasis, but it was not possible to carry out conclusive studies among different metastatic sites.

We never observed a statistically significant agreement between primary tumors and their metastases; the overall percentage of concordant sensitivities varied in the different tumor types, with a trend to be lower in testicular and higher in breast cancers. These results show that interlesion heterogeneity, in the analysis of clinical predictivity of in vitro assays for individual patients, may be a possible source of error in prediction, which is related to the

Table 8. Relationship between the sensitivity of the primary tumor and that of its metastasis in different tumor types

Tumor type	Sensitivity (S) or resistance (R) of primary tumor/S or R of metastasis				Agreement rate (%)	p
	S/S	S/R	R/R	R/S		
Testicular	4	6	5	8	39	NS
Gastrointestinal	0	2	5	5	42	NS
Breast	6	9	17	8	58	NS
Overall	10	17	27	21	49	NS

The association between the sensitivity of the primary and that of its metastasis was assessed by means of the kappa test

site from which tumor material was derived for the in vitro assay. This possibility was confirmed by the analysis of the clinical predictivity in relation to tumor site (reported elsewhere in this volume).

Discussion

As reported by other authors involved in similar assays, the evaluation of drug effects through the interference of labeled nucleotide precursor incorporation is a potentially useful approach to drug sensitivity testing on human tumors (Volm et al. 1979; Group for sensitivity testing of tumors 1981). Our effort, partly described in this chapter, was to verify the correctness of the methodology and some basic assumptions we made in setting up the assay (such as the processing of tumor material, criteria for the definition of drug concentration, and the cutoff values of in vitro sensitivity), by means of as many analyses as possible of in vitro results to substantiate the evidence of specificity and clinical usefulness of in vitro chemosensitivity.

The analysis of dose-effect plots and of the relationship between the results obtained on the same tumor material with the anitmetabolic assay and the TSCA, which evaluates the effect of the drugs on colony growth, proved the general reliability and specificity of the use of the interference on nucleotide precursor incorporation as a biological marker of drug action. Our comparative study between the antimetabolic and the clonogenic assay was not performed to evaluate the two systems, since they appear very different in their biological characteristics and clinical applicability. In fact, the clonogenic assay is very complex and may be much more reliable for basic biological studies, whereas it may be less appropriate for immediate clinical use because of its actual low feasibility, as reported by Von Hoff et al. (1981). In addition, the material used for the clonogenic assay may not be representative of the original tumor, because of the multiple manipulations needed to obtain single-cell suspensions, which may strongly affect the overall composition of the original tumor cell population, and because of the possibility of selective growth of some tumor cell populations as a consequence of the particular culture conditions.

The evaluation of the clinical relevance of the antimetabolic assay involved different steps, such as the analysis of the actual feasibility of the assay, and its ability specifically to predict the sensitivity to individual drugs and reproduce the complex patterns of sensitivity of the different tumor types to conventional drugs. Moreover, the findings of our test showed no significant correlation between sensitivities to drugs, even when they had similar mechanism of action, suggesting that the sensitivity or resistance to individual drugs cannot be indicative of a general sensitivity or resistance. Only some tumor types sensitive to a large number of drugs, such as testicular tumors, showed concomitant sensitivity to many drugs. In vitro testing of all the drugs planned for use in clinical treatment is thus of the utmost importance for a correct prediction of clinical response. Moreover, the in vitro sensitivity of the different tumor types to conventional drugs was very similar to the patterns of clinical sensitivity, thus showing the potentiality of the assay for its application to the screening of the more active drugs for each tumor type. The very short time needed to obtain results and the high percentage of tumors in which a successful assay was possible make the assay potentially suitable for the clinical application of drug screening on individual tumors.

Even if intralesion heterogeneity seems to be an expected important feature in human tumors, it was partly overcome by an accurate tumor sampling of different areas of the tumor, so that the overall in vitro response represents a mediated response of the pooled tumor cell population. The study of interlesion heterogeneity (Zaffaroni et al. 1983), the in vitro evidence for which has been reported here, points out the importance of caution when evaluating data derived from primary tumors to predict chemosensitivity of metastases.

References

Alberts DS, Chen HSG (1980) Tabular summary of pharmacokinetic parameters relevant to in vitro drug assay. In: Salmon SE (ed) Cloning of human tumor stem cells. Liss, New York, pp 351–359

Costa A, Piazza R, Sanfilippo O, Silvestrini R (1977) A quantitative test for chemosensitivity of short-term cultures of human lymphomas. Tumori 63: 237–247

Daidone MG, Silvestrini R, Sanfilippo O (1981) Clinical relevance of an in vitro antimetabolic assay for monitoring human tumor chemosensitivity. In: Salmon SE, Jones SE (eds) Adjuvant therapy of cancer III. Grune & Stratton, New York, pp 25–32

Daidone MG, Silvestrini R, Zaffaroni N, Sanfilippo O, Varini M (to be published) Reliability of a short-term antimetabolic assay in predicting drug sensitivity of breast cancer. Rev Endocrine-Related Cancer

Group for sensitivity testing of tumors (KSST) (1981) In vitro short-term test to determine the resistance of human tumors to chemotherapy. Cancer 48: 2127–2135

Hamburger AW, Salmon SE (1977) Primary bioassay of human tumor stem cells. Science 197: 461–463

Moon TE (1980) Quantitative and statistical analysis of the association between in vitro and in vivo studies. In: Salmon SE (ed) Cloning of human tumor stem cells. Liss, New York, pp 209–221

Salmon SE (1980) Cloning of human tumor stem cells. Liss, New York

Sanfilippo O, Daidone MG, Costa A, Canetta R, Silvestrini R (1981) Estimation of differential in vitro sensitivity of non-Hodgkin lymphomas to anticancer drugs. Eur J Cancer 17: 217–226

Sanfilippo O, Daidone MG, Zaffaroni N, Silvestrini R (1983a) Criteria for the definition of in vitro sensitivity in a short-term antimetabolic assay (abstract 1218). Proceedings, 74th Annual Meeting of the American Association for Cancer Research, San Diego, May 25–28

Sanfilippo O, Silvestrini R, Zaffaroni N, Piva L (1983b) Potentiality of an in vitro chemosensitivity assay for human germ cell testicular tumors (GCTT) (abstract C-145). Proceedings of 19th Annual Meeting of the American Society of Clinical Oncology, San Diego, May 25–28

Schmoll H (1982) Review of etoposide single-agent activity. Cancer Treat Rev (Suppl A) 9: 21–30

Silvestrini R, Sanfilippo O, Daidone MG (1983) An attempt to use incorporation of radioactive nucleic acid precursors to predict clinical response. In: Dendy PP, Hill BT (eds) Human tumour drug sensitivity testing in vitro. Academic Press, London, pp 281–290

Schmoll H (1982) Review of etoposide single-agent activity. Cancer Treat Rev (Suppl A) 9: 21–30

Sky-Peck HH (1971) Effects of chemotherapy on the incorporation of ^{3}H-thymidine into DNA of human neoplastic tissue. Natl Cancer Inst Monogr 34: 197–203

Tisman G, Herbert V, Edlis H (1973) Determination of therapeutic index of drugs by in vitro sensitivity tests using human host and tumor cell suspensions. Cancer Chemother Rep 57: 11–19

Volm M, Waiss K, Kaufmann M, Mattern J (1979) Pretherapeutic detection of tumor resistance and the results of tumor chemotherapy. Eur J Cancer 15: 983–993

Von Hoff DD, Casper J, Bradley E, Sandbach J, Jones D, Makuch R (1981) Association between human tumor colony-forming assay results and response of an individual patient's tumor to chemotherapy. Am J Med 70: 1027–1032

Weisenthal LM, Marsden JA, Dill PL, Macaluso CK (1983) A novel dye exclusion method for testing in vitro chemosensitivity of human tumors. Cancer Res 43: 749–757

Zaffaroni N, Silvestrini R, Sanfilippo O, Daidone MG, De Marco C (1983) Tumor heterogeneity: analysis of the chemosensitivity of different tumor sites of the same patient. Proceedings of the 13th International Congress of Chemotherapy, Vienna, Austria, August 28–September 2

*Predictive Relevance for Clinical Outcome of In Vitro Sensitivity Evaluated Through Antimetabolic Assay**

R. Silvestrini, O. Sanfilippo, M.G. Daidone, and N. Zaffaroni

Istituto Nazionale per lo Studio e la Cura dei Tumori, Oncologia Sperimentale C, Via Venezian 1, 20133 Milano, Italy

Introduction

Apart from resistance to drugs, which varies in the different human tumor types and still requires much basic research, experience arising from chemotherapeutic management of patients with cancers allows us to evidence some crucial points, the investigation and solution of which may improve clinical results in potentially chemoresponsive tumors. One of these is the variability in sensitivity to the same drugs of clinically similar tumors; another deals with the unforeseeable recurrence of tumors that have shown similar responses to first-line treatment.

To resolve the first problem, it is important to identify markers indicative of chemosensitivity to drugs as a whole, to subgroups of drugs with a common mechanism of action, or to individual drugs, or otherwise directly to assess at the preclinical level the sensitivity of individual tumors. The search for chemosensitivity markers can be performed more quickly if the findings from in vitro testing (once its reliability is checked) instead of those from clinical trials can be used. For the assessment of chemosensitivity on individual tumors at the preclinical level, the antimetabolic test on short-term cultures appears to be one of the most suitable systems among those available. In fact, it has been opportunely adapted for processing solid samples and cell suspensions (Sanfilippo et al. 1981; Daidone et al. 1983; Silvestrini et al. 1983) to make it feasible on most tumors, and is simple to perform and not time consuming.

To come to the second point, the different probability of relapse found for tumors clinically similar and homogeneously treated has to be researched at the biological level. Intertumor heterogeneity, selection of resistant subpopulations preexistent or induced as a mutagenic effect by the first treatment, or inadequacy of treatment for the most biologically aggressive tumors may be invoked. This last hypothesis may be investigated by analyzing to what degree the intrinsic biological aggressiveness of tumors is responsible for the lack of long-term response to chemotherapy. We have approached these problems by investigating and trying to answer three different questions:

1. Are morphological, kinetic, biochemical, or pathological features indicative of sensitivity to drugs?
2. Is the in vitro chemosensitivity predictive of clinical response?

* Supported in part by Grant PFCCN n° 83.00946.96

Recent Results in Cancer Research. Vol. 94

3. Can some biological features indicative of aggressiveness, such as cell kinetics (Meyer and Hixon 1979; Durie et al. 1980; Costa et al. 1981; Gentili et al. 1981; Tubiana et al. 1981), better define the long-term clinical outcome of chemoresponsive or chemoresistant tumors? All three points have been explored in non-Hodgkin lymphomas (NHL) and in breast cancers, and the first two in germ cell testicular tumors.

Methodological Approaches

In vitro chemosensitivity was always determined with an antimetabolic assay (Sanfilippo et al. 1981; Daidone et al. 1983; Silvestrini et al. 1983) by evaluating the interference of drugs with nucleic acid precursor incorporation in short-term cultures (3 h). The reproducibility of in vitro data was guaranteed by the optimization and standardization of technical conditions and result evaluation. The mean value of variations on nucleic acid precursor incorporation defined for each drug, on at least 25 cases for each tumor type, was used as the objective criterion to define tumor sensitivity or resistance. In most cases, all the drugs included in clinical trials were tested in vitro. Since the drugs were individually tested in vitro, whereas they were used in association in clinical trials, a tumor was defined as clinically sensitive on the basis of in vitro testing when it was sensitive to at least one drug used in clinical treatment. The first evidence of the specificity of the test, and of the reliability of the criteria of in vitro sensitivity used, derives from the satisfactory agreement, observed for each tumor type, between in vitro rate of sensitivity to individual drugs and rate of clinical response to the same drugs used in monochemotherapy regimens (Bonadonna 1974; Wasserman et al. 1975; Hoogstratten and Fabian 1979).
The cell kinetic characteristic was determined on in vitro fresh fragments (Silvestrini et al. 1979) or cell suspensions (Silvestrini et al. 1977) at the time of in vitro chemosensitivity assay and evaluated as ^3H-thymidine labeling index (LI). By considering the tumor heterogeneity, the LI value for each tumor was defined by scoring different areas of the tumor. The median LI value determined for each tumor type on a large series of patients was used as cutoff to define slowly or fast-proliferating tumors (Costa et al. 1981; Gentili et al. 1981).

Non-Hodgkin Lymphomas

Biological Features and In Vitro Chemosensitivity

The sensitivity to some drugs more frequently used in clinical trials was analyzed for subsets of tumors defined according to the variables considered of utmost clinical relevance or important in characterizing tumor biology (Table 1). Morphological features were defined according to the Rappaport classification (Rappaport 1966). The distinction of slowly or fast-proliferating tumors was established according to the median LI value of 4% obtained on a series of 88 tumors, which had already proved to have prognostic relevance (Costa et al. 1981). The overall frequency of in vitro sensitivity to different drugs was similar to the clinical response rates of NHL to the same drugs used in monochemotherapy (Bonadonna 1974). No significant differences in sensitivity rates were found between the two distinctly analyzed morphological or kinetic subsets. Only a trend of a higher sensitivity of fast-proliferating or diffuse tumors than of slowly proliferating or nodular tumors to 4-hydroperoxycyclophosphamide was observed. Again, the subsets of tumors at different pathological stages according to the Ann Arbor classification did not show any relevant difference in sensitivity to the four tested drugs.

Table 1. Relationship between in vitro sensitivity and biological features in non-Hodgkin lymphomas

	In vitro response rates (%)			
	Doxorubicin hydrochloride	4-Hydroperoxy-cyclophosphamide	Vincristine	Bleomycin
Overall	52 (103)	60 (35)	26 (70)	25 (59)
Morphology				
Nodular	58 (44)	50 (14)	23 (26)	32 (22)
Diffuse	49 (59)	71 (21)	27 (44)	21 (37)
Cell kinetics				
Slow-proliferating	57 (59)	54 (22)	25 (40)	30 (33)
Fast-proliferating	45 (44)	69 (13)	23 (30)	19 (26)

In parenthesis, number of case

Table 2. Association between in vitro sensitivity[a] and clinical response[b] in 65 patients with non-Hodgkin lymphomas

In vitro sensitive (S) or resistant (R)/clinical S or R				p
S/S	S/R	R/R	R/S	
38	9	10	8	< 0.0006
True positive 81%		True negative 56%		

[a] Inhibition of ^3H-thymidine incorporation superior to the mean value, by at least one drug used for the clinical treatment

[b] Complete remission after ABP ~ CVP or BACOP

The association between in vitro and clinical sensitivity was assessed by means of the kappa test (Fleiss 1973)

Correlation Between In Vitro Sensitivity and Short-Term Clinical Response of Individual Tumors

The analysis was performed on 65 patients (Table 2); most of them were treated with alternative courses of ABP ~ CVP, and a minority with BACOP or other drug combinations[1]. All the drugs used in clinical treatment were tested in vitro, except for cyclophosphamide, since at the beginning of the study the active metabolite, 4-hydroperoxycyclophosphamide, was not available. In vitro sensitivity to at least one drug predicted the clinical response in terms of complete remission in 81% of the cases, whereas in vitro resistance to all drugs corresponded to an actual lack of short-term response in only 56% of the cases. The poorer predictive accuracy of resistance could be partially due to the lack of data in more than 50% of the cases on the in vitro effect of cyclophosphamide, which is known to be very active against NHL.

1 *ABP ~ CVP*, doxorubicin, bleomycin, prednisone ~ cyclophosphamide, vincristine, prednisone; *BACOP*, bleomycin, doxorubicin, cyclophosphamide, vincristine, prednisone

Table 3. Relevance of in vitro sensitivity on long-term clinical outcome in patients with non-Hodgkin lymphomas

	Disease-free survival (%)[a]	Survival (%)[b]
In vitro sensitive	45	59
	$p < 0.025$	$p < 0.005$
In vitro resistant	20	0

[a] At 3 years
[b] At 3.5 years

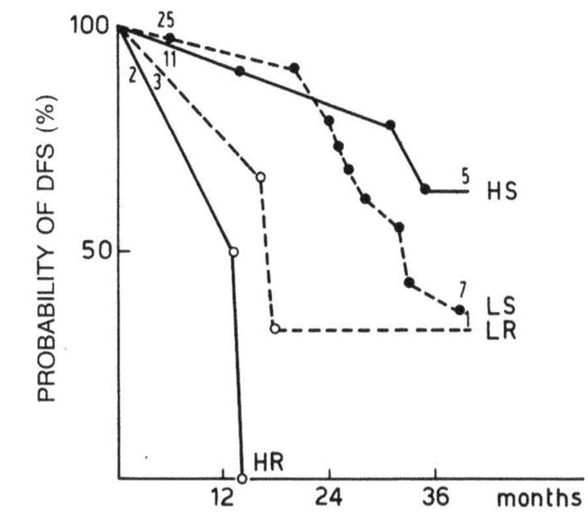

Fig. 1. Relevance of cell kinetics and in vitro sensitivity on disease-free survival (DFS) of patients with non-Hodgkin lymphomas. (●– – –●), low labeling index (LI), sensitive, *LS*; (○– – –○), low LI, resistant, *LR;* (●———●), high LI, sensitive, *HS;* (○———○) high LI, resistant, *HR*. HS versus HR, $p < 0.005$. The actuarial life table method was used to summarize disease-free survival distribution; the statistical significance of differences observed was assessed by the log rank test, and the p value quoted represents the comparison of the entire plot (Peto et al. 1977)

In Vitro Chemosensitivity and Cell Kinetics as Discriminants of Long-Term Clinical Outcome

The study, performed on 41 patients who had reached complete remission (Table 3), showed a significantly higher probability of 3-year disease-free survival (DFS) in patients whose tumors proved to be sensitive in vitro (45%) than in those with in vitro resistant tumors (20%). Again, analysis of a series of 58 patients who had or had not reached complete remission indicated that in vitro sensitivity is also indicative of survival. In fact, the 3.5-year probability of survival was 59% for patients with sensitive tumors, whereas no patient with a resistant tumor survived for that length of time. When the pretreatment cell proliferative activity was analyzed in association with the in vitro response to the drugs, the DFS curves (Fig. 1) were not statistically different for the two kinetic subgroups, for patients with sensitive or for those with resistant tumors. However, for the latter ones, a somewhat different clinical outcome was observed, consisting of relapse within 14 months of all patients with fast-proliferating tumors. The subset of patients with fast-proliferating, in vitro resistant tumors thus appears to be the group at the highest risk of relapse. Moreover, a significant difference ($p < 0.005$) in DFS was observed between the subgroups of patients with fast-proliferating sensitive tumors (64%) and fast-proliferating resistant tumors (0%). Conversely, the probability of survival (Fig. 2) was still similar, regardless of proliferative rate in patients with resistant tumors, and all patients died within 40 months, whereas the survival curves appeared significantly ($p < 0.05$) displayed for the two kinetic subgroups within the group of sensitive tumors (69% vs 44%). The proliferative activity of the tumor thus appears to be a further discriminant of survival in patients with chemosensitive tumors.

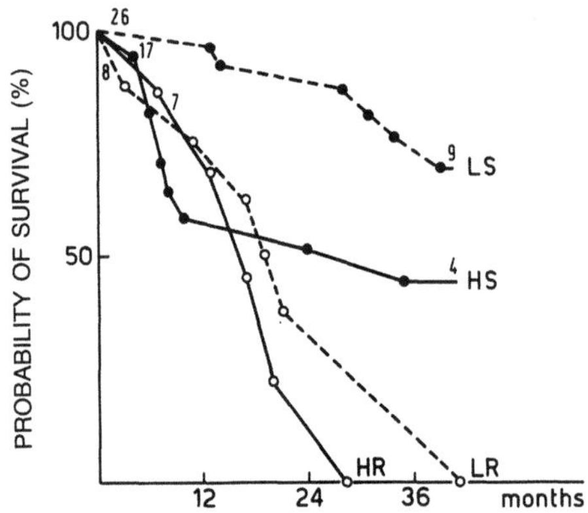

Fig. 2. Relevance of cell kinetics and in vitro sensitivity on survival of patients with non-Hodgkin lymphomas. (●- - -●), low labeling index (LI), sensitive, *LS;* (○- - -○), low LI, resistant, *LR;* (●——●), high LI, sensitive, *HS;* (○——○), high LI, resistant, *HR*. LS versus HS, $p < 0.05$. The actuarial life table method was used to summarize survival distribution; the statistical significance of differences observed was assessed by the log rank test, and the p value quoted represents the comparison of the entire plot (Peto et al. 1977)

Breast Cancer

Biological Features and In Vitro Chemosensitivity

The study was mainly performed on locally advanced disease because of the larger amount of biological material available for this stage than for operable cancers. In vitro sensitivity to drugs frequently used (doxorubicin hydrochloride [Adriamycin] and 4-hydroperoxy-cyclophosphamide) was analyzed in relation to estrogen receptor (ER) status and cell kinetics (Table 4), which are considered to be among the most characteristic variables of breast cancer biology (Meyer et al. 1977, 1983; Silvestrini et al. 1979). ER status was determined with the charcoal absorption technique, and 5 fmol/mg protein or an association constant of more than 1.5×10^{-9} M was used as the cutoff of positivity (Di Fronzo et al. 1978). The discrimination of slowly and fast-proliferating tumors was done by using the median LI value of 2.8% defined on 541 cases (Gentili et al. 1981). The overall responses were similar to those reported from monochemotherapy regimens using the same drugs (Hoogstratten and Fabian 1979). Only a trend of a higher frequency of sensitivity to doxorubicin hydrochloride and 4-hydroperoxycyclophosphamide was observed for ER-positive tumors than for ER-negative ones, whereas similar response rates to both drugs were observed for the two kinetic subgroups. Again, the stage, operable versus locally advanced tumors, did not influence sensitivity to the tested drugs.

Correlation Between In Vitro Sensitivity and Short-Term Clinical Response of Individual Tumors

The analysis was performed on 41 patients with locally advanced breast cancers. The patients were submitted to three cycles of induction chemotherapy with doxorubicin and vincristine, then to surgery or radiotherapy, and to seven subsequent cycles of maintenance therapy with the same drugs. Doxorubicin was always tested in vitro and vincristine on about half of the patients. However, vincristine was only occasionally active in vitro. Moreover, tumors sensitive to vincristine were also sensitive to doxorubicin. The retrospective analysis of in vitro response to doxorubicin and tumor volume reduction after induction chemotherapy showed an overall agreement of 78% between in vitro and clinical sensitivity and in vitro and clinical resistance, with a predictive accuracy of sensitivity and resistance of 75% and 81% respectively (Table 5).

Table 4. Relationship between in vitro sensitivity and biological features in breast cancers

	In vitro response rates (%)	
	Doxorubicin hydrochloride	4-Hydroperoxy-cyclophosphamide
Overall	28 (119)	30 (63)
Receptor status		
ER+	30 (60)	33 (33)
ER−	17 (30)	10 (10)
Cell kinetics		
Slow-proliferating	39 (31)	25 (16)
Fast-proliferating	31 (49)	32 (22)

In parentheses, number of cases
ER, estrogen receptors

Table 5. Association between in vitro sensitivity[a] and clinical response[b] in 41 patients with locally advanced breast cancer

In vitro sensitive (S) or resistant (R)/clinical S or R				p
S/S	S/R	R/R	R/S	
15	5	17	4	< 0.00016
True positive 75%		True negative 81%		

[a] Inhibition of ^{3}H-uridine incorporation superior to the mean value by doxorubicin hydrochloride
[b] Complete + partial remission > 50% after induction chemotherapy with doxorubicin hydrochloride and vincristine
The association between in vitro and clinical sensitivity was assessed by means of the kappa test (Fleiss 1973)

Table 6. Relevance of in vitro sensitivity on long-term clinical outcome in patients with locally advanced breast cancer

	2.5-year probability of survival (%)
In vitro sensitive	78
In vitro resistant	74

In Vitro Chemosensitivity and Cell Kinetics as Discriminants of Long-Term Clinical Outcome

In vitro sensitivity proved not to be discriminant of survival (Table 6). In fact, a similar 2.5-year probability of survival was found for patients with sensitive (78%) or resistant (74%) tumors. Conversely, when pretreatment proliferative activity was analyzed in association with in vitro chemosensitivity, a probability of 100% survival was found for

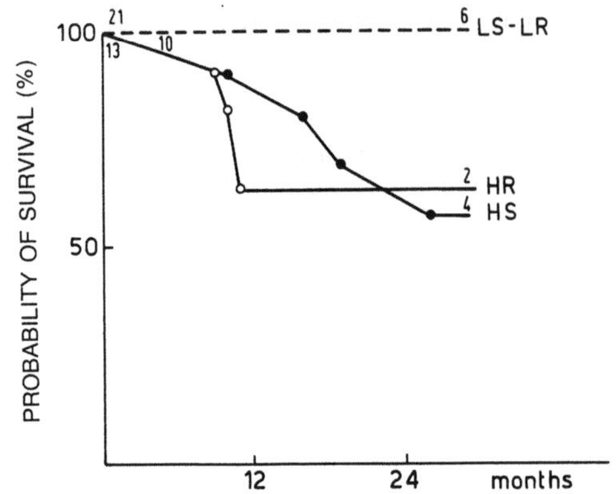

Fig. 3. Relevance of cell kinetics and in vitro sensitivity on survival of patients with locally advanced breast cancer. (– – –), low labeling index (LI), sensitive and resistant, *LS-LR;* (●———●), high LI, sensitive, *HS;* (○———○), high LI, resistant, *HR*. The actuarial life table method was used to summarize survival distribution; the statistical significance of differences observed was assessed by the log rank test

slowly proliferating tumors, regardless of sensitivity to drugs, and a significantly lower survival (about 60%) was observed for chemosensitive or chemoresistant fast-proliferating tumors (Fig. 3).

Germ Cell Testicular Tumors

Biological Features and In Vitro Chemosensitivity

The study on this tumor type is hampered by the complexity of the histological findings, which consist of the presence of combined seminomatous and nonseminomatous histological features, as well as different nonseminomatous histological subtypes and the frequent occurrence of different histological features in the various tumor sites of the same patient. Sensitivity to *cis*-platinum, vinblastine, and bleomycin was analyzed in relation to histomorphology in 66 nonseminomatous tumors and ten seminomas from primary or metastatic sites (Table 7). The overall frequency of in vitro sensitivity to different drugs was in agreement with the frequency of clinical responses reported for the same drugs used in monochemotherapy (Wasserman et al. 1975). No significant difference was observed between seminomas and nonseminomatous tumors, but the former showed a higher frequency of sensitivity to *cis*-platinum and bleomycin. Among nonseminomatous tumors, embryonal carcinomas in pure or mixed forms showed very similar sensitivities to the three tested drugs, whereas mature teratoma appeared less sensitive to bleomycin.

Correlation Between In Vitro Sensitivity and Short-Term Clinical Response of Individual Tumors

A retrospective analysis was performed on 36 samples from 32 patients with advanced disease: 26 of these, who had been previously untreated, were subjected to induction chemotherapy with three cycles of *cis*-platinum, vinblastine, and bleomycin; the six pretreated patients were given other drugs (Table 8). All the drugs used in clinical protocols were tested in vitro, and the assay was performed on 16 primary tumors and 20 metastatic sites (lymph node or lung). In four patients, both primary and metastatic sites were tested. The overall analysis showed an accuracy of 96% and 36% of in vitro data to

Table 7. Relationship between in vitro sensitivity and morphological features in germ cell testicular tumors

	In vitro response rates (%)		
	Cis-platinum	Vinblastine	Bleomycin
Seminomas	60 (10)	37 (11)	54 (11)
Nonseminomatous			
Overall	38 (66)	39 (64)	33 (58)
Embryonal ca pure	41 (27)	43 (30)	37 (27)
Embryonal ca mixed	35 (23)	36 (25)	33 (24)
Mixed teratoma	37 (16)	33 (9)	14 (7)

In parentheses, number of cases

Table 8. Association between in vitro sensitivity[a] and clinical response[b] in 34 patients with germ cell testicular tumors

	In vitro sensitive (S) or resistant (R)/clinical S or R				p
	S/S	S/R	R/R	R/S	
Primary tumor	10			6	NS
Metastasis	14	1	4	1	0.001
Overall	24	1	4	7	0.0094
	True positive 96%		True negative 36%		

[a] Inhibition of ^{3}H-thymidine incorporation superior to the mean value, by at least one drug used for the clinical treatment
[b] Complete + partial response > 50% after induction chemotherapy with *cis*-platinum, vinblastine, bleomycin (PVB) or other drugs
The association between in vitro and clinical sensitivity was assessed by means of the kappa test (Fleiss 1973)

predict clinical sensitivity or resistance, respectively. However, when the in vitro sensitivity of the primary tumor and that of metastases were separately analyzed, it was found that the sensitivity of metastases but not that of the primary tumor was significantly indicative of clinical response. In vitro sensitivity of the primary tumor was consistently indicative of clinical response, but the six cases which had proved to be resistant in vitro also showed a clinical response, Conversely, the in vitro sensitivity or resistance of metastases predicted the clinical response or a lack of response in 93% and 80% respectively.

Conclusions

The morphological, biological, and pathological features of the human tumors studied are not indicative of chemosensitivity, at least for the drugs considered. Pretreatment cell kinetics, which has proved to be directly related to chemosensitivity in experimental tumors, also does not appear to be indicative of sensitivity to the tested drugs in NHL, germ cell testicular tumors, or breast cancers. The lack of a correlation among the different tumor features and in vitro sensitivity to drugs is in agreement with the findings reported about the lack of predicitve relevance of some of these features on clinical response to chemotherapy (Kiang et al. 1978; Lippman et al. 1978; Bonadonna et al. 1980; Daidone et al. 1982) and indicates that sensitivity to drugs is an intrinsic and unforeseeable peculiarity of individual tumors.

The reliability of this analysis, which is based on the in vitro antimetabolic effect of drugs, is supported by the evidence of the correlation between in vitro findings and short-term clinical response to the same drugs and in the same patient. Such a correlation makes this in vitro test an important and valid means of predicting clinical response to drugs in individual patients as far as short-term response is concerned. A peculiarity of the antimetabolic assay we used is the high accuracy in recognizing active drugs. The lower accuracy in defining inactive drugs, i.e., the lower frequency of true negatives, could be due, in addition to the general reasons linked to in vitro testing and common to all in vitro systems, to the insufficient time (3 h) allowed for the drugs, particularly if phase-specific, to express their antimetabolic effect to an evaluable degree.

The main limitation for a definite verification of the predictive relevance of the antimetabolic assay is the present impossibility of comparing in vitro activity with clinical response to the same drugs used in monochemotherapy regimens. However, in the study on locally advanced breast cancers, this limitation is almost eliminated. In fact, clinical studies have shown that vincristine does not significantly improve the clinical result obtained with doxorubicin alone (Steiner et al. 1981). This clinical finding is also in agreement with in vitro finding that showed a sensitivity to vincristine of tumors sensitive also to doxorubicin.

The results from the study on germ cell testicular tumors also indicate the opportunity of assessing, when possible, the chemosensitivity of tumor metastases. The heterogeneity of tumors and selection of cell subpopulations during metastatic spread could make the primary tumor not adequately representative of the biological behavior of metastases and thus not indicative of the clinical course of the disease.

Moreover, the predictive relevance of in vitro sensitivity observed for short-term clinical response was not consistently found for long-term clinical outcome. In NHL, in vitro sensitivity is indicative of both DFS and survival, even though the intrinsic biological aggressiveness evaluated as the fraction of proliferating cells represents a further discriminant of survival within in vitro sensitive tumors. In locally advanced breast cancers, in vitro chemosensitivity was not indicative of survival, which instead was correlated with tumor cell kinetics. This disagreement could be due to biological differences between the two tumor types as well as to the different clinical protocols.

It may be supposed that NHL, in spite of their morphological complexity, are biologically homogeneous cell populations, as cytophotometric DNA determinations have demonstrated (Diamond and Braylan 1980; Diamond et al. 1982; Silvestrini et al. 1982), and that clinical protocols including at least five drugs are adequate to control tumor growth once the tumor has been shown to be chemosensitive.

In contrast, in breast cancers it may be supposed that the heterogeneity shown by the analysis of different biological aspects (Auer and Tribukait 1980; Bichee et al. 1982) and the selection of resistant subpopulations, or the inadequacy of a bichemotherapy regimen, or both these reasons, deny the long-term predictivity of in vitro sensitivity and indicate the biological aggressiveness as the only discriminant of long-term clinical outcome.

All these findings clearly show the difficulty of planning the optimal treatment schedule but also indicate the approaches to be followed if the best clinical results are to be achieved.

References

Auer G, Tribukait B (1980) Comparative single cell and flow DNA analysis in aspiration biopsies from breast carcinomas. Acta Pathol Microbiol Scand 88: 355–358

Bichee P, Poulsen HS, Andersen J (1982) Estrogen receptor and ploidy of human mammary carcinoma. Cancer 50: 1771–1774

Bonadonna G (1974) Chemoterapia dei linfomi linfocitico e istiocitico. In: Bucalossi P, Veronesi U, Bonadonna G, Emanuelli H (eds) I linfomi maligni. Casa Editrice Ambrosiana, Milan, pp 269–275

Bonadonna G, Valagussa P, Tancini G, Di Fronzo G (1980) Estrogen-receptor status and response to chemotherapy in early and advanced breast cancer. Cancer Chemother Pharmacol 4: 37–41

Costa A, Bonadonna G, Villa E, Valagussa P, Silvestrini R (1981) Labeling index as a prognostic marker in non-Hodgkin's lymphomas. J Natl Cancer Inst 66: 1–5

Daidone MG, Silvestrini R, Di Fronzo G (1982) Prognostic relevance of cell kinetics, alone or in association with estrogen receptor (ER) status, in operable and advanced breast cancers. Breast Cancer Res Treat 2: 277

Daidone MG, Silvestrini R, Zaffaroni N, Sanfilippo O, Varini M (1983) Reliability of a short-term antimetabolic assay in predicting drug sensitivity of breast cancers. In: Stool BA (ed) Reviews on endocrine-related cancer. Stuart Pharmaceuticals, Division of ICI American Inc.

Diamond LW, Braylan RC (1980) Flow analysis of DNA content and cell size in non-Hodgkin's lymphoma. Cancer Res 40: 703–712

Diamond LW, Nathwani BN, Rappaport H (1982) Flow cytometry in the diagnosis and classification of malignant lymphoma and leukemia. Cancer 50: 1122–1135

Di Fronzo G, Bertuzzi A, Ronchi E (1978) An improved criterion for the evaluation of estrogen receptor binding data in human breast cancer. Tumori 64: 259–266

Durie BG, Salmon SE, Moon TE (1980) Pretreatment tumor mass, cell kinetics, and prognosis in multiple myeloma. Blood 55: 364–372

Fleiss JL (1973) Statistical methods for rates and proportions. Wiley series in probability and mathematical statistics. Wiley, New York

Gentili C, Sanfilippo O, Silvestrini R (1981) Cell proliferation and its relationship to clinical features and relapse in breast cancer. Cancer 48: 974–979

Hoogstratten B, Fabian C (1979) A reappraisal of single drugs in advanced breast cancer. Cancer Clin Trials, Summer: 101–109

Kiang DT, Frenning DH, Goldman AI, Ascensao VF, Kennedy BJ (1978) Estrogen receptors and responses to chemotherapy and hormonal therapy in advanced breast cancer. N Engl J Med 299: 1330–1334

Lippman ME, Allegra JC, Thompson EB, Simon R, Barlock A, Green L, Huff KK, Do HMT, Aitken SC, Warren R (1978) The relation between estrogen receptors and response rate to cytotoxic chemotherapy in metastatic breast cancer. N Engl J Med 298: 1223–1228

Meyer JS, Hixon B (1979) Advanced stage and early relapse of breast carcinomas associated with high thymidine labeling indices. Cancer Res 39: 4042–4047

Meyer JS, Rao BR, Stevens SC, White WL (1977) Low incidence of estrogen receptor in breast carcinomas with rapid rates of cellular replication. Cancer 40: 2290–2298

Meyer JS, Friedman E, McCrate MM, Bauer WC (1983) Prediction of early course of breast carcinoma by thymidine labeling. Cancer 51: 1879–1886

Peto R, Pike MC, Armitage P, Breslow NE, Cox DR, Howard SV, Mantel N, McPherson K, Peto J, Smith PG (1977) Design and analysis of randomized clinical trials requiring prolonged observation of each patient. Br J Cancer 35: 1–39

Rappaport H (1966) Tumors of the hematopoietic system. In: Atlas of tumor pathology, sect III, fasc 8. Armed Forces Institute of Pathology, Washington DC, pp 91–166

Sanfilippo O, Daidone MG, Costa A, Canetta R, Silvestrini R (1981) Estimation of differential in vitro sensitivity of non-Hodgkin lymphomas to anticancer drugs. Eur J Cancer 17: 217–226

Silvestrini R, Piazza R, Riccardi A, Rilke F (1977) Correlation of cell kinetic findings with morphology of non-Hodgkin's malignant lymphomas. J Natl Cancer Inst 58: 499–504

Silvestrini R, Daidone MG, Di Fronzo G (1979) Relationship between proliferative activity and estrogen receptors in breast cancer. Cancer 44: 665–670

Silvestrini R, Costa A, Del Bino G, Mazzini G (1982) Flow cytofluorometric determination of DNA characteristics in non-Hodgkin's lymphomas. Proceedings, 13th International Cancer Congress, Seattle, September 8–15

Silvestrini R, Sanfilippo O, Daidone MG (1983) An attempt to use incorporation of radioactive nucleic acid precursors to predict clinical response. In: Dendy PP, Hill BT (eds) Human tumour drug sensitivity testing in vitro. Techniques and clinical applications. Academic Press, London, pp 281–290

Steiner R, Stewart JF, Knight RK, Rubens RD (1981) The contribution of vincristine to Adriamycin in the treatment of advanced breast cancer (abstract). UICC Conference on clinical oncology, Lausanne, October 28–31

Tubiana M, Pejovich MJ, Renaud A, Contesso G, Chavaudra N, Gioanni J, Malaise EP (1981) Kinetic parameters and the course of the disease in breast cancer. Cancer 47: 937–943

Wasserman TH, Comis RL, Goldsmith M, Handelsman H, Penta JS, Slavik M, Soper WT, Carter SK (1975) Tabular analysis of the clinical chemotherapy of solid tumors. Cancer Chemother Rep 6: 399–419

Biochemical Short-Term Predictive Assay: Results of Correlative Trials in Comparison to Other Assays

M. Kaufmann

Universität Heidelberg, Frauenklinik, Vosstraße 9,
6900 Heidelberg, FRG

Introduction

Various prediction test systems for human tumors have been developed during the past decades (Dendy 1976; Kaufmann 1980, 1983; Possinger and Ehrhardt 1983; Kaufmann and Kubli 1983; Salmon 1980; Volm and Mattern 1982). In cancer chemotherapy the availability of an assay which makes an appropriate selection of treatment for the individual patient feasible has long been dreamed of. General informations on an in vitro short-term predictive assay (STPA) will be presented. Besides these basic data, results of correlative trials of this assay are compared to results of other assays involving human tumors.

Short-Term Predicitive Assay

Principles and In Vitro Test Results

The STPA was developed by Volm, Kaufmann, Mattern, and Wayss about 12 years ago (Volm et al. 1979). They began to evaluate a short and simple method demonstrated in animal transplantation tumors, and standardized the assay for solid human tissues as well as for tumor cells from body effusions. The easily reproducible method has already been described in detail in various papers [Group for Sensitivity Testing of Tumors (KSST) 1981; Kaufmann 1980; Pfleiderer 1980; Volm et al. 1979].

Standard anticancer drugs can be tested in freshly mechanically prepared single tumor cell suspensions or cell clumps, and results are obtained within 1 day. In this 3-h assay, the binding of a radioactive nucleic acid precursor of proliferating cells is measured.

One major factor limiting the application of all predictive assays is that the prepared tumor cells must be sufficient in quantity and above all in viability for test. A second factor is that growth of the tumor cells and therefore test success rates are limited.

Table 1 shows the success rates for various human tumor types tested using our STPA or comparable antimetabolic STPAs. Best results were yielded in ovarian cancer, which at present is the best studied human tumor, with success rates of more than 90%. The in vitro inhibitory effect of a drug is commonly determined in percent compared to control values. Many laboratories use different arbitrary levels in their assays to distinguish between sensitive and resistant tumors. Most investigators use sensitivity indices, which are derived from dose-response curves and defined as the area under the curve, as shown in Fig. 1 as F_1.

Recent Results in Cancer Research. Vol. 94
© Springer-Verlag Berlin · Heidelberg 1984

Table 1. Success rates (%) for testing of human tumor types using STPAs

Reference	Year	Tumor type	*n*	%
KSST	1981	Ovarian	333	95
Kaufmann	1982	Ovarian	78[a]	96
Pfleiderer	1982	Ovarian	154[a]	95
Sanfilippo et al.	1983	Testicular germ cell	227	75
Kaufmann	1982	Breast (primary)	235	69
KSST	1981	Cervical	111	82

[a] FIGO stages III and IV

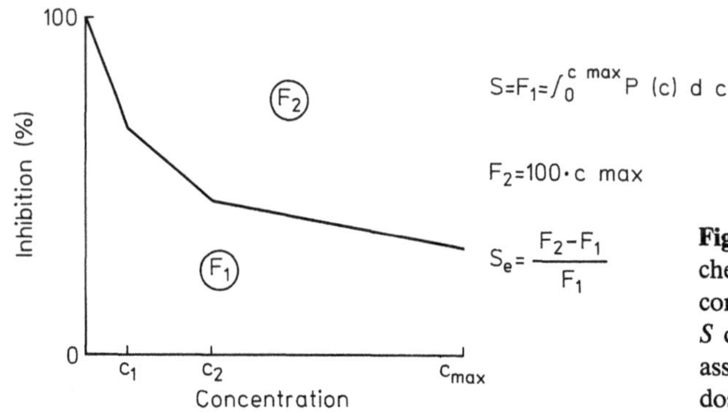

Fig. 1. Definition of an improved chemosensitivity index, S_e in comparison to the sensitivity index S commonly used in predictive assays. $S = F_1$ = area below dose-response curve

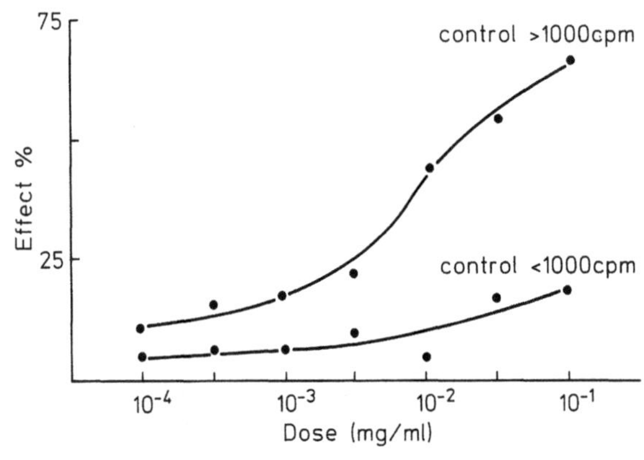

Fig. 2. Dose-effect curves for doxorubicin in two ovarian carcinomas tested by the STPA, showing different effects depending on the control incorporation (> or < 1,000 cpm) of tritiated uridine

In our recent studies we have used an improved modified chemosensitivity index, which measures relative rather than absolute drug effects and therefore allows different drug effects to be compared (Abel and Kaufmann 1982).

In vitro investigations of human tumors are easy to perform after successful tumor disaggregation using standard anticancer drugs and their corresponding nucleic acid precursor according to their mode of action (Kaufmann 1982; Volm et al. 1979). Usually, drugs are tested in three to four log concentrations. Test curves and therefore sensitivity indices differ from each single tumor to the next. The behavior of tumor cells in this in vitro STPA seems to be an individual, tumor-characteristic phenomenon. Differences exist

Table 2. Differential sensitivity to doxorubicin (10 µg/ml) in the STPA

Tumor type (solid)	*n*	% sensitive
Breast (primary)	162	50[a]
Ovarian (nonpretreated)	78	61[b]

[a] Sensitive = effect of doxorubicin > 25% compared to control values
[b] Sensitive = effect of doxorubicin > 35% compared to control values

Table 3. Chemosensitivity profiles of ovarian cancer (body effusions) in the STPA

Treatment	No. of patients	Sensitive to doxorubicin[a]
Cyclophosphamide – first-line CT	32	14 (44%)
Doxorubicin – second-line CT	21	3 (14%)

$p < 0.05$ exact Fisher test
[a] Effect of 10 µg/ml > 50% compared with control values
CT, combination chemotherapy

between median incorporation rates of and drug effects on different human tumor types, for example ovarian and breast cancer, as well as in different cell manifestations, for example tumor cells from solid tissues and those from body fluids. Positive interrelations exist between the incorporation rates of thymidine, uridine, or deoxyuridine as different nucleic acid precursors used in this assay.

Figure 2 shows dose-effect curves for the effect of doxorubicin on tritiated uridine incorporation in two ovarian carcinomas which differ in having control incorporation rates higher than and less than 1,000 cpm. It can easily be seen that, e.g., the effects of 10^{-2} mg doxorubicin per milliliter show large differences in the single tumors. This means that thresholds between sensitivity and resistance depend not only on tumor type, tumor site, and the tested concentrations of various drugs, but also on the incorporation rate.

It is well known that rapidly proliferating human tumors generally respond to cytotoxic agents better than slowly proliferating tumors. In our experiments these proliferation-dependent inhibitory effects of a broad spectrum of cytotoxic drugs could be demonstrated in vitro. However, our usual finding in vitro was that if a proliferating tumor which had not been cytotoxically pretreated responded to doxorubicin, other drugs in most cases produced similar reactions. In only a few instances were we able to identify different effects in different tested drugs.

Differential chemosensitivity to doxorubicin, using one concentration, is shown in Table 2 for primary breast and nonpretreated advanced ovarian carcinomas. Because of the different median incorporation rates for breast and ovarian carcinomas, different thresholds between sensitivity and resistance apply. Metastatic ovarian cancer is more sensitive than primary breast cancer.

Table 3 shows different doxorubicin chemosensitivity profiles obtained with cells from body effusions in pretreated ovarian carcinomas. Compared to nonpretreated ovarian tumor cells, with 61% chemosensitivity to doxorubicin, only 44% of cells pretreated in vivo with cyclophosphamide/5-fluorouracil showed chemosensitivity to doxorubicin in vitro. Of

Table 4. Heterogeneity of human tumors with respect to sensitivity to doxorubicin: testing of multiple samples using STPA

Source	No. of combinations	Disagreement S/R or R/S
Ovarian		
Solid/solid	17	5 (29%)
Solid/ascites	28	9 (32%)
Breast		
Solid/solid (primary/axillary lymph nodes)	59	14 (24%)

S, chemosensitive; *R*, chemoresistant

these 32 pretreated patients, in 21 who were further treated with doxorubicin/vincristine the sensitivity to doxorubicin was reduced to only 14%. There is a clear drift from nonpretreated sensitive to pretreated resistant tumor cells in patients.

The following conclusions can be drawn from our results with this STPA. A tumor which has not been pretreated by chemotherapy mostly responds to all proliferative cytotoxic agents in a similar way depending on its proliferative status (primary sensitivity or resistance). With low tumor cell kinetics, primary resistance is probable; with high kinetics, primary chemosensitivity is probable. Therefore, with this assay it is possible to predict common primary chemosensitivity using only one cytotoxic drug, for example doxorubicin, as a common test drug; doxorubicin proved to be most sensitive in the uridine system.

A chemotherapeutically pretreated tumor which is clinically resistant mostly does not react to the drug, which has become ineffective in therapy or reacts more weakly than other drugs (drug-induced resistance on the cellular level). The recognition of drug-induced biochemical resistance, however, requires the investigation of various cytotoxic agents.

The heterogeneity of human tumors is another major factor limiting the clinical application of predictive assays, since it makes it difficult to compare in vitro with in vivo results. Table 4 demonstrates that tested tumor samples are often not representative for the total tumor. In ovarian carcinomas we found a discrepancy of 29% in the results of doxorubicin tests on solid specimens and a discrepancy of 32% between solid tissue cells and ascites cells. In breast carcinomas there was a difference of 24% between values for primary tumors and involved axillary lymph nodes.

In conclusion, the benefits of the STPA are as follows: the test is possible in all human tumor types where single-cell suspensions or cell clumps can be achieved; there is no selection of tumor cells; the proliferation of the original tumor is maintained; the test results can be completed within 1 day; and the cost of the assay is low.

The limiting factors are that not only tumor cells are tested; cells do not enter a new cell cycle; changes in the nucleotide pool can result in false nucleic acid synthesis estimations; and not all drugs, for example not antimitotics, can be tested.

Quality control studies are essential in order to ensure comparability of test results. Interlaboratory studies on animal transplantation tumors have shown that the standardized doxorubicin-uridine system is easily reproducible, but nevertheless continuous quality control studies are necessary.

Table 5. In vitro (STPA)/in vivo correlation of chemosensitivity (*S*) and chemoresistance (*R*)

Reference	Year	No. of				
		Tests	S/S	R/R	S/R	R/S
Volm et al.	1979	49	19	25	5	0
KSST	1981	149	40	72	35	2
Silvestrini	1981	94	42	31	5	16
Possinger et al.	1982	154	38	88	28	0
Pfleiderer	1982	83	23	37	20	3
Kaufmann	1982	58	17	28	10	3
Daidone et al.	1983	41	15	17	5	4

Table 6. Predictive accuracy (%) for chemosensitivity (*S*) and chemoresistance (*R*) and overall predictive accuracy of the STPAs

Reference	Year	No. of			
		Tests	S	R	Total
Volm et al.	1979	49	79	100	90
KSST	1981	149	53	97	75
Silvestrini	1981	94	88	71	79
Possinger et al.	1982	154	58	100	82
Pfleiderer	1982	83	54	93	72
Kaufmann	1982	58	63	90	78
Daidone et al.	1983	41	75	81	78

Prediction of Clinical Therapy Results in Advanced Cancer

The relevance of sensitivity testing in human tumors can be proved only in correlative trials. Such in vitro/in vivo correlations are shown in Table 5 for short-term assays. It is obvious that with the exception of the overall data of the Milan group (Silvestrini 1981) the constellation "resistant in vitro, but sensitive in vivo" (R/S) occurs in only few cases. The constellation "sensitive in vitro and in vivo" (S/S) occurs very often because in all of these trials most of the patients were not pretreated with cytotoxic drugs.

Table 6 shows the predictive accuracy for chemosensitivity and chemoresistance and the overall predictive accuracy of STPAs. Correct prediction of chemoresistance is possible over a range of 71%–100%, which is very high. However, prediction of chemosensitivity appears possible over a range of only 53%–88%.

Test Results and Prognosis in Operable Cancer

In further studies we examined the prognostic value of the short-term doxorubicin-uridine system in clinically nondisseminated human tumors; it was established as a simple proliferation test in advanced carcinomas.

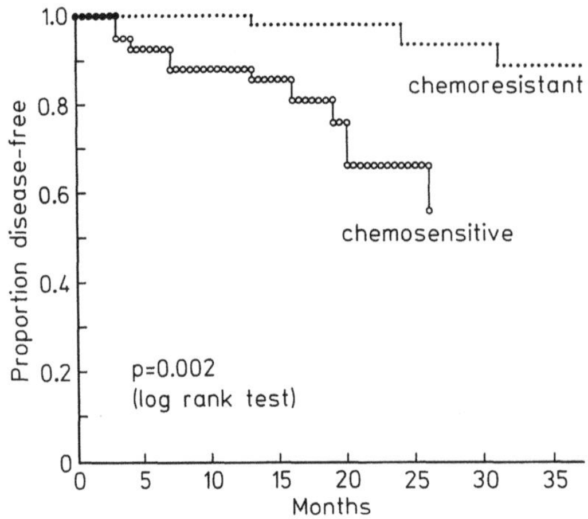

Fig. 3. Disease-free survival (life table analysis, Kaplan Meier estimation) in 60 patients with nodal negative primary breast carcinomas (T_{1-4}, N_0, M_0). Patients are divided according to the results of doxorubicin-uridine testing (chemoresistant-sensitive) of their primary tumors at the time of surgery (modified radical mastectomy)

In 60 nodal negative operable breast carcinomas (T_{1-4}, N_0, M_0), test results were analyzed for disease-free survival. Patients tumors were divided into resistant and sensitive tumors, according to slow and rapid proliferation. None of the patients received postoperative adjuvant cytotoxic treatment.

In Fig. 3 it is clearly shown that disease-free survival is significantly longer in patients with resistant tumors than in those with sensitive tumors. With this doxorubicin-uridine assay it is therefore possible to identify patients with sensitive rapidly proliferating primary breast tumors as patients at high risk of early recurrence.

These results are in contrast to the data obtained in metastatic carcinomas, where chemosensitive tumors respond better to chemotherapy. However, early data in patients with primary breast cancer who received aggressive postoperative adjuvant chemotherapy show that chemosensitive tumors involve the same disease-free survival rates as chemoresistant tumors.

Results of Correlative Trials in Various Predictive Assays

Predictive Accuracy Rates

Table 7 demonstrates the preditive accuracy for chemosensitivity and chemoresistance and the overall predictive accuracy of the human tumor colony assay (HTCA), first described by the Tucson group (Salmon 1980), and of the subrenal capsule assay (SRCA), first described by Bogden et al. (1981). In the case of the HTCA, correct prediction of drug resistance in 82%–100% can again be taken as the optimum attainable. For the SRCA, however, not enough correlative data exist as yet.

The divergence between true false positive and false negative prediction can be explained in purely theoretical terms: resistant tumor cells are more likely to determine the course of the patient's disease. The achievable drug concentrations differ according to the individual patient as a result of varying biotransformations or interactions of the applied medicaments, and even as a result of the variations in the vascularization of the individual tumors. Therefore, plasma level drug concentrations achievable in vivo are not absolutely comparable with in vitro ones.

Table 7. Predictive accuracy (%) for chemosensitivity (*S*) and chemoresistance (*R*) and overall predictive accuracy of the HTCA and the SRCA

Assay	Reference	Year	No. of			
			Tests	S	R	Total
HTCA	Salmon and von Hoff	1981	193	63	94	87
HTCA	Von Hoff et al.	1981	151	47	98	93
HTCA	Natale and Kushner	1981	28	77	93	86
HTCA	Bernheim et al.	1982	31	89	100	98
HTCA	Carney et al.	1982	147	84	92	81
HTCA	Mann et al.	1982	37	82	96	70
HTCA	Hogan et al.	1982	67	44	92	81
HTCA	Bertelsen et al.	1983	87	62	82	
HTCA	Parker et al.	1983	6	67	100	83
HTCA	Von Hoff et al.[a]	1983	246	60	85	80
SRCA	Griffin et al.	1982	41	91	78	88
SRCA	Stratton et al.	1983	25			63

[a] Prospective clinical trial with single agents

Table 8. Overall survival of tumor patients with chemosensitive (*S*) or chemoresistant (*R*) tumors according to the STPA and HTCA

Assay	Reference	Year	Tumor type	No. of patients	*p* value S vs R
STPA	KSST	1981	Lung	21	< 0.05
STPA	Sanfilippo et al.	1981	Non-Hodgkin's lymphoma	15[a]	< 0.005
STPA	KSST	1981	Ovarian	46	0.1
STPA	Kaufmann	1982	Ovarian	58	< 0.01
HTCA	Alberts et al.	1982	Ovarian	54	< 0.0001

[a] Only rapidly proliferating tumors

Prediction and Overall Survival Rates

Table 8 presents current data on overall survival of tumor patients based on test results obtained either with STPAs or with the HTCA. Because it is often difficult to establish objective response to chemotherapy, these data are of higher predictive value and even more convincing. Patients with treated advanced carcinomas whose tumors were found in the assays to be chemosensitive showed a longer survival in all studies.

Comparison of the STPA and the HTCA

In view of the similar predictive accuracy of all the different assays, it seemed to us of great interest to make a direct comparison between at least two test systems. In such a study we compared our STPA with the HTCA in 16 advanced ovarian carcinomas (Table 9).

Table 9. Comparison of the STPA and HTCA in ovarian carcinomas

Assay	Chemosensitive[a] (S)	Chemoresistant[a] (R)
STPA	5	11
HTCA	3	11

Correlation for S: 3/5 (60%); correlation for R: 11/11 (100%)
[a] Tested drug: doxorubicin

Table 10. Comparison of the HTCA and SRCA (Bogden et al. 1983)

Assay	Chemosensitive (S)	Chemoresistant (R)
HTCA	2	63
SRCA	7	76

Correlation for S: 2/7 (29%); correlation for R: 63/76 (83%)

Adequate test results were yielded in the STPA in 94% and in the HTCA in 57%. No correlation was found between the control values (tritiated thymidine or uridine incorporation rates) in the STPA and the number of colonies in the HTCA. Activity criteria for doxorubicin were established according to the commonly used thresholds between sensitivity and resistance for each test system. In the STPA (3 h drug exposure and also total test period) a dose of 10 μg/ml doxorubicin was used, whereas in the HTCA the dose was only 0.1 μg/ml for 1 h drug exposure to the cells, which were then kept in culture for 12 days.

For correct in vitro tumor response the correlation between these two assays was 60% (three out of five) for sensitivity and 100% (11 out of 11) for resistance. So far no clinical correlations are available. These in vitro data, however, confirm the high predictive accuracy of both test systems for chemoresistance.

Comparison of the HTCA and the SRCA

Table 10 represents the current available data for comparison of the HTCA and the SRCA (Bogden et al. 1983). Adequate test results were yielded in the HTCA in 42% and in the SRCA in 89%. The correlation between the two assays is greater for the prediction of chemoresistance (83%) than for the prediction of chemosensitivity (29%).

Conclusions

1. Predictive assays should make it possible to provide individualized and thereby improved cancer chemotherapy.
2. Problems of cell preparation and insufficient test rates are the major limiting factors inhibiting clinical application of predictive assays. Achievable test rates for the STPA

are about 60%–95% and those for the HTCA about 30%–70%. Here especially, more improvements are necessary in methods for tumor disaggregation and tumor culture techniques.

3. Heterogeneity of human tumors is a biological factor limiting predictive accuracy.
4. The STPA and HTCA are suited for clinical predictive trials but cannot yet be recommended for routine clinical work. For the SRCA too few data are at present available to allow any reliable statements.
5. Prediction of chemoresistance is possible with more than 90% accuracy.
6. Correlation between STPA and HTCA results (n = 16 ovarian carcinomas) is high only for the prediction of chemoresistance.
7. Quality control of predictive assays will improve the results of correlative trials.
8. Prospective randomized clinical trials are needed to establish the present encouraging results for clinical application. Only such trials can confirm whether chemotherapy tailored by predictive assay is superior to a treatment selected on the basis of the clinician's experience.

References

Abel U, Kaufmann M (1982) A methodologically improved definition of chemosensitivity indices. Cancer Res 42: 1610

Alberts DS, Surwit EA, Leigh S, Moon TE, Salmon SE (1982) Improved survival for relapsing ovarian cancer patients using the human tumor stem cell assay to select chemotherapy. Third conference on human tumor cloning, Tucson

Bernheim J, Naaktgeboren N, Roobol C, Thennissen J, Sips H (1982) Efficiency of oncograms by the human tumor stem cell clonogenic assay. Experience after 56 assays. Third conference on human tumor cloning, Tucson

Bertelsen CA, Korn EL, Kern DH (1983) Clinical correlations of the clonogenic assay (CA) with human solid tumors. ASCO Abstract C-26

Bogden AE, Cobb WR, Lepage DJ, Haskell PM, Gulkin TA, Ward A, Kelton DE, Esber HG (1981) Chemotherapy responsiveness of human tumors as first transplant generation xenografts in the normal mouse. Cancer 48: 10

Bogden AE, von Hoff DD, Schwartz JH (1983) Comparison of the human tumor cloning (ATC) and subrenal capsule (SRC) assays. ASCO Abstract C-11

Carney DN, Gazdar AF, Minna JD (1982) In vitro chemosensitivity of clinical specimens and established cell lines of small cell lung cancer. Third conference on human tumor cloning, Tucson

Daidone MG, Silvestrini R, Zaffaroni N, Sanfilippo O, Varini M (1983) Reliability of a short-term antimetabolic assay in predicting drug sensitivity of breast cancer. In: (eds) Endocrine-related tumors. In press

Dendy PP (1976) Human tumors in short-term cultures. Academic Press, New York

Griffin TW, Bogden AE, Reich SD, Hunter RE, Greene HL, Antonelli DM, Constanza ME (1982) Clinical update on the subrenal capsule assay as a predictive test for response to chemotherapy. ASCO Abstract C-67

Group for sensitivity testing of tumors (KSST) (1981) In vitro short-term test to determine the resistance of human tumors to chemotherapy. Cancer 48: 2127

Hogan WM, Ozols RF, Willson JKV, Foster BJ, Grotzinger KJ, McKoy W, Young RC (1982) Application of the human tumor stem cell assay to the study of epithelial ovarian cancer. 13th Annual Meeting of the Society of Gynecologic Oncologists, Marco Island

Kaufmann M (1980) Clinical applications of in vitro chemosensitivity testing. In: Newman CE, Ford CHJ, Jordan JA (eds) Ovarian cancer. Advances in the biosciences, vol 26. Pergamon, Oxford, p 189

Kaufmann M (1982) Nucleic acid precursor incorporation assay for testing tumour sensitivity and clinical application. Drugs. Exp Clin Res 8: 345

Kaufmann M (1983) Clinical aspects of chemosensitivity testing modalities in ovarian cancer. In: Grundmann E (ed) Cancer campain, vol 7. Carcinoma of the ovary. Fischer, Stuttgart

Kaufmann M, Kubli F (1983) Gegenwärtiger Stand der Chemosensibilitätstestung von Tumoren. Dtsch Med Wochenschr 108: 150–154

Mann BD, Kern DH, Giuliano AE, Burk MW, Campbell MA, Kaiser LR, Morton DL (1982) Clinical correlations with drug sensitivities in the clonogenic assay. Arch Surg 117: 33

Natale RB, Kushner B (1981) Applications of the human tumor cloning assay to ovarian cancer. AACR Abstract 617

Parker RL, Welander CE, Homesley HD, Jabson VW (1983) Chemotherapy sensitivity of carcinomas of the cervix (CACX) determined by the human tumor stem cell assay (HTSCA). ASCO Abstract C-87

Pfleiderer A (1980) Is chemosensitivity testing worth-while? Report of a cooperative study. Excerpta Medica, Amsterdam, International Congress Series, p 1143

Pfleiderer A (1982) Six years of experience with Volm's chemosensitivity testing of ovarian cancer. EORTC symposium on ovarian cancer, Stockholm

Possinger K, Ehrhardt H (1983) Prädiktive Tumorteste im chemotherapeutischen Behandlungskonzept maligner Erkrankungen. Klin Wochenschr 61: 77–84

Possinger K, Hartenstein R, Ehrhart H, Wilmanns W (1982) In-vitro-Erkennung von Tumorzellresistenz gegen antiproliferative Substanzen. Deutscher Krebskongreß, München

Salmon SE (ed) (1980) Cloning of human tumor stem cells. Prog Clin Biol Res 48

Salmon SE, von Hoff DD (1981) In vitro evaluation of anticancer drugs with the human tumor stem cell assay. Semin Oncol 8: 377

Sanfilippo O, Daidone MG, Costa A, Canetta R, Silvestrini R (1981) Estimation of differential in vitro sensitivity of non-Hodgkin lymphomas to anticancer drugs. Eur J Cancer 17: 217–226

Sanfilippo O, Silvestrini RS, Zaffaroni N, Piva L (1983) Potentiality of an in vitro chemosensitivity assay for human germ cell testicular tumors (GCTT). ASCO Abstract C-145

Silvestrini R (1981) Experimental approaches to improve the clinical management of tumors. Chemother Oncol 5: 14

Stratton JA, Micha JP, Braly PS, Rettenmaier MA, DiSaia PJ (1983) Correlation between chemotherapeutic sensitivity measured by the subrenal capsule tumor implant assay and the clinical response of patients with gynecologic tumors. ASCO Abstract C-125

Volm M, Mattern J (1982) Prätherapeutischer Nachweis der Resistenz von Tumoren gegen Zytostatika. Wien Klin Wochenschr 94: 599–604

Volm M, Wayss K, Kaufmann M, Mattern J (1979) Pretherapeutic detection of tumour resistance and the results of tumour chemotherapy. Eur J Cancer 15: 983

von Hoff DD, Casper J, Bradley E, Sandbach J, Jones D, Makuch R (1981) Asociation between human tumor colony-forming assay results and response of an individual patient's tumor to chemotherapy. Am J Med 70: 1027

von Hoff DD, Clark GM, Stogdill BJ, Sarosdy MF, O'Brien MT, Casper TJ, Mattox DE, Page CP, Cruz AB, Sandbach JF (1983) Prospective clinical trial of a human tumor cloning system. Cancer Res 43: 1926–1931

*In Vitro Chemosensitivity Assay Based on the Concept of Total Tumor Cell Kill**

L.M. Weisenthal, R.H. Shoemaker, J.A. Marsden, P.L. Dill, J.A. Baker, and E.M. Moran

Veterans Administration Medical Center, 5901 East Seventh Street, Long Beach, CA 90822, USA

Introduction

Attempts to develop in vitro assays to predict in vivo response to antineoplastic therapy in patients date back more than 40 years for radiotherapy and more than 25 years for chemotherapy (Von Hoff and Weisenthal 1980; Weisenthal 1981; Weisenthal and Lippman, to be published). More than 2,500 individual clinical correlations between assay results and results of therapy have been published. The combined average of the accuracy of "true positive" results of these tests is 66%, while the "true negative" accuracy average is 90%. The published data do not indicate that any one type of approach to in vitro chemosensitivity testing has superior validity over all other approaches.
Roper and Drewinko (1976) reported that the results of clonogenic assays in established cell lines did not agree with the results of nonclonogenic assays. Steel (1977) described evidence supporting a "stem cell" model of human tumor growth. Salmon et al. (1978) described positive clinical correlations with a clonogenic assay. These three publications stimulated enormous interest in using clonogenic assays for clinical drug testing and preclinical drug screening.
We tested the agar clonogenic assay described by Salmon et al. (1978). The first phase of the study was to determine assay success rates and optimize culture conditions. During the first 18 months, we cultured 143 specimens in soft agar. Four different types of culture media were tested in each assay, including the original "Hamburger-Salmon" medium (Hamburger and Salmon 1977), omitting the mouse spleen cell conditioned medium. From specimens with cytology or histology positive for cancer, more than 30 colonies per plate grew in 31 of 89 specimens. From cytologically negative specimens, colony growth was obtained in only five of 54 specimens ($p < 0.01$). Two of the five "false positive" cultures were from patients with proven metastatic carcinoma. There was no medium tested which was superior to RPMI-1640 plus 15% FBS plus 5% horse serum. Some researchers have reported that a simple RPMI-1640 medium was optimum for lung cancer specimens (Carney et al. 1980, 1981), while others have reported that simplified media were optimal in other tumors (Tveit et al. 1981; Kern et al. 1982).
Our culture success rates were disappointing, though consistent with those reported by Von Hoff et al. (1981), Callahan and Von Hoff (1982), Carney et al. (1983), and Lieber and Kovach (1982). Particularly discouraging was the finding that only 25 of the first 71 lung

* Supported by the U.S. Veterans Administration

cancer specimens (35% of those received) yielded enough cells to set up assays for three or more drugs. If we presume that with extensive further work a 50% culture success rate might be attained, the limited number of cells derived from typical biopsy specimens indicated that we could expect to test three drugs in only 18% of lung cancer specimens (50% of 35%). It was clear that an assay technique was needed which had a higher success rate and which required fewer cells.

Development of a Practical Nonclonogenic Assay System

Assays measuring damage in the total cell population (rather than strictly in the clonogenic cell population) have been used to predict responsiveness to radiotherapy in cervical cancer since the middle 1940s (Glücksmann 1974; Trott 1980). The impressive results (71% true positives and 88% true negatives in more than 1,000 clinical trials) indicate that assays

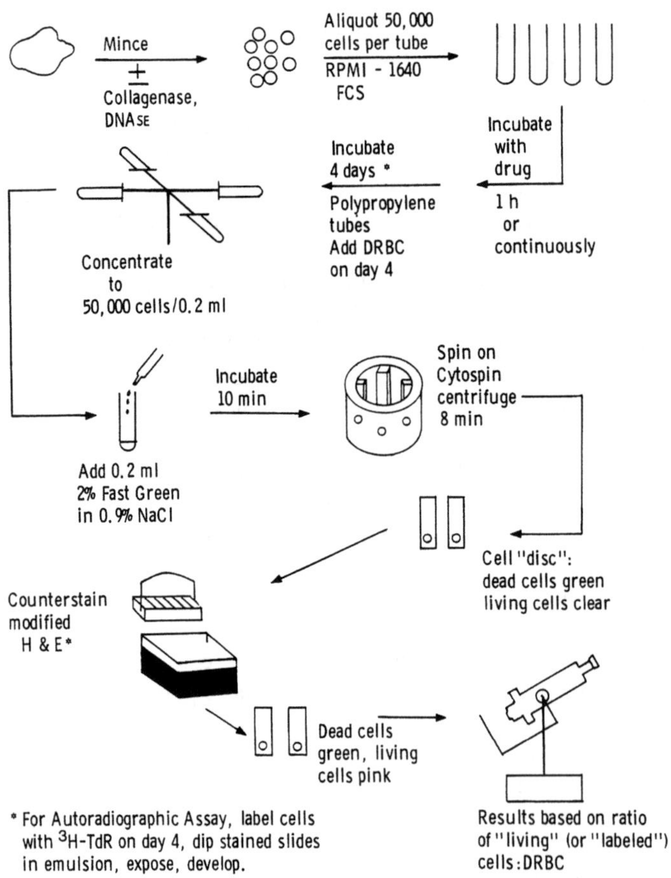

Fig. 1. Schematic depiction of novel dye exclusion (DE) and autoradiographic (AR) assays. *DRBC*, duck red blood cells (added as an internal standard). Both DE and AR assays may be performed on the same microscope slide. The DE assay is counted first and then the slides are dipped in emulsion for autoradiography. *FCS*, fetal calf serum; ^{3}*H-TdR*, tritiated thymidine; *H* and *E*, hematoxylin-eosin

based on the concept of total tumor cell kill may have clinical validity. We attempted to develop a practical in vitro assay based upon this concept.

Dye exclusion assays have long been used to quantify cell viability (Shrek 1936). Recent authors have questioned the value of dye exclusion assays (Roper and Drewinko 1976, 1979). However, many key technical problems have not been addressed by previous investigators. These problems include: (a) insufficient time passage following drug treatment to allow lethally damaged cells to lose their membrane integrity; (b) division of the surviving cell fraction that may continue during the time required for the lethally damaged cells to lose membrane integrity; (c) recognition of early disintegration of lethally damaged cells; (d) prevention of fibroblast growth; and (e) formulation of methods to distinguish tumor cells from nontumor cells present in cultures. We attempted to circumvent the above problems by developing a novel dye exclusion assay system (Weisenthal et al. 1983a–c; Bosanquet et al. 1983). Our methodology is illustrated in Fig. 1. Results will now be described.

Success Rates

Of 482 cytologically positive specimens received (excluding specimens from fiberoptic bronchoscopy), 335 (70%) were successfully assayed with three or more drugs (median ten drugs per successful assay for solid tumors). Success rates for the most commonly tested neoplasms are listed in Table 1. Solid tumors are preferred to pleural effusions, as the latter specimens may contain mesothelial cells, which may be difficult to distinguish from tumor cells. Notwithstanding this potential problem, more than 60% of malignant effusions may be successfully assayed with our technique.

Validation of the Assay

The validity of the assay was examined by:

1. Comparing the results of the assay with the results of the standard agar clonogenic assay in two different established cell lines.
2. Participating in a blind screening trial in which we were supplied with "unknown" drugs by the National Cancer Institute and attempted to identify active agents. These unknown drugs were simultaneously supplied to four different clonogenic assay laboratories and the results were compared.

Table 1. Novel dye exclusion assay: success rates

All specimens received	335/482	(70%)
Small-cell lung cancer	19/21	(90%)
Non-small-cell lung cancer	48/69	(70%)
Bladder cancer	48/59	(81%)
Acute leukemia	43/53	(81%)
Chronic lymphatic leukemia	30/35	(86%)
Non-Hodgkin's lymphoma	10/15	(66%)
Multiple myeloma	15/18	(83%)

Not all tumor types tested are listed in this table

3. Carrying out in vitro phase II trials to determine whether in vitro chemosensitivity patterns corresponded with known in vivo chemosensitivity patterns.
4. Attempting to determine whether previously treated patients were more or less sensitive in vitro than previously untreated patients.
5. Attempting to determine whether in vitro sensitivity or resistance correlated with clinical responsiveness in individual patients.

Comparison with Clonogenic Assay in Established Cell Lines

We compared the effects of eight drugs at three to four concentrations in two different established tumor cell lines, using both our dye exclusion assay and an agar clonogenic assay (Weisenthal et al. 1983a). Our dye exclusion assay was in general qualitative agreement with the clonogenic assay, but the clonogenic assay often gave higher estimates of cell kill at a given drug concentration. We discussed reasons why clonogenic assays may tend to give misleadingly high estimates of cell kill (Weisenthal et al. 1983a). It is possible that different assays may require different in vitro drug concentrations to be predictive for in vivo response.

National Cancer Institute Sponsored "Blind" Screening Trial

Twenty-five compounds were supplied by the National Cancer Institute for drug-screening studies. These compounds were identified only by a coded number and represented a variety of agents in different stages of evaluation as candidate anticancer drugs. Each compound was assigned for testing with ten different human tumors from fresh clinical biopsies using the dye exclusion assay. The same compounds were simultaneously shipped to four other laboratories for testing in the human tumor clonogenic assay as part of the first phase of a National Cancer Institute program for application of the human tumor clonogenic assay to new drug screening (Shoemaker et al. 1983). An in vitro response in the dye exclusion assay was defined as reduction of the viable tumor cell number to 30% or less of the control value. For the clonogenic assay, response was defined as reduction of colony formation to 30% or less of the control colony counts. Nine different tumor types were used for testing in the dye exclusion assay. More than 30 different tumor types were utilized in the clonogenic assay. All drug testing was by "continuous exposure" at a final concentration of 10 μg/ml. The laboratories were not given any information concerning the solubility of the drugs. In some cases, different solvents were used and possibly different final concentrations of solubilized drugs were achieved. These differing conditions reflect the actual situation encountered in the screening of new drugs, but they make direct comparisons more difficult.

Results of testing in both systems is summarized in Table 2 in terms of in vitro response rates. An overall qualitative similarity is apparent in the activity rankings of the various compounds in the two assay systems. Spearman's rank order correlation analysis was performed giving a correlation coefficient of 0.74 ($p < 0.0001$). If a 20% in vitro response rate is applied to these data as a minimum criterion for selection of compounds for further development, each assay yields eight compounds. For six out of eight compounds, the two assays are in agreement. Four of these compounds (Spirogermanium; homoharringtonine; dihydroxyanthracenedione; and CBDCA, a new platinum analog) have recently been selected for development for clinical trials. Sangivamycin was tested in the clinic in the

Table 2. National Cancer Institute "blind" drug screening trial: comparison of dye exclusion and clonogenic assays: percentage of specimens sensitive to tested drug

Drug	Dye exclusion assay	Clonogenic assay
Spirogermanium	100% (10/10)	41% (21/51)
Sangivamycin	100% (14/14)	36% (20/56)
Homoharringtonine	92% (12/13)	33% (25/75)
Henkel compound	91% (10/11)	17% (9/54)
Dihydroxyanthracenedione	64% (9/14)	40% (21/53)
Colchicine	46% (6/13)	19% (10/52)
CBDCA (platinum analog)	46% (5/11)	21% (11/52)
NSC 268562	36% (4/11)	24% (14/59)
Acer negundo	18% (2/11)	20% (10/51)
DON	17% (2/12)	18% (9/49)
NSC 295453	9% (1/11)	18% (9/50)
Aphidicolin	0% (0/10)	25% (13/51)
NSC 240419	0% (0/10)	22% (11/50)
Gallium nitrate	0% (0/13)	18% (10/55)
O,P-DDD	0% (0/9)	14% (6/44)
PALA	0% (0/12)	13% (7/53)
Hadacidin	0% (0/12)	10% (6/57)
Taxol	0% (0/15)	10% (5/50)
Butyric acid	0% (0/8)	9% (4/44)
Hydroxyurea	0% (0/13)	9% (4/47)
Dimethylsulfoxide	0% (0/13)	8% (4/49)
NSC 265595	0% (0/10)	7% (3/44)
Compound X[a]	0% (0/12)	7% (4/58)
NSC 276374	0% (0/12)	6% (3/51)
N-Methylformamide	0% (0/12)	5% (2/38)

[a] Drug supplied under commercial discrete agreement, prohibiting identification

1960s, but was dropped because of toxicity problems. Currently, there is renewed interest in this compound and a new phase I study has been started. Of interest is that one of the new compounds, NSC 268562, showed good activity in both systems and had only limited activity in the National Cancer Institute in vivo screening models and had not previously been considered to be a promising compound. In addition to these six compounds, the dye exclusion assay selected a "Henkel compound" which is currently undergoing clinical trial and colchicine, a potent microtubule inhibitor which has not received extensive consideration as an anticancer drug. The clonogenic assay selected aphidicolin and NSC 240419, two compounds which have shown significant activity in the National Cancer Institute in vivo screening models and are currently undergoing development. Using the 20% response rate threshold, a number of compounds are "missed" which either have activity in one or more animal tumor models or have shown clinical activity. Notable among the missed compounds is hydroxyurea. It was not unexpected that this noncytolytic compound should be negative in the relatively short-term dye exclusion assay. Hydroxyurea was also found to lack activity in the National Cancer Institute clonogenic assay program. It is speculated that reduced nucleotides present in the culture system interfere with the action of hydroxyurea.

The 20% response rate was used here as an activity criterion only for purposes of discussion. Additional data would be required in order to establish definitive criteria for

Table 3. Novel dye exclusion assay: in vitro phase II trials: specimens sensitive/specimens tested

Tumor type	DOX	HN2	LPAM	BCNU	DDP	MIT-C	5-FU	VBL	VCR	VP-16
Small cell	11/15	9/14	2/7	1/12	6/8	3/6	3/8	5/6	2/5	1/13
Non-small cell lung	2/42	5/38	2/25	2/30	5/25	3/18	4/31	2/20	2/13	1/15
CLL	9/16	11/16	12/17	5/13	NT	0/1	2/2	2/5	2/7	8/14
Multiple myeloma	2/9	4/8	6/9	3/10	NT	0/1	0/3	2/3	1/2	1/6
Non-Hodgkin's lymphoma	4/7	6/7	4/7	1/4	NT	1/1	NT	0/3	2/5	0/3
GI adeno-carcinoma	0/14	1/7	0/5	0/5	1/9	1/7	2/12	0/4	0/2	0/3
Bladder	2/14	NT	NT	NT	8/20	NT	NT	NT	NT	NT
Ovary	4/30	6/22	2/17	2/5	5/17	0/4	3/18	2/6	0/11	2/16
Breast	3/10	3/8	3/7	1/3	1/3	0/1	0/8	1/5	0/2	0/3
Melanoma	0/7	1/6	1/6	1/6	0/3	NT	0/5	NT	0/1	0/3

Drug concentrations tested were as follows (μg/ml × 1 h, unless indicated as continuous exposure): doxorubicin (*DOX*), 1.2; nitrogen mustard (*HN2*), 3.5; melphalan (*LPAM*), 12.5; carmustine (*BCNU*), 11; Cisplatin (*DDP*), 3.3 (continuous exposure); mitomycin C (*MIT-C*), 0.5 (continuous); 5-fluorouracil (*5-FU*), 20 (continuous); vinblastine (*VBL*), 0.53 (continuous); vincristine (*VCR*), 0.26 (continuous); etoposide (*VP-16*), 125
CLL, chronic lymphatic leukemia; *GI*, gastrointestinal; *NT*, not tested

use with either assay in a new drug-screening mode. The qualitative similarity of the results with the two assays, however, is evidence for the biological validity of the dye exclusion assay in primary human tumor cultures.

In Vitro Phase II Trials

The results of our phase II trials, in which in vitro drug sensitivity was correlated with known in vivo response patterns, have been reported (Weisenthal et al. 1983b). Our updated data are presented in Tables 3 and 4, which suggest the biological validity of the assay. Our results with a phase II in vitro trial with dexamethasone are summarized in Table 5. A similarity between our in vitro results and the expected clinical results with glucocorticoid therapy in the individual disease classifications was noted.

In another phase II in vitro trial carried out in transitional cell bladder cancer (Weisenthal et al. 1983c) we tested three drugs in concentrations which are achievable by intravesical administration. When given by the intravesical route, thiotepa is commonly administered at a dose of 30–60 mg/ml for 1 h. Considering dilution from continued urine production and absorption of the drug from the bladder, the drug concentration in vivo will range from 2 mg/ml at the time of instillation to considerably less at the end of 1 h. The in vitro response rates to thiotepa in our assay ranged from four out of 27 at 0.5 mg/ml to 11 out of 12 at 1.5 mg/ml. The known in vivo response rate to thiotepa is 60% (Koontz et al. 1981). We also tested doxorubicin in vitro at 100 μg/ml per hour. This concentration is easily achieved by intravesical administration (Eksborg et al. 1980). In our assay, 33 out of 33

Table 4. Novel dye exclusion assay: in vitro phase II trials: summation of specimens sensitive/specimens tested with single agents

GI adenocarcinoma	5/68	7%
Melanoma	3/37	8%
Non-small cell lung cancer	28/257	11%
Ovarian adenocarcinoma	26/146	18%
Breast adenocarcinoma	12/50	24%
Multiple myeloma	19/51	37%
Small-cell lung cancer	43/94	46%
Non-Hodgkin's lymphoma	18/37	49%
Chronic lymphatic leukemia	50/91	55%

GI, gastrointestinal

Table 5. Glucocorticoid sensitivity of human neoplasms

Neoplasm	Dexamethasone 10^{-6} *M*	
	Number sensitive	Number resistant
ALL	16	6
ANLL	2	10
CML (blast crisis, TdT+)	1	0
CML (blast crisis, TdT−)	0	4
MM + NHL + CLL (untreated)	6	8
MM + NHL + CLL (treated)	1	17
MM + NHL + CLL (documented prednisone resistance)	0	11
Solid tumors	0	23

ALL, acute lymphatic leukemia; *ANLL*, acute nonlymphatic leukemia; *CML*, chronic myelogenous leukemia; *TdT*, terminal deoxynucleotidyltransferase; *MM*, multiple myeloma; *NHL*, non-Hodgkin's lymphoma; *CLL*, chronic lymphatic leukemia

specimens were sensitive to this concentration of doxorubicin. This 100% response rate exceeds that reported in the literature. We noted evidence that treatment failures with intravesical doxorubicin may result from inadequate drug penetration to the tumor cells, rather than from intrinsic resistance of the cells to the drug (Weisenthal 1983c).

Previously Treated Versus Previously Untreated Patients

Further validation of the assay was provided by the comparison of previously treated and previously untreated patients. We did not have large numbers of patients with a given type of cancer studied who were sensitive to a given drug in vitro and were treated with the same drug in vivo. Therefore, meaningful comparisons can only be made between treated and untreated patients in certain groups. If patients with chronic lymphocytic leukemia (CLL), multiple myeloma, and lymphoma are grouped together for the purpose of our study, some

Table 6. Novel dye exclusion assay: treated versus untreated patients (multiple myeloma + CLL + lymphocytic lymphoma). In vitro assay results[a]

Clinical status	HN2(0.7)		LPAM(2.5)		VP-16(125)		DOX(1.2)		DEX(0.4C)	
	S	R	S	R	S	R	S	R	S	R
Never treated	8	5	8	3	6	2	9	4	6	8
Previously treated	2	17	2	17	3	12	6	13	1	17
Failed alkylating agents + prednisone	0	12	0	13	0	9	3	8	0	11

CLL, chronic lymphatic leukemia; *S*, sensitive; *R*, resistant; *DEX*, dexamethasone
[a] Cells were exposed to the drugs at the indicated concentrations (μg/ml) either for 1 h or continuously (*C*)

comparisons may be made. The treatment of these neoplasms is similar, in that the drugs used are alkylating agents and prednisone. We have divided the patients studied into three categories: (a) never treated, (b) previously treated with alkylating agents and prednisone, and (c) documented as nonresponsive to alkylating agents and prednisone.
The data summarized in Table 6 provide support for the validity of our assay in lymphatic neoplasms. It is of interest that previous therapy with alkylating agents and prednisone is associated in the assay with cross-resistance to VP-16 and, to a lesser extent, to doxorubicin.

In Vitro/In Vivo Correlations in Individual Patients

Valid correlations are exceedingly difficult to make, because most patients received drug combinations rather than single agents. We have analyzed our results in several ways. By accepting patients treated with drugs that showed mixed in vitro sensitivity (Von Hoff et al. 1981) and clinical results reported retrospectively and prospectively by experienced oncologists at outside institutions, we had 20 true positive (TP), 11 false positive (FP), and 31 true negative (TN) correlations and 1 false negative (FN) correlation. If we included retrospective and prospective correlations from patients treated only with drugs which were either all active or all inactive in vitro and included only patients who could be independently verfied at our institution, we had 8 TP, 8 FP, 22 TN, and 1 FN. Finally, if we used the latter criteria but included only prospective correlations, we had 8 TP, 8 FP, 10 TN, and 1 FN.
The above clinical correlations are obviously very limited and preliminary. We would certainly not claim that the assay has any given TP or TN predictive accuracy. Considering all our data summarized in the above subsections, it is justifiable to conclude that the assay is feasible and practical, and does have biological validity in human cancer.

Autoradiographic Assay

Our novel dye exclusion assay measures cell kill in the entire population of tumor cells. We attempted to determine the labeling index of 93 primary human tumor cell cultures to determine what percentage of the tumor cells were actually cycling after 4 days in liquid

Table 7. Theoretical considerations concerning dividing and nondividing cells in chemosensitivity assays

1. Cell kinetics do not completely explain chemosensitivity differences between different tumors. However, the most radioresistant and chemoresistant cells in any given tumor are likely to be the nondividing cells.
2. "Recruitment" of nondividing cells in vivo is primarily a matter of tumor angiogenesis. This takes place over a matter of days to weeks.
3. Recruitment in a clonogenic assay results from supplying nutrients and oxygen, which takes minutes to hours. Thus clonogenic assays measure a cell population with a growth fraction of 100%.
4. Tumor control and cure will ultimately correlate with sterilization of the most resistant cells rather than with sterilization of the most sensitive cells.
5. Differences between tumor stem cells and tumor nonstem cells may be much less important than differences between dividing stem cells and nondividing stem cells.
6. Kill of nondividing nonstem cells may predict the kill of nondividing stem cells for some tumors and some drugs.

culture. The labeling index [percentage of tumor cells labeled with tritiated thymidine (^{3}H-TdR)] typically ranged between 1% and 5% on day 4. Only occasional cultures had labeling indices greater than 10%. Thus we assayed a population of tumor cells which was mostly nondividing, at least after 4 days in culture. This is quite different from the clonogenic assay, where the population of cells being tested is 100% dividing (when true colonies and not clumps are counted). Colonies are, by definition, formed only from dividing cells.

The ability to measure cell kill in nondividing tumor cells has major potential advantages. One reason why human tumors are resistant to drugs is thought to be that they contain cells which are reversibly nondividing (Drewinko et al. 1981; Gavosto 1973). It is possible that long-term disease control may correlate more closely with the kill of nondividing tumor cells than with the kill of dividing tumor cells (Table 7). We were, however, concerned that the activity of strictly cycle-specific agents would be underestimated if the predominant cells being tested were noncycling.

In order to determine whether the kill of cycling or noncycling cells was more important for tumor control for different drugs, an autoradiographic assay was developed (Weisenthal et al. 1982; schematically depicted in Fig. 1). The classic type of ^{3}H-TdR autoradiographic assay is based upon the labeling index (Sky-Peck 1971; Livingston et al. 1980). This method may give erroneous results because of the following problems:

1. "Early" labeling shortly following drug exposure is affected by artifacts related to changes in nucleoside pool sizes and the relative activities of the salvage versus de novo thymidylate synthesis pathways (Livingston et al. 1980).
2. If labeling is delayed to allow for a reduction in these artifacts, some killed tumor cells may disintegrate and some surviving tumor cells could continue to proliferate, which would artifactually raise the labeling index and lower the estimate of cell kill. Our autoradiographic assay appeared to be more reliable in detecting cell kill than the standard technique (unpublished data).

To date, we have attempted a simultaneous comparison of the dye exclusion assay with the autoradiographic assay in 93 specimens. The dye exclusion assay was successful in 82 of 93

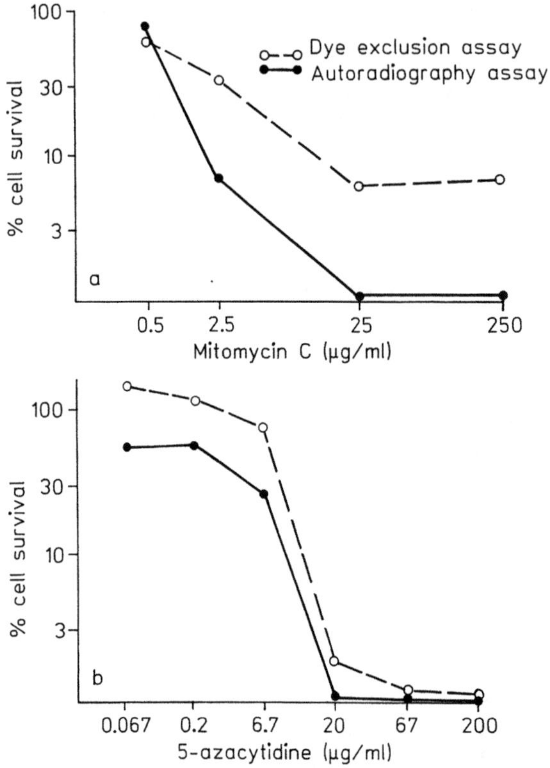

Fig. 2a, b. Comparison between dye exclusion and autoradiography assays in two human neoplasms. **a** Assay performed on cells from a patient with a previously untreated lung adenocarcinoma. Durg exposure was for 1 h. The patient was subsequently treated with single-agent mitomycin C. He achieved at most a transient disease stabilization before dying from progressive disease. **b** Assay performed on cells from a patient with acute nonlymphatic leukemia in relapse after treatment with cytarabine and doxorubicin. The patient was subsequently treated with single-agent 5-azacytidine and had a 99% reduction in his peripheral blast count (100,000 prior to treatment), but did not achieve a bone marrow remission

specimens, while the autoradiographic assay was successful in only 30 of 93 specimens, generally because of poor labeling (11 of 11 specimens of CLL failed to label, but all were successfully assayed with the dye exclusion technique). Our success rate with the autoradiographic assay was about the same as with the clonogenic assay, but with one important exception. Because of the small number of cells required (50,000 per assay tube), we were able to test ten times as many drugs per successful assay with the autoradiographic technique as with the clonogenic technique. We found that the autoradiographic technique was superior to ^{3}H-TdR labeling and scintillation counting for two reasons. First, we were able to test many more drugs per assay because ten times fewer cells were required for autoradiography than for statistically meaningful scintillation counting. Secondly, occasional cultures contained large numbers of transformed lymphocytes, the labeling of which could exceed the labeling of the tumor cells. This could obviously cause erroneous results in a scintillation counting assay, but labeled transformed lymphocytes can be identified as such in an autoradiographic assay.

When we compared the results of the dye exclusion assay with the autoradiographic assay, we found that the two assays agreed closely in established tumor cell lines (Weisenthal et al. 1982; Weisenthal and Marsden, to be published). The former assay measures all tumor cells, while the latter assay measures only dividing tumor cells. Since nearly all cells in established lines are dividing, it is not surprising that the two assays agreed. In primary human tumor cultures, however, the autoradiographic assay has often demonstrated a greater degree of cell kill at a given drug concentration than the dye exclusion assay (Fig. 2). Further experience should permit us to answer the question of whether kill of dividing or nondividing cells correlates better with short- and long-term responses to individual agents in different tumor types.

Theoretical Considerations of the Predictive Accuracy of Chemosensitivity Assays

The mathematical expressions that describe the intrinsic validity of a predictive assay are the "sensitivity" and the "specificity". For chemosensitivity assays, sensitivity reflects the probability of identifying a truly active drug as being active. Specificity reflects the probability of identifying a truly inactive drug as being inactive.
Sensitivity and specificity are quantified by the following expressions:

Sensitivity = TP/TP + FN
Specificity = TN/TN + FP.

Another expression which describes the performance of a test is the "predictive accuracy". Compared to the above expressions, predictive accuracy addresses a different question, for it reflects the probability that a given test result is correct in a given patient population.

Predictive accuracy for a positive test = TP/TP + FP
Predictive accuracy for a negative test = TN/TN + FN.

Predictive accuracy is profoundly affected by the characteristics of the patient population, and the same test can have large differences in predictive accuracy, depending on the patient population being studied. The relationship has been mathematically described (Makuch 1982) and practical examples have been given (Redwood et al. 1976).
Let us consider two different patient populations tested with an assay shown to have an intrinsic sensitivity of 0.86 and an intrinsic specificity of 0.76 (these vaules represent the accumulated results of more than 2,000 published clinical correlations with various types of chemosensitivity assays):

(A) One hundred patients with non-small-cell lung cancer: expected rate of response to cisplatin = 10%
(B) One hundred patients with testicular cancer: expected rate of response to cisplatin = 80%.

In patient population A, it can be mathematically demonstrated (Redwood et al. 1976) that the true positive predictive accuracy will be 29% and the true negative predictive accuracy will be 99%. With patient population B, the true positive accuracy will be 93% and the true negative accuracy will be 58%.
There are two important points which must be made concerning the above calculations. First, the true negative predictive accuracy of a test will be much higher in patients unlikely to respond to therapy than in patients likely to respond to therapy. Secondly, the most promising use of predictive assays is in choosing between drugs which are known clinically to have an equal chance of producing a response. In the above examples, a drug predicted to be active in patient population A (29% response probability) would be 29 times more likely to produce a response than a drug predicted to be inactive (1% response probability), provided that both drugs were known to have a 10% response rate in that particular patient population. In patient population B, an active drug in the assay (93% response probability) would be about twice as likely to produce a response as an assay-inactive drug (42% response probability), provided that both drugs were known to produce an 80% response

rate. However, an assay-inactive drug with a known 80% response rate would be more likely to produce a response than an assay-active drug with a known 10% response rate. Therefore, it would be very dangerous to forsake a drug with a known high response rate on the basis of a negative chemosensitivity assay result.

Conclusions

1. The novel dye exclusion assay is a rapid and practical test. It may be successfully used in a wide variety of neoplasms, including those which do not form colonies in soft agar and do not incorporate ^{3}H-TdR. A major advantage of the test is that small numbers of cells are required.
2. Indirect evidence strongly suggests that the test will prove useful in determining the sensitivity of lymphatic neoplasms to glucocorticoids and alkylating agents. Evidence for the validity of the assay in other tumor types and for other drugs in encouraging, but preliminary.
3. Screening for drug activity in nondividing cell populations is a worthwhile goal of chemosensitivity testing.
4. The dye exclusion assay may be a useful tool in drug screening.
5. Predictive chemosensitivity assays will be most useful in facilitating a choice from a group of drugs known on the basis of clinical experience to be equally likely to produce a response. It is dangerous to forsake a known highly active drug in favor of a known poorly active drug on the basis of a chemosensitivity assay result.

References

Bertoncello I, Bradley TR, Campbell JJ et al. (1982) Limitations of the clonal assay for the assessment of primary human ovarian tumor biopsies. Br J Cancer 45: 803

Bosanquet AG, Bird MC, Price WJP, Gilbey ED (1983) An assessment of a short term tumour chemosensitivity assay in chronic lymphocytic leukaemia. Br J Cancer 47: 781

Callahan SK, Von Hoff DD (1982) Growth of oat cell carcinoma of the lung in the human tumor cloning assay. Stem Cells 1: 295

Carney DN, Gazdar AF, Minna JD (1980) Positive correlation between histologic tumor involvement and generation of tumor cell colonies in agarose in specimens taken directly from patients with small cell carcinoma of the lung. Cancer Res 40: 1820

Carney DN, Gazdar AF, Bunn PA, Guccion JG (1981) Demonstration of the stem cell nature of clonogenic tumor cells from lung cancer patients. Stem Cells 1: 149

Carney DN, Broder L, Edelstein M et al. (1983) Experimental studies of the biology of human small cell lung cancer. Cancer Treat Rep 67: 27

Drewinko B, Patchen M, Yang LY, Barlogie B (1981) Differential killing efficacy of twenty antitumor drugs on proliferating and non-proliferating human tumor cells. Cancer Res 41: 2328

Eksborg S, Nilsson SO, Edsmyr R (1980) Intravesical instillation of Adriamycin: a model for the standardization of chemotherapy. Eur Urol 6: 218

Gavosto F (1973) An outline of the objectives of the study of leukemic cell kinetics. In: Dutcher RM, Chieco-Bianchi G (eds) Unifying concepts of leukemia. Bibl Haemat, no 39, Karger, Basel, pp 968–977

Glücksmann A (1974) Histological features in the local radiocurability of carcinomas. In: Friedman M (ed) The biological and clinical basis of radiosensitivity. Thomas, Springfield

Hamburger AW, Salmon SE (1977) Primary bioassay of human tumor stem cells. Science 197: 461

Kern DH, Campbell MA, Cochran AJ et al. (1982) Cloning of human solid tumors in soft agar. Int J Cancer 30: 725
Koontz WW, Prout GR, Smith W et al. (1981) The use of intravesical thiotepa in the management of non-invasive carcinoma of the bladder. J Urol 125: 307
Lieber MM, Kovach JS (1982) Soft agar colony formation assay for chemotherapy sensitivity testing of human solid tumors. Mayo Clin Proc 57: 527
Livingston RB, Titus GA, Heilbrun LK (1980) In vitro effects on DNA synthesis as a predictor of biologic effect from chemotherapy. Cancer Res 40: 2209
Makuch RW (1982) Interpreting clonogenic assay results. Lancet 2: 438
Redwood DR, Borer JS, Epstein SE (1976) Wither the ST segment during exercise? Circulation 54: 703–706
Roper PR, Drewinki B (1976) Comparison of in vitro methods to determine drug-induced cell lethality. Cancer Res 36: 2182
Roper PR, Drewinko B (1979) Cell survival following treatment with antitumor drugs. Cancer Res 39: 1428
Salmon SE, Hamburger AW, Soehnlen BJ, Durie BGM et al. (1978) Quantitation of differential sensitivities of human tumor stem cells to anticancer drugs. N Engl J Med 298: 1321
Selby P, Buick RN, Tannock I (1983) A critical appraisal of the human tumor stem cell assay. N Engl J Med 308: 129
Shoemaker RH, Wolpert-DeFilippes MK, Makuch RW, Venditti JM (1983) Use of the human tumor clonogenic assay for new drug screening. Proch Am Assoc Cancer Res 24: 311
Shrek R (1936) A method for counting the viable cells in normal and in malignant cell suspensions. Am J Cancer 28: 389
Sky-Peck HH (1971) Effects of chemotherapy in the incorporation of 3*H*-thymidine into DNA of human neoplastic tissue. Nat Cancer Inst Monogr 34: 197
Steel GG (1977) Growth and survival of tumor stem cells. In: Steel GG (ed) Growth kinetics of tumors. Clarendon, Oxford, pp 217–267
Trott KR (1980) Can tumor response be assessed from a biopsy? Br J Cancer (Suppl IV) 41: 163
Tveit KM, Endresen L, Rugstad HE et al. (1981) Comparison of two soft agar methods for assaying chemosensitivity of human tumors in vitro: malignant melanomas. Br J Cancer 44: 539
Von Hoff DD, Weisenthal LM (1980) In vitro methods to predict for patient response to chemotherapy. Adv Pharmacol Chemother 17: 133
Von Hoff DD, Casper J, Bradley E et al. (1981) Association between human tumor colony-forming assay results and response of an individual patient's tumor to chemotherapy. Am J Med 70: 1027
Weisenthal LM (1981) In vitro assays in preclinical antineoplastic drug screening. Semin Oncol 8: 362
Weisenthal LM, Lippman ME (to be published) Clonogenic and non-clonogenic in vitro chemosensitivity assays
Weisenthal LM, Marsden JA (to be published) Comparison of 3*H*-thymidine autoradiography and cloning in soft agar for determining cancer chemosensitivity in vitro
Weisenthal LM, Dill PL, Marsden JA, Campbell CK (1982) Comparison of 3 in vitro chemosensitivity assays: dye exclusion, autoradiography, and agar cloning. (Abs.) Proc Am Assoc Cancer Res 23: 184
Weisenthal LM, Dill PL, Kurnick NB, Lippman ME (1983a) Comparison of dye exclusion assays with a clonogenic assay in the determination of drug-induced cytotoxicity. Cancer Res 43: 258
Weisenthal LM, Marsden JA, Dill PL, Macaluso CK (1983b) A novel dye exclusion method for testing in vitro chemosensitivity of human tumors. Cancer Res 43: 749
Weisenthal LM, Lalude AO, Miller JB (1983c) In vitro chemosensitivity of human bladder cancer. Cancer 51: 1490

Evaluation of In Vitro Results

*Predictive Tests and Infrequent Events in Cancer Chemotherapy**

M. Rozencweig and M. Staquet**

Bristol-Myers Company, Pharmaceutical Research and Development Division,
P.O. Box 4755, Syracuse, NY 13221–4755, USA

Introduction

Peritoneoscopy is widely used as a diagnostic procedure to detect hepatic metastases. Liver involvement may be demonstrated by biopsy and microscopic examination. In contrast, normal biopsy is not a proof that the liver is free of invasion. The risk of false negative findings at peritoneoscopy is evidently related to the selection of patients (Bleiberg et al. 1978). Thus the risk may be noticeably lower if the procedure is performed to confirm that a malignant disease is still operable than if it is used because of altered liver function tests at the advanced stage of the same disease. These effects of the prevalence in diagnostic tests must also be taken into consideration for predictive tests.

Various models have been developed to predict for activity and toxicity of cancer chemotherapy. In this paper, we will analyze the effect of the prevalence on the predictive value of such models in relation to specific characteristics of currently available cytotoxic agents. In new drug development programs, predictive tests are usually selected because of favorable correlations between experimental and clinical findings. It is then assumed that these correlations will remain favorable with new compounds that have not yet been tested in humans. In this analysis, it is also assumed that the test result and the clinical effect of the treatment may be expressed dichotomously as positive or negative. Predictive tests for efficacy and toxicity will be dealt with separately. In addition, it will be shown that limitations related to the occurrence of rare events also apply to results of clinical trials.

Predictive Tests for Drug Efficacy

Prospective trials have indicated that drug testing in a cloning assay might be successfully used for defining optimal chemotherapy regimens in individual cancer patients (Alberts et al. 1980; Von Hoff et al. 1983). The routine application of this assay is hindered by a number of difficulties (Selby et al. 1983). Other tests have been utilized with variable

* This work was supported in part by grant 3.4535.79 of the Fonds de la Recherche Scientifique Medicale (F.N.R.S. Belgium)

** The authors acknowledge the assistance of Ms. Geneviève Decoster in the preparation of this manuscript results. Our purpose is not to discuss technical aspects or to compare the relative merits of these procedures but to underline general limitations inherent in the prediction of drug efficacy in individual patients. The relevance of conventional animal screening for new antitumor agents will also be addressed.

Sensitivity, Specificity, and Predictive Value

The reliability of a predictive test may be defined by its sensitivity and its specificity (Chiang et al. 1956; Vecchio 1966). Sensitivity is the ability of the test to identify patients whose disease does respond clinically to chemotherapy, whereas specificity is the ability of the test to identify patients whose disease does not respond clinically. These indexes characterize the test with one regimen and one tumor type for fixed criteria of experimental and clinical response. They are determined during preliminary evaluation of the test and are calculated as follows:

$$\text{Sensitivity} = \frac{\text{Patients with clinically responsive disease and positive test}}{\text{All patients with clinically responsive disease}} \times 100$$

$$\text{Specificity} = \frac{\text{Patients with clinically refractory disease and negative test}}{\text{All patients with clinically refractory disease}} \times 100$$

Sensitivity and specificity may be readily converted to the rate of false negative (β error = 1 − sensitivity) and false positive tests (α error = 1 − specificity) respectively.
If the test is then used to predict clinical results of chemotherapy, it is essential to know its predictive value. The predictive value of a positive test (PV pos) gives the probability that a patient with a positive test will achieve clinical response, whereas the predictive value of a negative test (PV neg) gives the probability that a patient with a negative test will fail to achieve clinical response. These indexes are calculated as follows:

$$\text{PV pos} = \frac{\text{Patients with clinically responsive disease and positive test}}{\text{All patients with positive test}} \times 100$$

$$\text{PV neg} = \frac{\text{Patients with clinically refractory disease and negative test}}{\text{All patients with negative test}} \times 100$$

The respective meaning of sensitivity, specificity, and predictive value is illustrated in the following example. A total of 100 patients with ovarian cancer are treated with cisplatin 100 mg/m^2 i.v. repeated every 3 weeks. A cloning assay is performed in all these patients with in vitro cell exposure to cisplatin for 1 h at a concentration of 0.1 μg/ml. A positive test is defined as a reduction by 70% or more in the number of colonies relative to the controls. Forty patients exhibit shrinkage by 50% or more of all measurable disease; among these, 24 have a positive test (sensitivity = 60%) and 16 a negative test (false negative tests = 40%). Treatment failure is seen in 60 patients and, among these, 48 have a negative test (specificity = 80%) and 12 a positive test (false positive tests = 20%). Under these conditions, the predictive value of the test is obtained by simple arithmetic. The likelihood of clinical response to cisplatin in patients with a positive test is 67%:

$$\text{PV pos} = \frac{24}{24 + 12}\,.$$

Conversely, the likelihood of clinical failure to respond to cisplatin in patients with a negative test is 75%:

$$\text{PV neg} = \frac{48}{48 + 16}.$$

Relation of Predictive Value to Sensitivity and Specificity

The predictive value is enhanced with increasing values of sensitivity and specificity. This relationship is documented in Tables 1 and 2 for a test performed with one agent in a population in whom 40% of the patients do respond clinically to this agent. The predictive value of a positive test is noticeably more influenced by its specificity than by its sensitivity. The reverse is true, although to a lesser extent, for the predictive value of a negative test.

In the previous example of a study of cisplatin in ovarian cancer, it was shown that with a sensitivity of 60% and a specificity of 80%, lack of in vitro drug effect was confirmed in vivo in 75% of the cases. The predictive value of these negative tests may be substantially improved by changing the experimental conditions toward increasing sensitivity. Modifications that could improve the reliability of a negative test include in vitro exposure of cells to higher concentrations of drugs (e.g., 10 µg per milliliter cisplatin) and less stringent criteria for in vitro drug efficacy (e.g., reduction by 50% in the number of colonies relative to controls).

Table 1. Predictive value of a positive test (PV pos) over a range of sensitivities and specificities when actual response rate in the entire population under study is 40% (example cisplatin in ovarian cancer)

Specificity (%)	Sensitivity (%)			
	60	70	80	90
60	50	54	57	60
70	57	61	64	67
80	67	70	73	75
90	80	82	84	86

Table 2. Predictive value of a negative test (PV neg) over a range of sensitivities and specificities when actual response rate in the entire population under study is 40% (example cisplatin in ovarian cancer)

Specificity (%)	Sensitivity (%)			
	60	70	80	90
60	69	75	81	90
70	72	78	84	91
80	75	80	86	92
90	77	82	87	93

Relation of Predictive Value to the Expected Response Rate in the Entire Population Under Study

The predictive value cannot be determined directly from the sensitivity and the specificity, since it is also related to the probability of response in the entire population under study (Chiang et al. 1956; Vecchio 1966). This probability may be estimated from past experience providing a historical response rate in this population. A lower expected response rate results in a reduced predictive value of a positive test and an increased predictive value of a negative test despite unchanged sensitivity and specificity. This effect of response rate is demonstrated by comparing data in Tables 1 and 3.

If the expected clinical response rate in the entire population under study is 10%, the predictive value of a positive test is 25% only when sensitivity is 60% and specificity is 80% (Table 3). Table 1 shows a corresponding figure of 66% using a test with identical reliability in a population with an expected response rate of 40%. The lowest predictive value in Table 1 becomes the highest achievable one in Table 3. Of note is that specificity must be at least 80% to increase the expected response rate of 10% in the entire population under study to a response rate higher than 25% in a fraction of patients selected through a positive test.

In contrast, the predictive value of a negative test is extremely high in a population of patients with a disease highly resistant clinically to the therapy given. With a response rate of 10% in the entire population, the predictive value of a negative test will vary from 93% to 99% when sensitivity and specificity vary between 60% and 90%.

The relative contribution of a predictive test, expressed by the ratio of the predictive value to the expected response rate in the entire population under study, varies widely according to this expected response rate. Thus, if the response rate in an unselected population is 40%, a doubling of this result may be obtained among patients selected through a predictive test characterized by a sensitivity of 60% and a specificity of 90% (Table 1). A fourfold increase in response rate would be achieved with the same test if responses were seen in 10% of unselected patients (Table 3).

A low response rate of 10% in an unselected patient population may be observed with conventional chemotherapy, for example, when treating advanced malignant melanoma with cisplatin. If drug efficacy is found in a predictive test carried out in a freshly sampled tumor obtained from such a patient, this patient has a likelihood of clinical response to cisplatin that does not exceed 50% even with a highly reliable test.

Table 3. Predictive value of a positive test (PV pos) over a range of sensitivities and specificities when actual response rate in the entire population under study is 10% (example cisplatin in malignant melanoma)

Specificity (%)	Sensitivity (%)			
	60	70	80	90
60	14	16	18	20
70	18	21	23	25
80	25	28	31	33
90	40	44	47	50

This situation might dramatically worsen with investigational agents that have still unestablished pharmacokinetics in humans. This information seems crucial for testing a compound at relevant concentrations in the cloning assay. In order not to miss antitumor activity, one might elect to investigate these new drugs at obviously high concentrations in vitro with resulting increase in the sensitivity of the test and a simulaneous decrease in specificity. Then, if a test predicts efficacy for a patient with a drug that yields a 10% response rate in an unselected population, the probability that this patient might respond clinically would not exceed 20% (sensitivity = 90% and specificity = 60%).
The relationship between predictive value and sensitivity and specificity is expressed mathematically by the following equations (Chiang et al. 1956):

$$\text{PV pos} = \frac{p \times \text{sensitivity}}{p \times \text{sensitivity} + (1\text{-}p)(1\text{-specificity})} \times 100$$

$$\text{PV neg} = \frac{(1\text{-}p) \times \text{specificity}}{(1\text{-}p) \times \text{specificity} + p\,(1\text{-sensitivity})} \times 100$$

In these equations, p represents the probability of response for a patient in the population under study. It is readily apparent that, when p tends to zero, the predictive value of a positive test tends to zero and the predictive value of a negative test tends to one, and vice versa. These equations should allow the reader to compute easily predictive values for variable sensitivities, specificities, and clinical response rates.

Validation of a Predictive Test

A prime requirement for a predictive test is that it improves results achievable with empirical chemotherapy. This requirement may be expressed by PV pos > prevalence or PV neg > 1 − prevalence or sensitivity + specificity > 1.
These considerations are also essential when the values of different predictive tests are compared (Staquet et al. 1981). Various reports have appeared on predictive tests for the response of human cancer to chemotherapy. None, however, have provided relevant information allowing estimates of sensitivity and specificity with sufficient accuracy. To obtain this crucial piece of information it would be necessary to evaluate the test in consecutive patients from homogeneous populations treated with the same chemotherapy regimens regardless of the test results.

Therapeutic Implications

The striking relationship between the predictive value of a test and the expected response rate in the entire population under study has major implications. A predictive test may identify a group of patients in whom response rate will be greater than in the entire population. However, even with a most reliable test, this improved response rate will always be dependent upon the degree of therapeutic effectiveness that may be clinically expected with currently available drugs in unselected patients. The predictive value of the same test will be higher for effective agents or tumors responsive to chemotherapy than for ineffective agents or refractory malignancies. The lower the efficacy of a drug given as empirical chemotherapy, the higher the reliability needed to keep the same predictive value of a positive test.

The response rate to single-agent anticancer treatment varies between 0 and 30% in most diseases. A significantly increased response rate may be observed in a fraction of patients in whom activity is detected in a test provided that the test has remarkable specificity. This increase, however, will remain necessarily limited and, in general, unless predictive tests with high reliability are used, patients with very resistant diseases will barely benefit from predictive tests as compared to empirical chemotherapy. In any event, the predictivity of the test will always be higher for clinical failure than for clinical response.

A greater impact on cancer treatment could be expected with predictive systems for combination chemotherapy regimens, which generally yield higher response rates than single-agent treatment in unselected patients. Little experience has been accumulated in this exciting area of research which might be hampered by specific additional methodological and technical difficulties. Moreover, a superiority over empirical chemotherapy might be difficult to detect with very active regimens since the relative contribution of predictive tests (PV pos/*p*) is lower when the expected response rate is higher in the entire population under study.

Screening for New Anticancer Agents

Currently available anticancer agents were introduced into clinical trials mainly on the basis of their activity against experimental tumors in mice. Advances in cancer chemotherapy with cures in an increasing number of human malignancies are encouraging for further refining screening strategies. A recent analysis of the experimental screening data for 1949 compounds and a comparison of murine and human data for 69 drugs evaluated clinically against solid tumors indicated that the correlation between screening and clinical results was low and that most drugs active in man would have been identified by four murine models, i.e., the P388 and the L1210 leukemias, the B16 melanoma, and the MX-1 human mammary tumor xenograft (Staquet et al. 1983).

The assessment of a screening method is restricted evidently to drugs evaluated clinically and correlations are possible only with human malignancies against which adequate numbers of active and inactive drugs are known. In tumor types highly refractory to chemotherapy, the predictivity of a screening system cannot be substantiated. For such tumors, whether current screening methodology is inadequate or whether few or no active drugs actually exist is unknown.

The potential of a cloning assay for the screening of new anticancer agents appears promising (Shoemaker et al. 1983). Its value relative to conventional screening in transplantable tumor systems would be difficult to ascertain, since compounds lacking

Table 4. Contingency table for relating screening results in mice to clinical activity

	Clinical findings		
	Active	Inactive	Total
Active in the screening	a	b	a + b
Inactive in the screening	?	?	c
	a + ?	b + ?	a + b + c

PV pos = a/a + b
Yield = a + b/a + b + c

anticancer properties in mice are not investigated clinically (Table 4). The effects of this biased sampling were extensively discussed elsewhere (Staquet et al. 1983). Comparisons between screening systems may be made, however, in terms of predictive value of positive tests and ratio of active drugs identified in a test to the total number of drugs tested.

Prediction of Toxicity

Animal toxicology studies with new anticancer agents are carried out to define appropriate starting doses for phase I clinical trials and to predict qualitatively for toxic effects in humans. A retrospective analysis on quantitative relationships between animal and human toxicology data indicated that the starting dose in phase I clinical trials could be safely and efficiently based on one-tenth LD_{10} (the lethal dose for 10% of the group) in the mouse, expressed in milligrams per square meter, with prior verification that the resulting dose is not lethal or life-threatening in the dog (Rozencweig et al. 1981). The rationale for this procedure currently recommended by the United States National Cancer Institute and the European Organization for Research on Treatment of Cancer was described in a publication that also focused on the limitations of qualitative predictions for toxicology (Rozencweig et al. 1981). These latter aspects may be summarized as follows.

The predictive value of a positive finding for a particular toxic effect may be defined as the percentage of drugs inducing this effect in man among the total number of drugs inducing this effect in animals. The predictive value of a negative finding for a particular toxic effect is the percentage of drugs that do not induce this effect in man among the total number of drugs that do not induce this effect in animals.

The qualitative prediction of toxic effects in man, based on animal experiments, cannot be correctly assessed without taking into account the prevalence of these toxic effects in man. Prevalence is defined here as the proportion of drugs inducing this effect in man among the total number of drugs investigated. Since the predictive value of a positive test tends to zero as prevalence tends to zero, whereas the predictive value of a negative test tends to zero when prevalence tends to 1 (see above), the probability of observing a particular toxic effect in humans with one chemotherapeutic agent inducing this effect in animals decreases if this toxic effect is rarely encountered in man with chemotherapeutic agents.

Table 5. Predictive value in man of qualitative toxicologic findings by organ systems in the dog

Organ system	Prevalence in man (%)	PV pos (%)	PV neg (%)
Gastrointestinal	92	92	–
Bone marrow	88	87	0
Liver	52	54	100
Renal	40	36	33
Cardiovascular	40	54	75
Neuromuscular	28	29	75
Integument	28	27	71
Injection site	24	31	83
Respiratory	20	20	80
Lymphoid	4	5	100

These concepts are illustrated by a retrospective analysis of clinical, chemical, and pathological toxicology data by organ systems with 25 anticancer agents in humans and in dogs (Table 5). The relatively small number of drugs available for analysis allows only a rough estimate of the relevance to the clinical setting of toxicologic findings in animals. Nevertheless, a number of striking observations may be made.

Results in Table 5 document the variability of the predictive value of a toxic effect, especially the predictive value of a positive test, according to its prevalence. Animal toxicology studies are carried out to alert the clinician to potential hazards of the drugs. However, by definition, the predictive value of a predictive test is dependent upon the prevalence, so that the predictability will be relatively high for common toxicities to which all investigators are a priori already alerted, whereas this predictability will be relatively low for rare toxic effects.

Moreover, it appears that, generally, the predictive values provide minimally additive information as compared to what may be expected from the prevalence of the respective toxic effects. For most of these, there are no clear trends indicating a superiority of the animal findings over the simple knowledge of prevalence. This situation is likely to persist as long as screening methods for selecting new drugs for cancer therapy remain unchanged.

Clinical Trials

Clinical trials identifying new active or more active treatments are rare events. Despite statistical significance, results may be obscured by the interference of variables related to the conventional α and β errors, as well as the probability that new active or more active treatments actually do exist (Staquet et al. 1979).

Suppose that 1,000 trials comparing two equivalent treatments are performed with random samples from a given population (Table 6). The decision rule for significance is set in such a way that 5% of the trials will show a significant conclusion. A type II error, with β equal to 50%, for example, would lead to 500 studies with incorrect (false negative) conclusions in 1,000 randomized trials with two treatments of different effectiveness.

Table 6 shows that two other errors may be easily computed by reading the data vertically. The error in positivity (δ error) is defined by the number of false positive trials over the total number of positive trials (50/550 or 9%). The error in negativity (ε error) is found by

Table 6. Theoretical outcome of a series of randomized trials

	Conclusion of the trial	
	Positive: significant difference in effectiveness	Negative: no significant difference in effectiveness
1,000 trials with 2 treatments of equivalent effectiveness	50 (false positive) (α = 0.50)	950
1,000 trials with treatments of different effectiveness	500	500 (false negative) (β = 0.50)
Total	550	1,450

calculating the ratio between the number of false negative trials and the total number of negative trials (500/1450 or 34%). δ and ε are related to α and β in the following way:

$$\delta = \frac{(1\text{-TER})\alpha}{(1\text{-TER})\alpha + \text{TER}(1\text{-}\beta)}$$

$$\varepsilon = \frac{(\text{TER})\beta}{(\text{TER})\beta + (1\text{-TER})(1\text{-}\alpha)}$$

where TER (true effectiveness ratio) is the ratio of true positive results to the total number of situations investigated (1,000/2,000 in Table 6). In randomized trials, the TER provides the a priori probability that the new treatment is better than the control, whereas in phase II trials, TER is the a priori probability that an investigational agent is active. Both δ and ε increase with increasing α and/or β and decrease with decreasing α and/or β. In addition, slight variations in TER may result in enormous changes in δ and ε errors even when α and β errors are kept constant. A reduction in TER augments δ and lowers ε, and vice versa, with most significant effect on the δ error when TER is < 0.5 (Table 7). In randomized trials, ethical considerations are not consistent with comparisons of treatments of apparently very different effectiveness, resulting in low TER in most cases. A critical analysis of published data in a variety of tumor types (Staquet 1978) indicates that the value of TER in randomized trials can reasonably be assumed to fall into a range of 1%–20%. Under these conditions, when a therapeutic difference is reported to be statistically significant, there is at least a 17% chance that this difference does not exist with α at 5% and a power $(1-\beta)$ of 95% (Table 7). This figure is further but moderately worsened when the power of the test declines as exemplified with $\beta = 50\%$ (Table 7). It can be easily calculated that, in the favorable case of TER = 20%, the α error must be, in fact, set at 1% at most, for a $\beta = 5\%$, in order to keep the δ error below the 5% limit. This requirement would necessitate a doubling of the total patient accrual as compared to a design using α and $\beta = 5\%$.

This analysis reemphasizes the need for solid rationales before embarking on clinical trials. Clearly, the highest priority must be given to testing the major principles of therapy and the investigation of minor questions probably should merely be avoided. The puzzling magnitude of the δ error also calls for the greatest care when designing clinical trials. Adequate questions and proper design not only are more consistent with large accrual of patients, but they also are instrumental in increasing the TER.

The δ error may also be extremely high in conventional phase II studies, but this observation is of much less importance than it is in the case of phase II trials. Phase II trials

Table 7. δ and ε errors at selected levels of true effectiveness ratio (TER) and α and β errors

TER (%)	$\alpha = \beta = 5\%$		$\alpha = 5\%$, $\beta = 50\%$	
	δ (%)	ε (%)	δ (%)	ε (%)
20	17	1	29	12
10	32	0.60	47	6
1	84	0.05	91	0.53

are conducted in small numbers of patients, the high rate of the α error in these trials is commonly recognized, and detection of activity usually prompts rapid activation of confirmatory clinical trials.

The Bayesian approach described in this analysis is not widely accepted in clinical research, due to the subjectivity in determining the a priori probability of success. Although other considerations than TER undoubtedly play a role in designing clinical trials and analyzing clinical data, this method gives one possible explanation for non reproducible results and underlines the importance of conducting confirmatory trials before embarking on large-scale use of new treatment modalities.

References

Alberts DS, Chen HSG, Soehnlen B, Salmon SE, Surwit EA, Young L, Moon TE (1980) In vitro clonogenic assay for predicting response of ovarian cancer to chemotherapy. Lancet 2: 340–342

Bleiberg H, Rozencweig M, Mathieu M, Beyens M, Gompel C, Gérard A (1978) The use of peritoneoscopy in the detection of liver metastases. Cancer 41: 863–867

Chiang CL, Hodges JL Jr, Yerushalmy J (1956) Statistical problems in medical diagnoses. In: Neyman J (ed) Proceedings of the third Berkeley symposium on mathematical statistics and probability, vol 4. University of California Press, Berkeley, pp 121–148

Rozencweig M, Von Hoff DD, Staquet MJ, Schein PS, Penta JS, Goldin A, Muggia FM, Freireich EJ, DeVita VT Jr (1981) Animal toxicology for early clinical trials with anticancer agents. Cancer Clin Trials 4: 21–28

Selby P. Buick RN, Tannock I (1983) A critical appraisal of the "human tumor stem cell assay". N Engl J Med 308: 129–134

Shoemaker RH, Wolpert-De Filippes MK, Venditti JM (1983) Application of a human tumor clonogenic assay to screening for new antitumor drugs. In: Spitzy KH, Korrer K (eds) Proceedings of the 13th international congress on chemotherapy, Vienna, 223: 14–19

Staquet MJ (ed) (1978) Randomized trials in cancer: a critical review by sites, vol 4. Raven, New York

Staquet MJ, Rozencweig M, Von Hoff DD, Muggia FM (1979) The delta and epsilon errors in the assessment of cancer clinical trials. Cancer Treat Rep 63: 1917–1921

Staquet M, Rozencweig M, Lee YJ, Muggia FM (1981) Methodology for the assessment of new dichotomous diagnostic tests. J Cronic Dis 34: 599–610

Staquet MJ, Byar DP, Green SB, Rozencweig M (1983) The clinical predictivity of transplantable tumor systems in the selection of new drugs for solid tumors – rationale for a three-stage strategy. Cancer Treat Rep 67: 753–765

Vecchio TJ (1966) Predictive value of a single diagnostic test in unselected populations. N Engl J Med 264: 1171–1173

Von Hoff DD, Clark GM, Stogdill BJ, Sarosdy MF, O'Brien MT, Casper JT, Mattox DE, Page CP, Cruz AB, Sandbach JF (1983) Prospective clinical trial of a human cloning system. Cancer Res 43: 1926–1931

*Pharmacologic Pitfalls in the Human Tumor Clonogenic Assay**

D.S. Alberts, J. Einspahr, R. Ludwig, and S.E. Salmon

The University of Arizona, Health Sciences Center,
Section of Hematology and Oncology, Tucson, AZ 85724, USA

Introduction

The human tumor clonogenic assay (HTCA) (Hamburger and Salmon 1977; Salmon et al. 1978) has proved to be a useful system for screening new chemical compounds for anticancer activity (Salmon 1980; Shoemaker et al. 1983), for testing new clinical agents for phase II activity (Salmon et al. 1981; Von Hoff et al. 1981; Alberts et al. 1981a), and for predicting clinical response to cytotoxic therapy in cancer patients (Salmon et al. 1978; Alberts et al. 1980a, 1981a; Salmon et al. 1980; Salmon and Von Hoff 1981; Von Hoff et al. 1983). Despite its widespread use, additional biological and pharmacologic studies are essential before the HTCA can be considered optimized for routine clinical and laboratory research applications (Alberts et al. 1981b; Selby et al. 1983). We have previously reported suggestions for improving drug sensitivity assay conditions (Alberts et al. 1980b, 1981b), and in this presentation will discuss more recent data concerning the evaluation of (a) in vitro drug stability; (b) drug combinations which are additive in their antitumor activity; and (c) cyclophosphamide biotransformation and the relative activity of its major metabolites and a new oxazaphosphorine analog (Asta-Werke Z7557).

In Vitro Drug Stability Testing

In the standard HTCA fresh human tumor cells are exposed in vitro to anticancer drugs either for 1 h prior to plating or continuously in the culture dish (Alberts et al. 1980b, 1981b; Salmon and Von Hoff 1981; Ludwig et al., to be published). In order to interpret the assay results accurately, it is essential to determine the chemical and biological stability of these agents during 1-h and continuous (i.e., up to 10-day) exposures. We have quantitated the chemical and biological stability of various cell-cycle-specific (e.g., bleomycin, etoposide, and vinblastine) and cell-cycle-nonspecific (e.g., doxorubicin, actinomycin D, and bisantrene) drugs commonly used in our laboratory for chemosensitivity testing (Ludwig and Alberts, to be published).

In these drug stability studies anticancer drugs were added to enriched CMRL 1066 medium (Hamburger et al. 1978) plus 15% horse serum, which is used for the upper layer

* This work was supported in part by grants CA 17094, CA 21839, and CA 23074 from The National Institutes of Health, Bethesda, MD 20205, USA

in the HTCA and incubated at 37° C for up to 10 days. The drugs were also added to standard CMRL 1066 without horse serum. Aliquots of the culture media were obtained immediately following addition of the drug and after 10 days of incubation, and then frozen in liquid nitrogen (−120° C) for later chemical and biological assay in the HTCA.

For biological evaluation of drug stability, cell lines (i.e., HEC-1A endometrial carcinoma and HL-A melanoma) rather than fresh tumors were utilized. A single-cell suspension from each human tumor cell line was incubated with the freshly prepared drugs for 1 h to serve as a control. Simultaneously, the thawed aliquots of drug incubated in culture media were exposed to the tumor cell suspensions for 1 h. Thereafter, the cells were washed twice and prepared for culture.

The survival of tumor-colony-forming units (TCFUs) after 1-h exposure to freshly prepared drug was compared to survival of TCFUs after 1-h exposure to drug preexposed at 37° C for different time intervals. The difference in survival was expressed as percentage loss of activity.

For chemical evaluation of drug stability, actinomycin D, bleomycin, and vinblastine concentrations in culture media were determined using radioimmunoassays. Etoposide (VP16-213) and bisantrene were quantitated by high-performance liquid chromatography (HPLC) assays.

Data from the biological assays showed that doxorubicin, actinomycin D, bisantrene, bleomycin, and vinblastine retained their biological activity over a period of up to 10 days. In contrast etoposide lost 60% of its activity against TCFUs over a 10-day period. For each of these drugs there was no difference in their activity against TCFUs whether incubations were in plain media or in enriched CMRL 1066 plus 15% horse serum. Also, for two of these agents (i.e., doxorubicin and bisantrene) there was no significant loss of anticancer activity (measured biologically) following incubation in CMRL 1066 plus 0.3% agar for 24 h or 4 days.

Similar results were obtained from the chemical assay studies. There was no loss in the chemically measured concentrations of actinomycin D, bisantrene, bleomycin, or vinblastine; however, etoposide concentration as measured by a HPLC assay (Dr. Donald Van Harken, Bristol Laboratories, Syracuse, N.Y.) disappeared completely over the 10-day incubation period.

Previous data support our findings of a high degree of in vitro chemical stability for actinomycin D, bisantrene, bleomycin, and vinblastine (Brotham et al. 1982; Peng et al. 1983; Root, personal communication) and the limited chemical stability of etoposide (Issell 1982). It was to be expected, although previously not shown, that the biological stability of these agents would parallel their chemical stabilities. Thus our data suggest that the HTCA can be used as a simple test for biological stability of new investigational agents prior to the development of adequate chemical assay methodology.

Preincubation of an anticancer drug in culture media for up to 10 days has proved a useful test of the bioavailability of both standard and experimental drugs to be used in the HTCA. For those drugs that undergo rapid loss of biological activity following incubation in the culture medium, in vitro dose adjustments must be made in order to guarantee an adequate concentration × time product for the drug when exposed to tumor cells. For example, if it is planned to maintain a drug concentration of 1 μg/ml throughout the 10-day continuous exposure period and it is determined that a drug like etoposide maintains only 40% of its biological activity at 10 days, etoposide concentrations must be refurbished daily in order to maintain the concentration × time product at the desired level.

We must caution that the bioavailability studies for standard and experimental drugs used in the HTCA do not take into consideration either metabolism to biologically active compounds or uptake into the intracellular space with resultant decrease in extracellular concentration. Although for most drugs the biological and chemical stability results are likely to be complementary, divergent results between these two assay methods would provide the basis for a search for both active and inactive metabolites.

Studies of the Additive Effects of Drug Combinations

For many years medical oncologists have combined two or more anticancer drugs for the clinical management of most solid and hematologic cancers. The drug combinations are almost always designed on an empirical basis. The HTCA could be used to develop a rational rather than empirical basis for combining two or more drugs for the treatment of solid tumors, and could help identify those drug combinations which might be additive or potentiating in the clinic.

In studies of fresh human ovarian cancers the HTCA has been used in this report to assess the additive effects of doxorubicin plus cisplatin and the three-drug combination of doxorubicin, cisplatin, and cyclophosphamide (i.e., 4-hydroperoxycyclophosphamide). These are two of the most commonly used drug combinations in the treatment of advanced ovarian cancer, being associated with complete plus partial response rates of about 70% (Wallach et al. 1980; Omura et al. 1982). It has been our objective in these studies to compare retrospectively the in vitro additivity of these compounds with their known clinical activities.

Each drug in the combination was tested for a 1-h incubation at two or three different drug concentrations (i.e., low, intermediate, and high). The two and three drugs were also combined simultaneously for 1-h incubations at one or two of the concentrations used for the single agents. Quantitation of in vitro drug effects was carried out according to previously described methods. (Drewinko et al. 1976; Alberts et al. 1981b, c). Thirty-nine fresh ovarian cancers (from patients who had not had prior chemotherapy), which yielded at least 30 colonies of greater than 60 μm diameter, were tested with a 1-h exposure to doxorubicin, cisplatin, and the two-drug combination at 0.01 and 0.1 μg of each drug per milliliter. Additive or synergistic effects from the two-drug combination were observed in 69% of the cancers. Thirty-eight percent showed additivity, 31% showed synergism, and 31% showed inhibitory effects. These results correlate with the findings of phase II and III clinical trials of doxorubicin plus cisplatin in the treatment of ovarian and other cancers, which have suggested at least additive effects from this drug combination in about 70% of patients (Higby et al. 1977; Bruckner et al. 1983).

In considerable contrast to the results of these two-drug combination studies have been our observations concerning the potential additive effects of doxorubicin plus cisplatin plus cyclophosphamide (i.e., 4-hydroperoxycyclophosphamide). Thirteen fresh ovarian cancers from previously untreated patients with advanced disease were tested with a 1-h exposure to each of these three drugs alone and in combination at concentrations of 0.01, 0.1, and 1.0 μg/ml. Surprisingly, no additive or synergistic activity was observed in any of the tumors for the three-drug combination. These results suggest either a true lack of additivity between these three clinically useful agents or, more likely, a problem of chemical compatibility between them under HTCA assay conditions.

Recently, Sondak et al. (1983) reported that only 6.3% of 332 solid tumors (not including ovarian cancer) showed evidence of "sensitivity" (i.e., $< 50\%$ TCFU survival) to any one of seven different drug combinations, but not to any single drug in the combination using continuous exposure wells. They concluded that their system demonstrated an absence of significant synergy and that in vitro chemosensitivity tests may not need to include combinations if each individual drug is tested. In contrast to those of Sondak et al. our present drug combination data were obtained using 1-h drug exposures of ovarian cancers followed by the standard Hamburg-Salmon assay methodology. Nevertheless, the lack of additivity of the doxorubicin, cisplatin, and cyclophosphamide combination even in a small number of patients is a troublesome finding and suggests the possibility of adverse in vitro chemical interactions between these drugs. One possible mutually inhibitory drug interaction could involve the chelation of cisplatin by the hydroxyquinone moiety of doxorubicin. Further, detailed evaluations of such potential in vitro interactions are indicated.

Biotransformation of Cyclophosphamide and the Activity of Major Metabolites and Asta-Werke Z7557

Several standard anticancer drugs (e.g., cyclophosphamide) may reguire bioactivation in vitro by microsomal enzymes to express their cytotoxicity in vitro (Conners et al. 1974). We (Alberts et al. 1981b) and Lieber et al. (1980) have presented evidence that the S-9 fraction of rat liver can be used to activate compounds with anticancer activity when combined with $MgCl_2$, KCl, glucose-6-phosphate, and nicotinamide adenine dinucleotide phosphate in phosphate buffer in the HTCA. Shown in Fig. 1 are dose-response curves against fresh, human ovarian TCFUs for chlorambucil, l-phenylalanine mustard (L-PAM), S-9 activated cyclophosphamide, phosphoramide mustard, and 4-hydroperoxycyclophosphamide (provided by Dr. Norbert Brock, Asta-Werke, Bielefeld, West Germany). Note that 4-hydroperoxycyclophosphamide was associated with less than 1% TCFU survival at 10 µg/ml (a concentration of active metabolites which may be achievable in patients at only the highest cyclophosphamide doses). The rat liver S-9 activated metabolites and phosphoramide mustard had intermediate activity (i.e., 35%–40% survival of TCFUs), and there appeared to be considerable resistance to both L-PAM and chlorambucil. Although this figure depicts the sensitivity of only one human tumor to a variety of bifunctional alkylating agents, it does demonstrate the heterogeneity of tumor response to anticancer drugs in the HTCA and the clear dose-response relationship associated with 4-hydroperoxycyclophosphamide.

A number of important research questions must be answered before a final decision can be made concerning the bioactivation of agents to be tested for cytotoxicity in the HTCA. Comparisons between rat liver slices, purified microsomes, and the S-9 mix should be

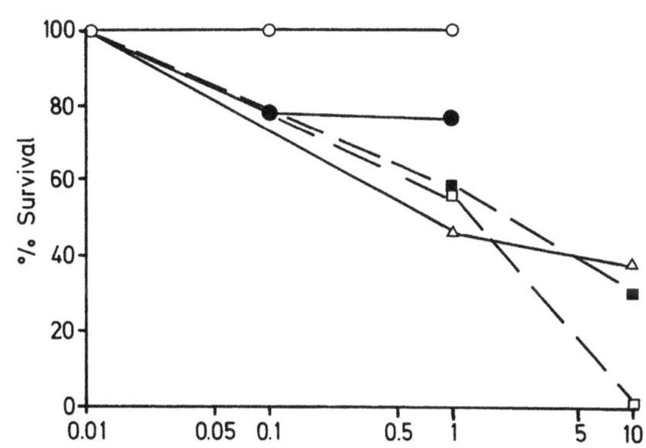

Fig. 1. Dose-response curves for chlorambucil (○——○) l-phenylalanine mustard (L-PAM) (●——●), phosphoramide mustard (■——■), rat liver S-9 fraction activated cyclophosphamide (△——△), and 4-hydroperoxycyclophosphamide (□——□) against tumor-colony-forming units from a previously untreated, fresh ovarian cancer in the human tumor clonogenic assay

Table 1. Cyclophosphamide metabolites and analogs versus fresh human tumor-colony-forming units (TCFUs) in the human tumor clonogenic assay (HTCA)

Tumor type	Median molar ID_{50} concentrations[a]			
	4-OOH cyclophosphamide	Asta-Werke Z7557	S-9 activated cyclophosphamide	Phosphoramide mustard
Ovarian (no. tested)	1.1×10^{-4} (45)	5.3×10^{-4} (12)	9.9×10^{-4} (23)	7.2×10^{-2} (21)
All tumors (no. tested)	5.7×10^{-5} (107)	6.7×10^{-5} (17)	9.9×10^{-4} (45)	2.5×10^{-4} (30)

4-OOH cyclophosphamide, 4 hydroperoxycyclophosphamide

[a] ID_{50} concentration = molar concentration reducing TCFU survival to 50% of control. All cells were exposed to drugs for 1 h prior to washing and plating in the HTCA

carried out to determine the optimal activating system for specific antineoplastic agents. Furthermore, isolated, active metabolites of these drugs (e.g., 4-hydroperoxycyclophosphamide or phosphoramide mustard) could prove more useful for in vitro cytotoxicity testing than any of the activating systems.

We have compared the activity of 4-hydroperoxycyclophosphamide, S-9 activated cyclophosphamide and phosphoramide mustard against advanced previously untreated ovarian, non-small-cell lung, and various other, genitourinary cancers. In addition we have evaluated the relative activity of a new oxazaphosphorine compound, Asta-Werke Z7557 {[4-(2-sulfonatoethylthio)-cyclophosphamide]}, which after dissolution undergoes rapid spontaneous hydrolyzation in vitro and in vivo with liberation of an activated 4-hydroxylated compound. Shown in Table 1 are the median molar concentrations of the cyclophosphamide metabolites and Z7557 (provided by Dr. Peter Hilgard, Asta-Werke, Bielefeld, West Germany) required to reduce survival of TCFUs to 50% of control (ID_{50}). It is of interest that 4-hydroperoxycyclophosphamide was associated with the lowest median molar ID_{50} concentrations for both the ovarian cancers and other solid tumors. Z7557 had the next lowest median ID_{50} concentrations, followed by S-9 activated cyclophosphamide and phosphoramide mustard, its ultimate intracellular metabolite. Consistent with these data are those showing that the 4-hydroperoxy compound achieved an ID_{50} concentration in the highest percentage of tumors of all types (i.e., 75%), whereas phosphoramide mustard again was shown to be the least active compound.

Further reason to use 4-hydroperoxycyclophosphamide as the marker for parent compound activity is the fact the both the S-9 activated drug and phosphoramide mustard degrade rapidly in vitro (i.e., within 1 and 7 days respectively), even when placed immediately in liquid nitrogen (i.e., at −120° F). As a result of these studies we have chosen routinely to use the 4-hydroperoxy compound for cyclophosphamide alkylating agent studies in the HTCA. On the basis of the high degree of in vitro activity demonstrated by Z7557 it appears to be an attractive compound for future clinical trials.

Conclusion

We have described our continued efforts pharmacologically to optimize the HTCA for use in the evaluation of chemotherapeutic agents for fresh, solid tumors. It is clear that there is need for continual updating of improved methods for drug sensitivity quantitation, scheduling, combination, and bioactivation. Such in vitro studies with the HTCA are likely to become a regular component of testing of new anticancer drugs.

References

Alberts DS, Chen HSG, Soehnlen B, Salmon SE, Surwit EA, Young L (1980a) In vitro clonogenic assay for predicting response of ovarian cancer to chemotherapy. Lancet 2: 340–342

Alberts DS, Chen HSG, Salmon SE (1980b) In vitro drug assay: pharmacologic considerations. In: Salmon SE (ed) Cloning of human tumour stem cells, chap 16. Liss, New York, pp 197–207

Alberts DS, Chen HSG, Salmon SE, Surwit EA, Young L, Moon TE, Meyskens FL Jr (1981a) Chemotherapy of ovarian cancer directed by the human tumour stem cell assay. Cancer Chemother Pharmacol 6: 279–285

Alberts DS, Salmon SE, Chen HSG, Moon TE, Young L, Surwit EA (1981b) Pharmacologic studies of anticancer drugs using the human tumour stem cell assay. Cancer Chemother Pharmacol 6: 253–264

Alberts DS, Salmon SE, Surwit EA, Chen HSG, Moon TE, Meyskens FL (1981c) Combination chemotherapy in vitro with the human tumour stem cell assay. Proc Am Assoc Cancer Res 22: 153

Brothman AR, Davis TP, Duffy JJ, Lindell TJ (1982) Development of an antibody to actinomycin D and its application for the detection of serum levels by radioimmunoassay. Cancer Res 42: 1184–1186

Bruckner HW, Cohen CJ, Goldberg J, Kabakow B, Wallach R, Holland JF (1983) Ovarian cancer: comparison of Adriamycin and cisplatin ± cyclophosphamide. Proc Am Soc Clin Oncol (abstract C-594)

Conners TA, Cox PJ, Farmer PB, Foster AB, Jarman M (1974) Some studies of the active intermediates formed in the microsomal metabolism of cyclophosphamide and isophosphamide. Biochem Pharmacol 23: 115–129

Drewinko B, Loo TL, Brown B et al. (1976) Combination chemotherapy in vitro with Adriamycin. Observations of additive, antagonistic and synergistic effects when used in two-drug combinations of cultured human lymphoma cells. Cancer Biochem Biophys 1: 187–195

Hamburger AW, Salmon SE (1977) Primary bioassay of human tumour stem cells. Science 197: 461–463

Hamburger AW, Salmon SE, Kim NB, Trent JM, Soehnlen BJ, Alberts DS, Schmidt HJ (1978) Direct cloning of human ovarian carcinoma cells in agar. Cancer Res 38: 3438–3444

Higby DJ, Wilbur D, Wallace HJ et al. (1977) Adriamycin-cyclophosphamide and Adriamycin-*cis*-dichlorodiammine-platinum II combination chemotherapy in patients with advanced cancer. Cancer Treat Rep 61: 869–873

Issell BF (1982) The podophyllotoxin derivatives VP-16 and VM 26. Cancer Chemother Pharmacol 7: 73

Lieber MM, Ames MM, Powis G et al. (1980) Anticancer drug testing in vitro: use of an activating system with the human tumour stem cell assay. Life Sci 28: 287–293

Ludwig R, Alberts DS (to be published) Chemical and biological stability of anticancer drugs in a human tumour clonogenic assay. Cancer Chemother Pharmacol

Ludwig R, Alberts DS, Miller TP, Salmon SE (to be published) Evaluation of anticancer drug schedule dependency using an in vitro human tumour clonogenic assay. Cancer Chemother Pharmacol

Omura GA, Ehrilch CE, Blessing JA (1982) A randomised trial of cyclophosphamide plus Adriamycin with or without cisplatinum in ovarian carcinoma. Proc Am Soc Clin Oncol 1: 104

Peng YM, Ormberg D, Alberts DS, Davis TP (1983) Improved high performance liquid chromatography of the new antineoplastic agents bisantrene and mitoxantrone. J Chromatograph Biomed Appl 233: 235–247
Salmon SE (1980) Applications of the human tumour stem cell assay to new drug evaluation and screening. In: Salmon SE (ed) Cloning of human tumour stem cells. Liss, New York, pp 291–314
Salmon SE, Von Hoff DD (1981) In vitro evaluation of anticancer drugs with the human tumour stem cell assay. Semin Oncol 8: 377–386
Salmon SE, Hamburger AW, Soehnlen B, Durie BGM, Alberts DS, Moon TE (1978) Quantitation of differential sensitivity of human tumour stem cells to anticancer drugs. N Engl J Med 298: 1321–1327
Salmon SE, Alberts DS Meyskens F et al. (1980) Clinical correlations of in vitro drug sensitivity. In: Salmon SE (ed) Cloning of human tumour stem cells. Liss, New York, pp 223–246
Salmon SE, Meyskens FL, Alberts DS et al. (1981) New drugs in ovarian cancer and malignant melanoma: in vitro phase II screening with the human tumour clonogenic cell assay. Cancer Treat Rep 65: 1–12
Selby P, Buick R, Tannock I (1983) A critical appraisal of the human stem cell assay. N Engl J Med 308: 129–134
Shoemaker RH, Wolpert-Defilippes MK, Makuch RW, Venditti JM (1983) Use of the human tumour clonogenic assay for new drug screening. Proc Am Assoc Cancer Res 24: 311
Sondak VK, Korn EI, Morton DL, Kern DH (1983) Absence of in vitro synergy for chemotherapeutic combinations tested in the clonogenic assay. Proc Am Assoc Cancer Res 24: 316
Von Hoff DD, Cottman CA, Porseth B (1981) Activity of mitoxantrone in a human tumor cloning system. Cancer Res 41: 1853–1855
Von Hoff DD, Clark GM, Stogdill MF et al. (1983) Prospective clinical trial of a human tumour cloning system. Cancer Res 43: 1926–1931
Wallach RG, Cohen C, Bruckner H (1980) Chemotherapy of recurrent ovarian carcinoma with cis-dichlorodiammine platinum II and adriamycin. Obstet Gynecol 55: 371–376

Heterogeneity and Variability of Test Results as Limiting Factors for Predictive Assays

P. Schlag and D. Flentje

Chirurgische Universitätsklinik, Sektion Chirurgische Onkologie,
Im Neuenheimer Feld 110, 6900 Heidelberg, FRG

A number of different human neoplasms have been postulated to arise from more than one single cell of origin (Ohno 1971; Petersen et al. 1979; Siracky 1979), whereas there is evidence that myelomas and some hematological malignancies may be monoclonal (Fialkow 1976). The question arises whether the wellknown morphologically different cell types found in many tumors point to a multicellular origin or whether these phenotypes only represent differentiation levels or directions of cells which have developed from a single stem cell (Nowell 1976).

Several investigators have approached this question experimentally. In fact, cellular diversity has been proved by antigen patterns (Fidler and Krupke 1980; Kerbel 1979), although antigenic diversity may only be related to differing functional stages of growth or immunologic modulation induced by interaction with other host cells. Fidler and colleagues have investigated the metastatic properties of cells derived from primary tumors and metastatic sites in experimental systems (Fidler et al. 1978).

They have been able to isolate clonal subpopulations of cells with differing metastatic patterns and varying criteria of malignant aggressiveness and phenotype expression. To what extent tumor progression and spontaneous mutations may account for the observed heterogeneity is difficult to evaluate.

Following the development of a clonal assay for the growth of primary human tumor cells (Salmon et al. 1978), there have been a number of reports describing various characteristics of heterogeneity in this system (Meyskens 1980; Epstein et al. 1980; Mackillop et al. 1982; Schlag et al. 1980). In this context, the presence of tumor subclones with differing sensitivity toward anticancer agents has been described (Rosenblum et al. 1981; Barranco et al. 1972; Heppner et al. 1978). Often, the finding of a plateau-type dose-response relationship in cell survival curves points to the presence of resistant subpopulations.

Heterogeneity in drug sensitivity is of utmost importance for the significance of the so-called predictive tests that have been developed during recent years to assess the individual tumor response to in vitro chemotherapeutic drugs. If such an in vitro chemosensitivity test is to be of use for individual treatment planning, the results of testing must be unequivocal, independent of the biopsy source.

With the different predictive test systems, discrepancies of in vitro and in vivo results have been observed. Besides pharmacokinetic in vitro/in vivo differences, this may in particular be due to heterogeneity of tumor cells in different tumor sites of one particular tumor-bearing individual.

A differing proliferative activity of tumor cells within one tumor lesion, as well as between

Recent Results in Cancer Research. Vol. 94

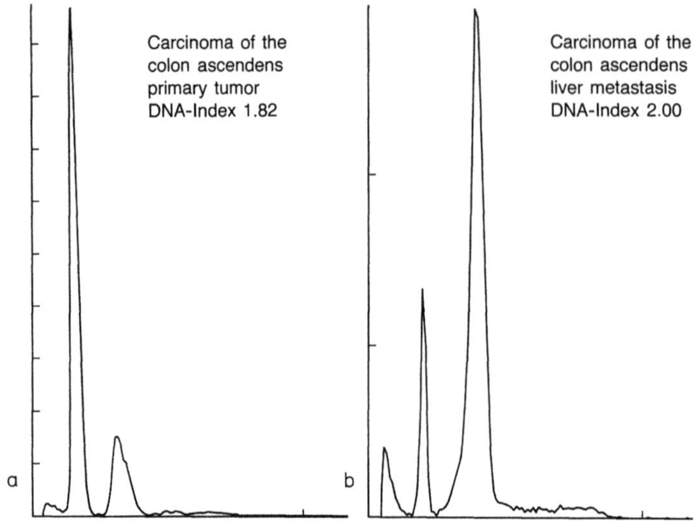

Fig. 1. Flow cytometric DNA-distribution of a primary colon carcinoma (**a**) and its liver metastasis (**b**). Reproduced with kind permission by Dr. G. Teichter, Section of Comparative and Experimental Pathology, University of Heidelberg

different metastases, could be observed in our own material and has also been seen by other investigators (Schlag and Schreml 1982; Epstein et al. 1980; Dexter et al. 1981; Vindelow et al. 1980). As an example, Fig. 1 shows the flow cytometric DNA analysis of a human colon carcinoma. When the primary lesion is compared to its synchronous liver metastasis, there are clear discrepancies concerning the pattern of proliferation. These proliferation data represent another explanation why some of the so-called drug sensitivity assays, which use pooled proliferation parameters as criteria for drug effects in vitro, must be evaluated with caution.

It was one objective in the development of the clonogenic assay to analyze chemosensitivity of a homogenous cell population. In this assay, it should at least theoretically be possible to test the effectiveness of different anticancer agents on clonogenic tumor cells only. It would thus be expected that chemosensitivity of tumor cell colonies from one individual tumor, but taken from different samples, would show only minor differences if any. This has so far not been confirmed satisfactorily.

In our own experience with simultaneously performed clonogenic assays of primary tumor lesions and their synchronous lymph node or visceral metastases, using the method described by Salmon et al. (1978), we observed no satisfactory consistency. The colonies grown from either the primary solid lesion or its synchronously assayed solid metastases differed with respect to colony size and other morphological criteria. Frequently, the colonies derived from metastatic lesions were larger and contained more cells than colonies derived from the primary tumors (Fig. 2).

Further on, in almost all cases studied, the number of tumor-colony-forming units (per 5×10^5 nucleated cells) was higher for the metastatic lesion than for its primary. As it is not possible to standardize the single-cell suspensions for the human tumor colony forming assay according to the number of viable tumor cells versus host cells in the compared cell suspensions, the cloning efficiency of primary and metastatic tissue from the same donor cannot be expected to be comparable, and in fact, that is what we observed.

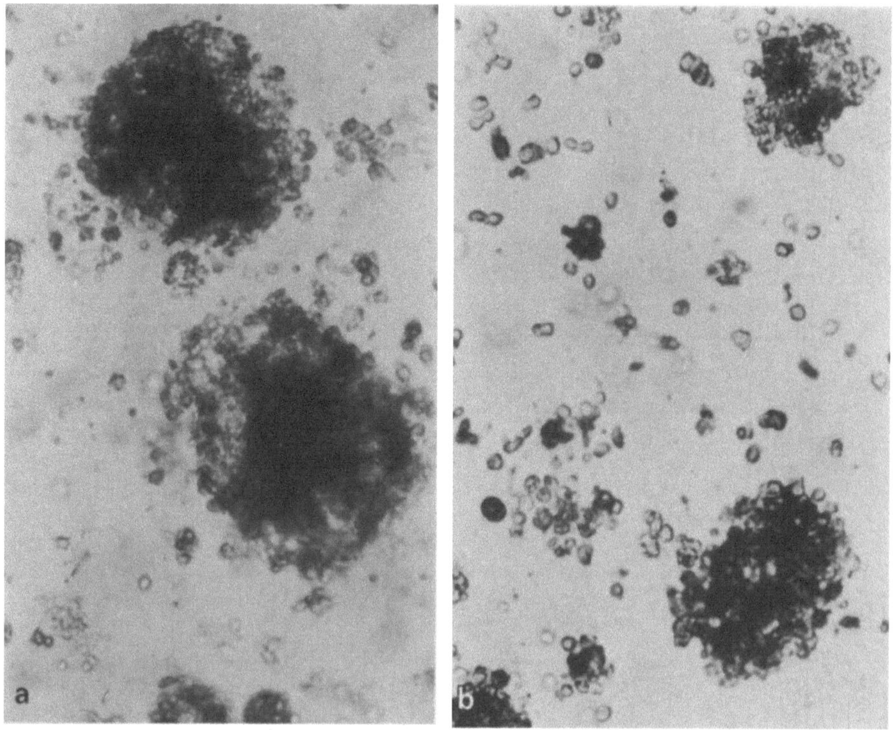

Fig. 2a, b. Comparison of colonies derived from a (breast carcinoma) metastasis and its corresponding primary tumor

In addition, the inhibition of colony growth from these paired samples by anticancer agents was investigated. In these experiments, the reduction of tumor colony growth by cytostatic drugs differed largely between primary tumor and metastatic lesions under the in vitro conditions used for the assay (Fig. 3). A regression analysis of sensitivity data for doxorubicin, 1,3-bis(2-chloroethyl)-1-nitrosourea (BCNU), 5-fluorouracil, and activated cylcophosphamide did not show any significant correlation, but the data suggest that metastasis-derived tumor colonies might be more sensitive in vitro to cytostatic agents than tumor cell colonies derived from the corresponding primary tumor.

Similar results for the simultaneous analysis of primary carcinomas and malignant peritoneal effusions have been reported by other investigators (Epstein et al. 1980). Better in vitro growth was observed for cells derived from the effusions than for cells from the solid specimens, whereas the contrary has been reported by Plasse. A comparison of sensitivity data from different solid lesions in the same patient, as described by Bertelsen, yielded differences in drug sensitivity in 28% of synchronously performed assays and in 40% of metachronous assays. Similar discrepancies of drug sensitivity between primary tumor and metastatic lesions have also been observed using nucleotide precursor assays (Sanfilippo et al. 1983) and xenograft models (Selby et al. 1979).

The observed differences in comparing results from different biopsy lesions may be due first of all to methodological problems. These problems may especially lie in the difficulties in standardizing the single-cell suspension. It has to be taken into consideration that the

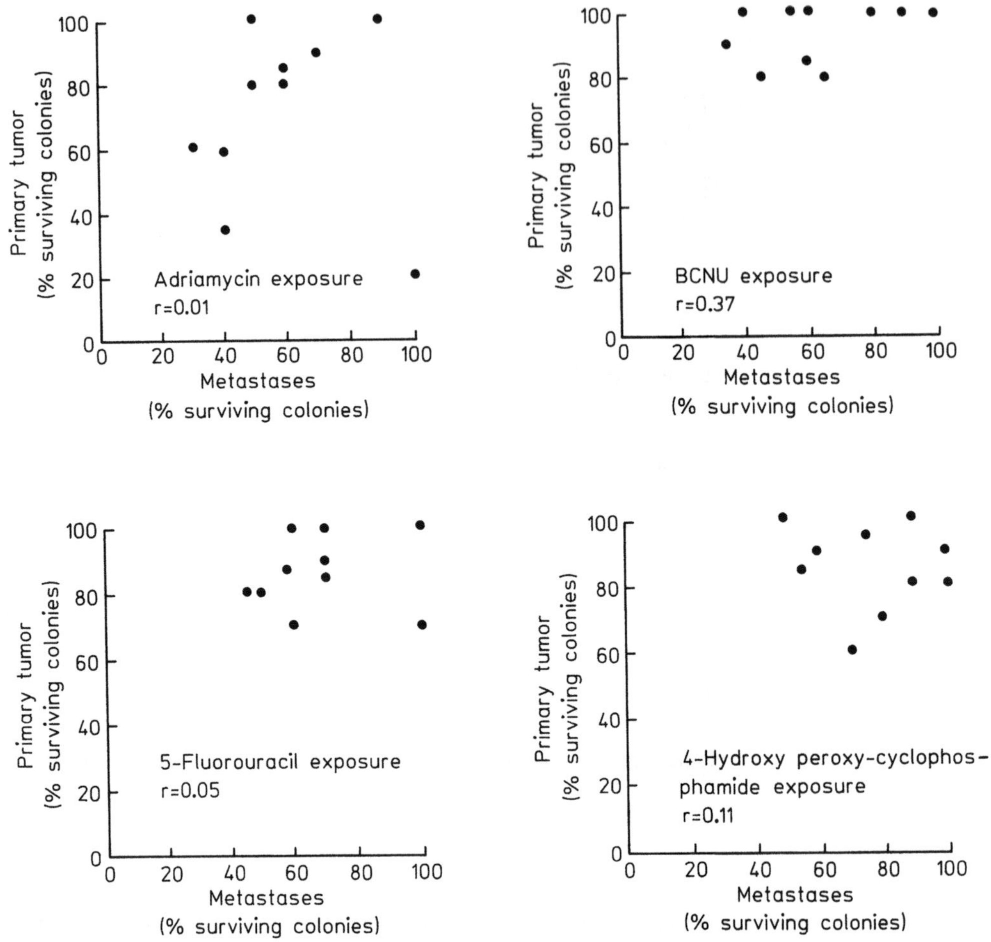

Fig. 3. Colony growth inhibition by Adriamycin, BCNU, 5-FU and cyclophosphamide in a set of primary tumors and their synchronously assayed metastasis

inoculum used for comparison is not necessarily identical. The specimen chosen for drug testing does not always represent the different subpopulations present in the whole tumor mass. The preparation of the single-cell suspension may damage one particular cell type and thus select for specific cells surviving the disaggregation. In addition, the colony assay may be particularly affected by interactions between malignant and nonmalignant host cells. As described, the ratio of tumor versus host cells differs widely in different lesions of the same neoplasm (Siracky 1979; Schlag and Schreml 1982). From the results of Buick et al. (1980) it is known that especially nonmalignant host cells may markedly influence cloning efficiency. Further, it is possible that the simultaneous killing of host cells during drug exposure differs from one sample to the other and may also be responsible for the observed phenomena. If the cell suspensions that are compared are composed of similar ratios of tumor and host cells, these differences may be concealed.

Before postulating the presence of heterogeneous clones to account for the observed differences, one has to exclude the possibility that the inherent variability of the assay system may in fact be responsible. Quality control studies have been able to define some of the problems that may influence these variations. It has become known that simple factors like the delay during sample transport, the freshness of media, the batch of fetal calf serum, or the heparin preservative may greatly influence the results obtained with clonogenic assays and may even select for the growth of subpopulations with differing chemosensitivity. Another factor might be that tumor cells from metastatic sites, especially ascites, possibly adapt more easily to growth in the culture system than cells derived from primary lesions (Epstein et al. 1980).

In addition to these technical problems and limitations, the results of the various assay systems may also be influenced by heterogeneity of tumor cells. Tumor cell heterogeneity concerns different proliferative activity as well as susceptibility to cytostatic drugs. It has been postulated that the cells which produce colonies in this assay represent the stem cells of a tumor. Yet the differences in morphological appearance and cloning efficiency of the colonies seen in our experiments may indicate heterogeneity also within the clonable tumor cell compartments of a tumor and its metastases. Thus this test exhibits similar difficulties known from other nonclonal predictive assays, which have not been shown to benefit patients studied in randomized controlled clinical trials.

Whatever the reasons for the variation in the test results may be, the results must be interpreted with caution. As past experience shows, the test results of predictive assays depend on a number of factors which we have not so far been able to interpret adequately; a more detailed analysis seems to be necessary to establish a firmer basis for the results obtained.

References

Barranco SC, Ho DHW, Drewinko B (1972) Differential sensitivities of human melanoma cells grown in vitro to arabinosyl-cytosine. Cancer Res 32:2733–2736

Buick RN, Fry SE, Salmon SE (1980) Effect of host-cell interactions on clonogenic carcinoma cells in human malignant effusions. Br J Cancer 41:695–704

Dexter DL, Spremulli EN, Fligid Z, Barbosa JA, Vogel R, Van Voorhees A, Calabresi P (1981) Heterogeneity of cancer cells from a single human colon carcinoma. Am J Med 71:949–956

Epstein LB, Jen-Ta Sehn, Abele JS, Reese CC (1980) Further experience in testing the sensitivity of human ovarian carcinoma cells to interferon in an in vitro semisolid agar culture system: comparison of solid and ascites forms of the tumor. In: Salmon SE (ed) Cloning of human tumor stem cells. Liss, New York, pp 277–290

Fialkow PJ (1976) Clonal origin of human tumors. Biochim Biophys Acta 458:283–291

Fidler IJ, Kripke ML (1980) Tumor cell antigenicity, host immunity and cancer metastasis. Cancer Immunol Immunother 7:201–205

Fidler IJ, Gersten DM, Budmen MB (1978) The biology of cancer invasion and metastasis. Adv Cancer Res 28:149–250

Heppner GH, Dexter DL, DeNucci T, Müller FR, Calabresi P (1978) Heterogeneity in drug sensitivity among tumor cell subpopulations of a single mammary tumor. Cancer Res 38:3758–3768

Kerbel RS (1979) Implications of immunological heterogeneity of tumors. Nature 280:358–360

Mackillop WJ, Stewart SS, Buick RN (1982) Density/volume analysis in the study of cellular heterogeneity in human ovarian carcinoma. Br J Cancer 45:812–820

Meyskens FL (1980) Human melanoma colony formation in soft agar. In: Salmon SE (ed) Cloning of human tumor stem cells. Liss, New York, pp 85–100

Nowell PC (1976) The clonal evolution of tumor cell populations. Science 194:23–28

Ohno S (1971) Genetic implication of karyologic instability of malignant somatic cells. Physiol Rev 51:496–526

Petersen SE, Bichel P, Lorentzen M (1979) Flow cytometric demonstration of tumor cell subpopulations with different DNA content in human colorectal carcinoma. Eur J Cancer 15: 383–386

Rosenblum ML, Ferosa MA, Dougherty DV (1981) Influence of tumor heterogeneity on chemosensitivity testing on human brain tumors. Stem Cells 1: 279

Salmon SE, Hamburger AW, Soehnlen BS, Durie BGM, Alberts DS, Moon TE (1978) Quantitation of differential sensitivity of human-tumor stem cells to anti-cancer drugs. N Engl J Med 298: 1321–1327

Sanfilippo O, Silvestrini RS, Zaffaroni N, Piva L (1983) Potentiality of an in vitro chemosensitivity assay for human germ cell testicular tumors (GCTT). Proceedings ASCO, 19th Annual Meeting 2: C–145

Schlag P, Veser J, Geier G, Breitig D, Herfarth Ch (1980) Cytostatic effects on nucleic acid synthesis in primary tumors and lymph node metastases of human breast cancer. Langenbecks Arch Chir [Suppl] Chr Forum 323–327

Schlag P, Schreml W (1982) Heterogeneity in growth pattern and drug sensitivity of primary tumor and metastases in the human tumor colony-forming assay. Cancer Res 42: 4086–4089

Selby PJ, Thomas JM, Peckham MJ (1979) A comparison of the chemosensitivity of a primary tumor and its metastases using a human tumor xenograft. Eur J Cancer 15: 1425–1429

Siracky J (1979) An approach to the problem of heterogeneity of human tumor cell subpopulations. Br J Cancer 39: 570–577

Vindelov LL, Hansen HH, Christensen IJ, Spang-Thomsen M, Hirsch FR, Hansen M, Nissen NJ (1980) Clonal heterogeneity of small-cell anaplastic carcinoma of the lung demonstrated by flow-cytometric DNA analysis. Cancer Res 40: 4295–4300

*Interlaboratory Comparison of In Vitro Cloning of Fresh Human Tumor Cells from Malignant Effusions**

V. Hofmann, E.E. Holdener, M. Müller, and U. Früh

Universitätsspital Zürich, Departement für Innere Medizin, Abteilung für Onkologie, Rämistrasse 100, 8091 Zürich, Switzerland

Introduction

The in vitro semisolid agar culture system recently developed by Hamburger and Salmon favors the cloning of human tumor cells (Salmon et al. 1978). Experience in several laboratories indicates that some tumor types, such as malignant melanomas, ovarian adenocarcinomas, and neuroblastomas, seem to grow better than, for instance, sarcomas and colorectal and non-small-cell bronchogenic carcinomas (Von Hoff 1981). Because successful in vitro cloning is critically dependent on fresh biopsies, experiments are performed only once. Due to these practical constraints, information concerning the reproducibility and interlaboratory performance of the assay is not available. Such comparisons would be possible if ongoing studies in other centers could demonstrate that cryopreserved tumor specimens retain the biological and pharmacologic properties of the fresh samples. Thus far, however, this information is not available.
In the present study, we have compared the performance of the Hamburger-Salmon assay on identical tumor samples in two Swiss laboratories. We have taken advantage of the fact that cell viability in malignant effusions remains high for several days after paracentesis and that subsequent clonogenic cultures can be initiated with high success rates. This has allowed us to perform culture studies on the same fresh material in two different centers.

Materials and Methods

Included in this study were 25 cytologically proven malignant effusions from 24 patients with the following neoplasias: breast (10), ovarian (4), pancreas (1), stomach (1), bronchogenic small-cell (1), bronchogenic non-small-cell (1), and adenocarcinomas of unknown primary origin (7). Patients had either to be untreated or to have received their last treatment course more than 4 weeks prior to paracentesis. The effusions were collected into sterile plastic bags containing 5–10 IU of preservative-free heparin per milliliter of effusion. The effusions were immediately separated in to two equal portions, one of which was sent at ambient temperature to the other laboratory. All samples were processed within 48 h.

---

* V.H. and U.F. were supported by the EMDO Foundation. E.E.H. was supported by grant SNF 3.873.0.79 and the Regional Cancer Leagues St. Gallen-Appenzell and Thurgau

Recent Results in Cancer Research. Vol. 94

Tumor Isolation and Culture

The two participating laboratories in Zürich and St. Gallen standardized their culture methods. Preparation of samples and culture media and evaluation of results were uniform. Effusions were centrifuged at 900 rpm for 20 min at room temperature. The cell sediments were washed twice using McCoy's 5A with 10% fetal calf serum. Red cells were eliminated by a short exposure to hypotonic solutions. Cells were counted in a hemocytometer and cell viability was determined by the trypan blue dye exclusion method. All cultures were performed in 35 × 10 mm Petri dishes with a 2-mm grid. The double agar layer was prepared as described by Salmon and Hamburger (Salmon et al. 1978) with the exceptions that conditioned medium and calcium were omitted. Subsequently, the Petri dishes were kept in an incubator at 37° and 5% CO_2 in humidified air. Quality of plates on day 0 and colony formation were checked using an inverted microscope. Number, shape, and size of clumps on day 0 were carefully registered. Growth was defined as de novo formation of a compact (round) cell aggregate of at least 80 µm in size. Plates were screened three times weekly and final scoring was usually performed between days 10 and 24.

Drug Assays

Stock solutions of melphalan (1 µg/ml), bleomycin (1 µg/ml), vinblastine (1 µg/ml), *cis*-platinum (1 µg/ml), 5-fluorouracil (1 µg/ml) and doxorubicin (0.1 µg/ml) were prepared in center A. The drugs were labeled from 1 to 6 and kept at appropriate temperature. The tumor cell suspensions were incubated with drugs 1–6 dilated 1:10 for 1 h at 37° C and then plated unwashed in the Petri dishes. Triplicates were set up for each drug and the control plates. Colonies were scored independently in each laboratory; the mean number of colonies of drug-exposed samples was expressed as percentage of control plates. Results were evaluated in each center without knowledge of the other's findings. Finally, after compilation of all data the two groups compared their results.

Labeling Index

Tumor cells were incubated with ^3H-thymidine for 1 h at 37° C as described by Durie and Salmon for myeloma cells. Unincorporated ^3H-thymidine was washed away and slides for autoradiography were prepared using a cytocentrifuge. Slides were developed and stained (Durie and Salmon 1975). A tumor cell with ⩾ 5 grains over the nucleus was considered labeled. The number of labeled tumor cells was related to the number of unlabeled tumor cells and expressed as labeling index (LI):

$$\mathrm{LI}(\%) = \frac{\text{No. of labeled tumor cells}}{\text{No. of tumor cells}} \times 100.$$

Results

The two laboratories found identical growth patterns in 20 out of 25 (80%) cytologically proven malignant effusions (Table 1). Tumor samples 1–11 failed to grow in both laboratories. Among these the most numerous were breast tumor samples. The median

Table 1. Characteristics of 25 malignant effusions

Sample no.	Tumor type	Origin	Preplating cell viability (%)		Labeling index (%)	Colony formation	
			Center A	Center B		Center A	Center B
1	Breast	P	54	96	ND	–	–
2	AUO	A	82	98	4.5	–	–
3	Breast	P	86	100	ND	–	–
4	Breast	A	79	97	ND	–	–
5	Breast	P	68	77	3.25	–	–
6	Breast	P	60	90	2.0	–	–
7	Pancreas	A	92	98	0	–	–
8	AUO	A	92	94	ND	–	–
9	Small cell br.c.	P	93	93	1.5	–	–
10	Breast	A	86	94	0.5	–	–
11	Stomach	A	81	98	13	–	–
12	Breast	A	92	95	ND	+	++
13	AUO	A	85	96	10	+++	+++
14[a]	Fallopian tube	A	93	97	11.5	+	+
15[a]	Fallopian tube	A	93	98	9.5	+	+
16	Breast	P	86	71	1.5	+++	+++
17	AUO	P	91	92	12.5	+++	+++
18	Ovary	A	88	97	21	+++	+++
19	Non small cell br.c.	P	90	76	9	+++	++
20	Ovary	A	80	90	0.75	+	++
21	AUO	A	94	98	1.5	–	+++
22	AUO	A	95	98	ND	–	+
23	Ovary	A	90	98	ND	–	+
24	Breast	P	60	98	5	++	–
25	Breast	P	73	92	0.25	+++	–

[a] Same patient, paracentesis 2 months apart
A, ascites; *P*, pleural effusion; *br.c.*, bronchogenic carcinoma; *AUO*, adenocarcinoma of unknown origin; +, 5–10 colonies; ++, 21–50 colonies; +++, > 50 colonies; *ND*, not done

labeling index of this group was 2.0% (range 0–13%). Tumor specimens 12–20 displayed growth of comparable magnitude in both centers. Colony formation was categorized into three groups: 5–20, 21–50, and above 50. The median labeling index in this instance was 9.5% (range 0.75%–21%), which was significantly higher ($p < 0.05$) than for the nongrowing tumors (Fig. 1). The two laboratories reported discordant findings for samples 21–25.

In addition, we compared the effect of six drugs on colony survival from identical samples. For this comparison, results derived from 82 cultures exposed to drugs are available. In a majority of cases [25 out of 41 (61%)], both groups detected sensitivity or resistance to the same drugs by the same samples. Sensitivity – defined as a reduction of colony survival by at least 70% – was observed in four instances. Resistance – defined as colony survival above 30% – was reported in 21 simultaneous experiments. Discordant results were observed in 16 out of 41 (39%) drug sensitivity studies (Table 2).

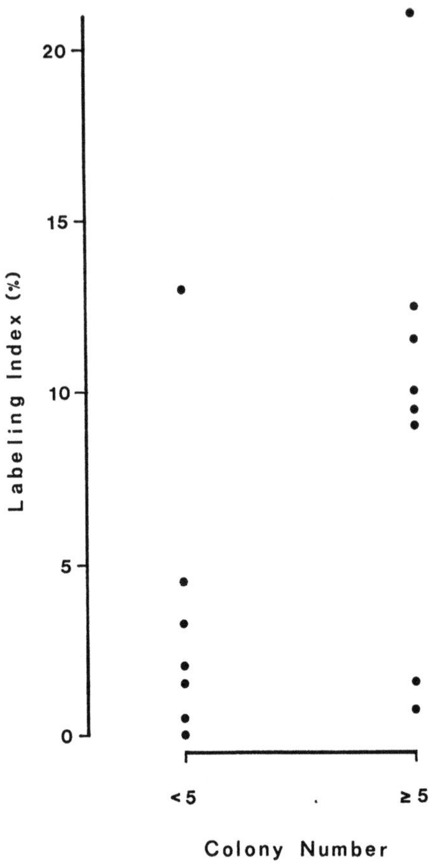

Fig. 1. Association of preplating labeling index with subsequent tumor colony formation

Table 2. Comparison of drug sensitivity (colony survival)

	No. of experiments
Sensitivity[a] predicted by both laboratories	4
Resistance[b] predicted by both laboratories	25
Discordant results reported by the two laboratories	16
Total	41

[a] Sensitivity is defined by a $\geq 70\%$ reduction in colony survival
[b] Resistance is defined by a $< 70\%$ reduction in colony survival

Discussion

The finding that a majority of malignant effusions display identical growth patterns in two different laboratories indicates that the soft agar tumor colony assay as described by Hamburger and Salmon favors the growth of some tumor samples (see chapter by Hofmann et al., to be published). This also indicates that growth or failure to grow is not a local laboratory artifact. Rather, this confirms that some tumors with yet poorly known biological properties are more prone to proliferate under the established in vitro conditions. Tumor histology certainly plays an important role. It is well known that, for instance, ovarian adenocarcinomas and neuroblastomas can be cloned with high success rates (Von Hoff et al. 1981; Hofmann et al., to be published). In this study, breast tumor specimens grew poorly. Other authors have also reported that mammary cancer cells are difficult to clone (Jones 1983). A finding of interest was that tumor cell suspensions that did grow had a significantly higher labeling index than the ones that failed to grow. We have recently confirmed this finding in a large cohort of tumor samples (see chapter by Hofmann et al., this volume). Durie and Salmon (1980) have made the same observation on myeloma cells. Since determination of ^{3}H-thymidine uptake is an easy and rapid procedure, it could be performed routinely to select samples with likelihood of subsequent in vitro cloning. Better growth conditions are necessary for tumor specimens in which ^{3}H-thymidine incorporation ist low or does not take place at all.

In the second part of the present investigation, we compared the drug survival curves obtained with the same drugs on the same tumor probes. The two laboratories found concordant results in a majority of cases. However, on about one-third of the samples the laboratories reported different results. Recently, Rozencweig (1983) has reported considerable interlaboratory differences for drug survival curves on the WiDR cell line. These results clearly demonstrate that in vitro results must be interpreted with great caution. Ultimately, each laboratory must demonstrate that its in vitro findings are correlated with in vivo response. In addition, it appears to be of the utmost importance to carry out interlaboratory quality controls before introducing the clonogenic assay into clinical routine practice.

References

Durie BGM, Salmon SE (1975) High speed scintillation autoradiography. Science 190: 1093–1095

Durie BGM, Salmon SE (1980) Cell kinetic analysis of human tumor stem cells. In: Salmon SE (ed) Cloning of human tumor stem cells, chap 13. Liss, New York, pp 153–163

Hofmann V, Berens M, Fruh U (to be published) Drug selection for perioperative chemotherapy. In: Metzger U (ed) Recent results in cancer research

Jones SE, Salmon SE, Dean JC (1983) In vitro cloning of human breast cancer. International Conference of Predictive Drug Testing on Human Tumor Cells, Zürich, July

Rozencweig M (1983) Quality control study of the human tumor clonogenic assay. Am Assoc Cancer Res 1239

Salmon SE, Hamburger AW, Soehnlen BJ, Durie BGM, Alberts DS, Moon TE (1978) Quantitation of differential sensitivity of human tumor cells to anticancer drugs. N Engl J Med 298: 1321–1327

Von Hoff DD, Casper J, Bradley E, Sandbach J, Jones D, Makuch R (1981) Association between human tumor colony-formation assay results and response of an individual patient's tumor to chemotherapy. Am J Med 70: 1027–1032

Pharmacology Phase II Studies and New Drug Development

*Evaluation of Schedule Dependency of Anticancer Drugs in the Human Tumor Clonogenic Assay**

R. Ludwig, D.S. Alberts, and S.E. Salmon

The University of Arizona, Health Science Center,
Section of Hematology and Oncology, Tucson, AZ 85724, USA

The cytotoxicity of anticancer drugs may depend on the schedule of administration as well as on the dose used. There are limited data concerning the evaluation of drug schedules using human tumor models. The purpose of our study was to evaluate schedule dependency of anticancer drugs by comparing 1-h to continuous drug exposures in the human tumor clonogenic assay (HTCA) (Salmon et al. 1978).

Fresh human tumor samples (e.g., ovarian, melanoma, breast, and lung) and human tumor cell lines (HEC-1A endometrial carcinoma, RPMI 8226 myeloma, and U266 myeloma) were used in these studies of drug schedule dependency. The following anticancer drugs were tested simultaneously for 1-h and continuous exposures: doxorubicin, actinomycin D, bleomycin, bisantrene (an antracene derivative), vinblastine, and etoposide (VP-16). For the 1-h exposure studies a single-cell suspension from fresh human tumors as well as from human tumor cell lines was incubated with the anticancer drug at concentrations between 0.001 and 0.1 μg/ml for 1 h, washed, and then plated. For the continuous exposure studies drugs were included in the upper layer for the entire incubation period at concentrations ranging between $^{1}/_{10}$ and $^{1}/_{1,000}$ of those used for 1 h.

By comparing the inhibitory effect on tumor colony formation after 1-h and continuous drug exposures, the HTCA results appeared to identify and separate in vitro schedule-dependent and schedule-independent drugs. Substances like vinblastine, bleomycin, and etoposide, which are known to have cell-cycle-specific characteristics (2), were associated with exponential reductions of tumor colony formation when used by continuous low-dose exposures, whereas after 1-h exposures each of these agents caused plateau-type dose-response curves (e.g., as shown for etoposide in Fig. 1). In contrast, doxorubicin, actinomycin D, and bisantrene resulted in similarly shaped dose-response curves following both 1-h and continuous exposures (e.g., see Fig. 2).

For statistical comparisons of the relative efficacy of the two dosing schedules, a ratio was calculated of the drug concentration which reduced growth of tumor colonies to 50% of controls: ID_{50} ratio = ID_{50} (1-h exposure)/ID_{50} (continuous exposure). Using fresh tumors ID_{50} ratios for doxorubicin, actinomycin D, and bisantrene ranged between 2 and 60, with a median of 14. However, for bleomycin, vinblastine, and etoposide the ID_{50} ratios ranged between 100 and 3,000 with a median of 600. A statistical comparison of the two medians

* This work was supported in part by grants CA-17094, 21839, and 23074 from the National Institutes of Health, Bethesda, MD, USA

Recent Results in Cancer Research. Vol. 94

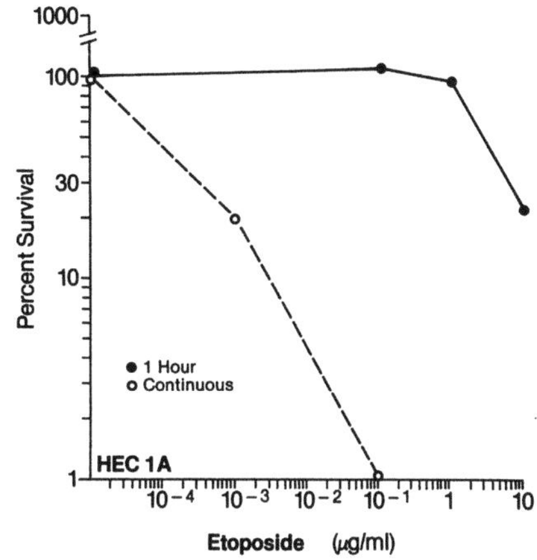

Fig. 1. Dose-survival curves of tumor-colony-forming units (HEC-1A cell line) following 1-h (——●——)or continuous exposure (- -O- -) for etoposide (VP-16)

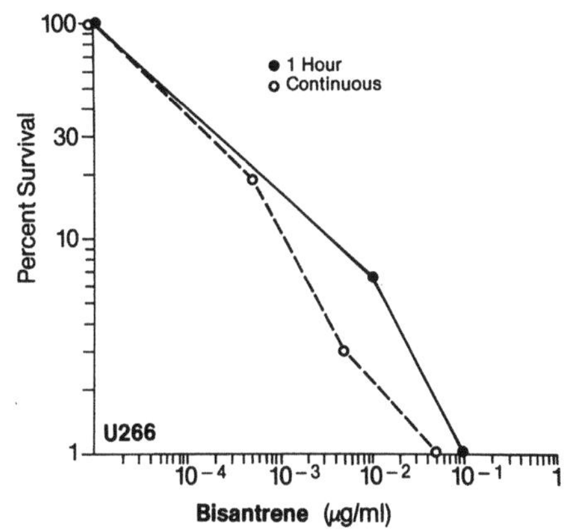

Fig. 2. Dose-survival curves of tumor-colony-forming units (U226 myeloma cell line) following 1-h (——●——) or continuous exposure (- - O - -) for bisantrene

showed a significant difference ($p < 0.01$). Similar results were achieved using the human tumor cell lines.

Thus, to summarize our results, we observed that drugs with cell-cycle-specific characteristics (Van Putten 1974) like bleomycin, vinblastine, and etoposide resulted in plateau-type dose-response curves after 1-h drug exposures, whereas continuous exposure resulted in exponential-type dose-survival curves. High ID_{50} ratios were found for these compounds. In contrast, drugs possessing cell-cycle-nonspecific characteristics, like doxorubicin, actinomycin D, and bisantrene resulted in similarly shaped dose-survival curves after 1-h and continuous exposure, as well as low ID_{50} ratios.

Before final conclusions can be drawn concerning the schedule dependency of an anticancer drug based on in vitro studies in the HTCA, it is essential to document drug stability under in vitro assay conditions. In a companion paper in this volume (Alberts et al.) we have shown that all of the drugs used in these studies except etoposide maintained chemical and biological stability for up to 10 days in the HTCA.

The HTCA may be used to select a specific drug-dosing schedule for clinical use. However, we must caution that the in vitro/in vivo correlation of continuous exposure studies has still to be established.

References

Salmon SE, Hamburger AW, Soehnlen B, Durie BGM, Alberts DS, Moon TE (1978) Quantitation of differential sensitivity of human tumor stem cells to anticancer drugs. N Engl J Med 29: 1321–1327

Van Putten LM (1974) Are cell kinetic data relevant for the design of tumor chemotherapy schedules? Cell Tissue Kinet 7: 493–504

*In Vitro Effect of Interferon-α on Human Granulocyte/Macrophage Progenitor Cells and Human Clonogenic Tumor Cells**

E.E. Holdener, P. Schnell, P. Spieler, and H. Senn

Division für Onkologie und Hämatologie, Abteilung für Medizin, Kantonsspital, 9007 St. Gallen, Switzerland

Introduction

Interferon (IFN), originally discovered by Isaacs and Lindenmann in 1957 (Isaacs and Lindenmann 1957; Isaacs et al. 1957) as an antiviral agent, has also been shown to be a potent inhibitor of normal and malignant cell growth (Gresser et al. 1972; Hilfenhaus and Karges 1974; McNeill and Gresser 1973; Strander et al. 1973). As early as 1962, Paucker et al. (1962a, b) found this antiproliferative effect of IFN to be dose and time dependent. Advances in the large-scale production (Bridgen et al. 1977; WHO Scientific Group 1982) and purification (Anfinsen 1981; Pestka et al. 1981; Staehelin et al. 1981) of the various interferons have only recently allowed numerous in vitro and in vivo studies to define the antiproliferative potential of these substances (De Clercq et al. 1982; Epstein and Marcis 1981; Gewert et al. 1981; Glasgow et al. 1978; Ito et al. 1980; Schlag et al. 1982; Verma et al. 1980, 1981).

Clinical studies have shown that neutropenia is a toxic side effect associated with partially purified human leukocyte interferon therapy, suggesting that these preparations of interferon were toxic to normal myeloid progenitor cells (Borden et al. 1982b; Horning et al. 1982; Rohatiner et al. 1982; Scott et al. 1981, 1982; Sherwin et al. 1982). These findings are consistent with in vitro studies which demonstrated that certain interferon preparations inhibited the formation of normal human granulocyte/macrophage colonies (GM-CFU-c) in soft agar (Bulkwill and Oliver 1977; Greenberg and Mosny 1977). To date, normal and malignant cells have exhibited varying degrees of sensitivity to the in vitro growth-inhibitory effects of the various interferons (Bradley and Ruscetti 1981; Van t'Hull et al. 1978; Verma et al. 1979; Von Hoff et al. 1982).

The purpose of this study was first to determine whether various human leukocyte interferon-α preparations can exert different growth-inhibitory effects on normal myeloid progenitor cells; secondly, to determine whether the growth-inhibitory effect of cloned interferon IFL-rA could be neutralized by a monoclonal antibody LI-1; and thirdly, to determine whether these interferons might preferentially inhibit the proliferation of GM-CFU-c in comparison to human tumor cell lines (CMF-7 and WiDr) or fresh clonogenic tumor cells, which would ultimately limit the therapeutic range of these substances in the treatment of cancer patients.

* Supported by grant SNF 3.873.0.79 from the Swiss National Foundation, by the Regional Cancer Leagues St. Gallen/Appenzell and Thurgau and the Eugen and Elisabeth Schellenberg Stiftung

Material and Methods

"Crude" leukocyte interferon was purchased from Imunoloski Zavod (Rockefellerova 2, Prof. D. Ikic, Zagreb, Yugoslavia). Dry net weight per ampule was 42 mg, protein content was 3.8 ± 0.2 mg (determined by Dr. H. Gehring, Institute of Biochemistry, University of Zurich); specific activity was 6.6×10^4 IU/mg protein; biological activity was determined by cytopathic effect (CPE) inhibition assay (poliovirus in BT 20 cells) and calibrated against the Hu IFN-Lc standard 69/19 (Dr. S. Arrenbrecht, Oncological Laboratories, University of Zurich).

Partially purified leukocyte interferon (Helveferon), prepared according to the technique by K. Cantell, was purchased from Helvepharm AG/SA, 3185 Schmitten, Switzerland (Lot No. 6974/6974-B). Biological specifications were determined by the Theodor Kocher Institute, University of Berne, Switzerland: biological activity 1×10^6 IU/ml (CPE inhibition assay), specific activity 1.13×10^6 IU/mg protein.

Bacterial cloned human leukocyte interferon (IFL-rA) (Ro 22.8181/601, PT 3414 Hl) was provided by Hoffmann-La Roche Company, Basel, Switzerland dry powder with a specific activity of $1.5-2.5 \times 10^8$ IU/mg protein. All IFN stock solutions were prepared with 0.9% saline and aliquots stored at −70° C.

Monoclonal antibody (LI-1) against cloned human leukocyte interferon (IFL-rA) (Staehelin et al. 1981) was kindly provided by Prof. T. Staehelin, Hoffmann-La Roche Company, Basel. LI-1 of the IgG class, in 0.9% phosphate-buffered saline containing 0.05% sodium azide, was diluted with 0.9% saline and aliquots stored at −70° C.

Tumor specimens were obtained during diagnostic and/or therapeutic surgical procedures. Mechanical disaggregation was performed according to Hamburger et al. (1978). To obtain optimal single-cell suspension, DNase (0.01%), collagenase (0.06%), and hyaluronidase (0.05%) were used if necessary (1 h incubation at 37° C).

Cells were washed three times using 10% fetal calf serum (FCS) in alpha medium. Viability was measured by trypan blue dye exclusion.

Malignant effusions were collected from patients using sterile transfer bags containing preservative-free heparin (10 U/ml effusion) in Hanks' balanced saft solution (HBSS). After centrifugation (225 *g*, 15 min, 22° C) the supernatant was removed, the remaining cells washed three times, and a sample stained (Papanicolaou) for cytological analysis. Cell number and viability were determined; aliquots of 2.5×10^5 cells were prepared for culture.

WiDr human colon adenocarcinoma cell line (ATCC CCL 218) was obtained from American-type culture collection (Rockville, Maryland). MCF-7 human breast adenocarcinoma cell line (Soule et al. 1973) was kindly provided by Dr. D. Hartmann (University of Basel, Switzerland). Cell lines were subcultered by trypsinization once a week.

Cells were cultured using a double-layer semisolid agar system as described by Hamburger and Salmon (1977) with the following modifications:

1. Mineral-oil-primed BALB/c mouse spleen conditioned media, diethylaminoethanol (DEAE) dextran, 2-mercaptoethanol, and horse serum were omitted.
2. Transferrin was added to the top layer (2 μg/ml).

Enriched McCoy's 5A medium with tryptic soy broth and asparagine (1%) were plated as bottom layer with agar 0.5%; 5×10^5 cells were suspendend in enriched CMRL 1066 medium with asparagine (1%) in agar 0.3% and used as top layer. Culture dishes were examined for cell clumps using an inverted microscope prior to incubation. The culture period was 14–21 days at 37° C, 7% CO_2 in air, and humified atmosphere.

Bone marrow aspirates for this study were obtained from hematologically normal adults undergoing staging for untreated solid tumors (normal marrow). After informed consent, aspirates (2 ml) were taken from the posterior iliac crest and added to 1.0 ml alpha medium containing 200 U preservative-free heparin. The buffer coat was obtained by centrifugation ($225 \times g$, 10 min at 22° C) and washed with alpha medium containing 10% FCS in the manner described previously (Holdener et al. 1983).

The agar culture technique was used with minor modifications as described previously (Pike and Robinson 1970). The cultures were incubated for 2 weeks at 37° C in an atmosphere of 5% CO_2 in air and humified atmosphere. Colonies of 50 or more cells were counted using an inverted microscope.

The effects of IFN were tested as follows:

1. Short time exposure: 5×10^5 nucleated bone marrow cells or tumor cells supended in alpha medium and McCoy's respectively, containing 10% FCS, were exposed to various concentrations of human leukocyte interferons for 1, 4, and 24 h at 37° C. The cells were washed with media as described above by centrifugation at 4° C. The control cultures were treated in an identical fashion but without interferon.
2. Continuous exposure: interferon (various concentrations: 10, 10^2, 10^3 and 10^4 U/ml) was added to the cell suspension prior to plating. Quadruplicate dishes were set up for each drug concentration. The mean number of colonies of the drug-exposed groups was expressed as a percentage of the control culture.

Interferon (10^4 U/ml) was preincubated with various concentrations of monoclonal antibody LI-1 in the presence of the target cells, e.g., bone marrow or tumor cells, for 1 h and brought into culture thereafter.

Results

Continuous exposure of human bone marrow cells to crude leukocyte interferon produced a dose-related inhibition of granulocyte/macrophage precursor colony growth (GM-CFU-c) (Fig. 1). In the presence of interferon 100 U/ml the number of colonies was reduced by 50% compared to the untreated control. At interferon concentrations of 1,000 U/ml or more, significant reduction ($p < 0.01$) of cluster formation (10–49 cells) could also be observed. With shorter exposure time, an increasing interferon concentration was necessary in order to produce the same growth-inhibitory effect (Fig. 2). Significant reduction of GM-CFU-c was obtained with either a continuous exposure to 100 U/ml ($p < 0.001$) or a 24-h exposure to 1,000 U/ml ($p < 0.01$). Following a short-term exposure (1 h or 4 h) a significant colony growth inhibition was seen only in the presence of high IFN concentration (10,000 U/ml). Occasional growth stimulation (15.4% of all bone marrow experiments) could be seen using low doses of interferon (10 U/ml interferon) ($p < 0.01$). Dose-related growth inhibition of GM-CFU-c was also found after a continuous exposure to partially purified interferon (IFN p.p.) or cloned interferon from *Escherichia coli* (IFL-rA). However, interferon with higher specific activities and comparable levels of antiviral activity (Fig. 3) consistently showed a lower antiproliferative effect. A strong suppression of GM-CFU-c by crude interferon was most prominent at very high concentrations of interferon (10^4 U/ml).

A monoclonal antibody LI-1 (1 μg/ml) of the IgG class against recombinant interferon IFL-rA was able to neutralize the growth-inhibitory effect of IFL-rA (10^4 U/ml, continuous exposure) by 62% (Fig. 4). The same antibody at the same concentration neutralized the antiproliferative effect of IFN p.p. and resulted in a 40% increase of GM-CFU-c. No

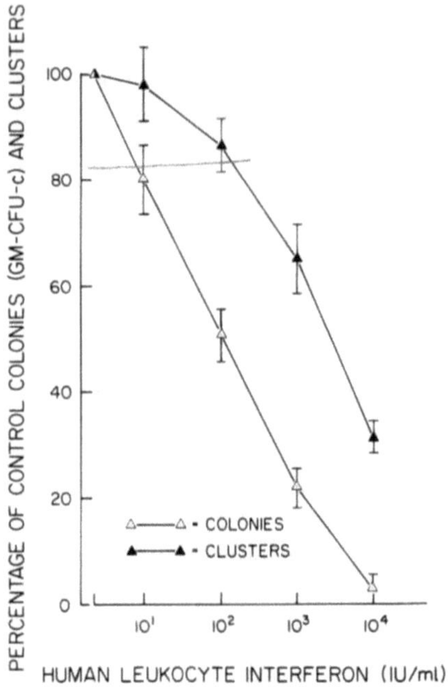

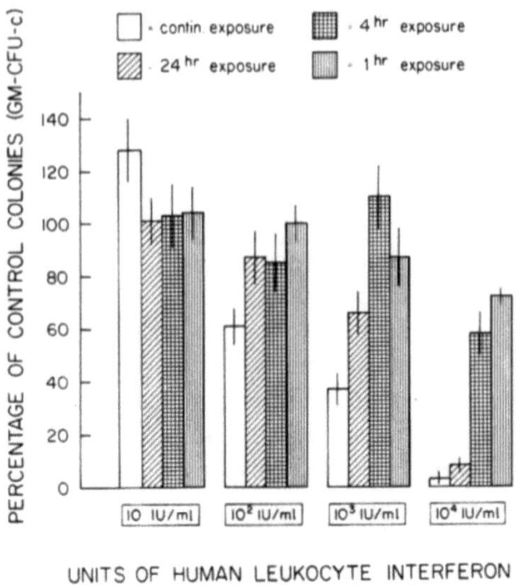

Fig. 1 (Left). Growth-inhibitory effect of human leukocyte interferon (crude) on human GM-CFU-c; 13 experiments, each set up in quadruplicate dishes. Bars, SD

Fig. 2 (Right). Time- and dose-related growth-inhibitory effect of human leukocyte interferon. Mean ± SD for quadruplicate dishes are shown

significant neutralization of the colony growth inhibition, however, was achieved with LI-1 against crude leukocyte interferon. GM-CFU-c was not reduced when LI-1 was added to the culture system in the absence of interferon.

Human leukocyte interferon (crude) was also tested against two human tumor cell lines; a mammary carcinoma (MCF-7) and a colon carcinoma (WiDr). Short-term exposure of MCF-7 did not result in tumor colony reduction even at 10^4 U interferon per milliliter (Fig. 5). However, continuous exposure with 10^4 U/ml reduced colony growth by more than 90% of the control. A similar, less than 50% growth inhibition was also found when WiDr colon carcinoma cells were exposed to graded concentrations of the various leukocyte interferons (crude, p.p., and recombinant). Continuous exposure to 10^4 U/ml, however, resulted in significant growth inhibition which could be neutralized with LI-1 by 54% and 100% for IFN p.p. and IFL-rA respectively. Crude leukocyte interferon showed nearly complete suppression which could not be neutralized by LI-1.

Seventy-two fresh human tumor specimens, 40 malignant effusions and 32 solid specimens, were obtained for in vitro culture. Only 12 specimens grew more than 30 colonies per control dish, which allowed multiple concentration tests of crude interferon. Two ovarian carcinomas (no. 1, no 4) showed slight but not significantly improved colony growth at a low interferon concentration (10 U/ml). At the same concentration, colony formation of a malignant melanoma (no. 9) was significantly increased. However, at 5×10^3 and 10^4 U/ml significant growth inhibition was found. Growth inhibition greater than 70% was only seen

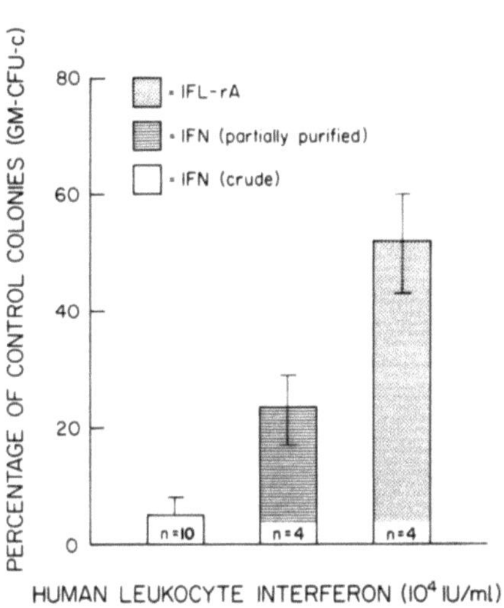

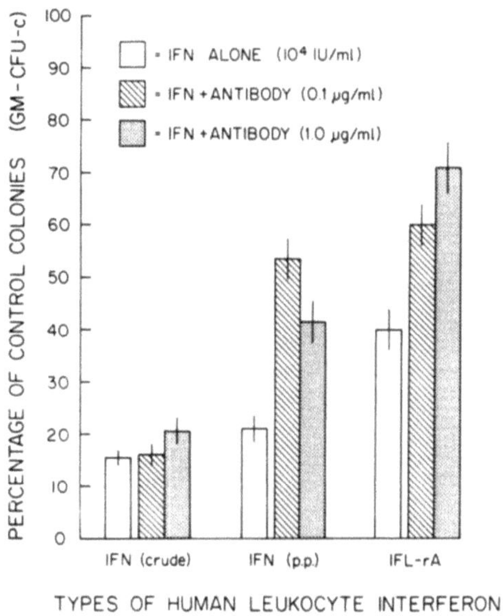

Fig. 3 (Left). Colony growth inhibition of three human leukocyte interferons of various specific activities; IFN (crude): 6.6×10^4 U/mg protein; IFN (partially purified): 1.13×10^6 U/mg protein; IFL-rA: $1.5-2.5 \times 10^8$ U/mg protein. *Bars,* SD; *n*, number of experiments set up in quadruplicate

Fig. 4 (Right). Neutralization of the antiproliferative effect of various human leukocyte interferons by a monoclonal antibody LI-1 against IFL-rA. Median of two experiments (each in quadruplicate) and SD are shown. *p.p.*, partially purified

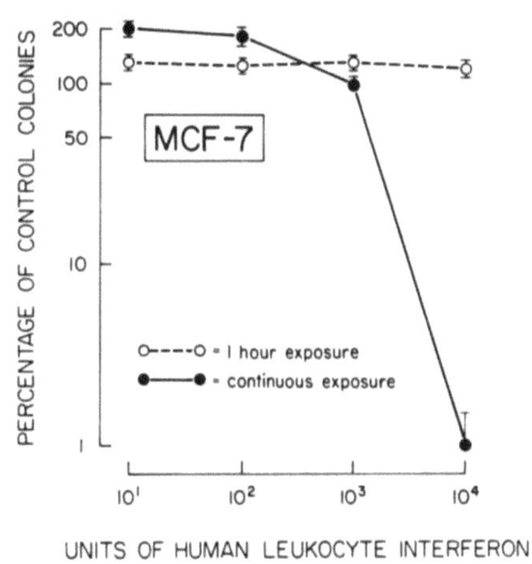

Fig. 5. Human mammary carcinoma (MCF-7) exposed to human leukocyte interferon (crude). Experiments set up in quadruplicate. *Bars,* SD

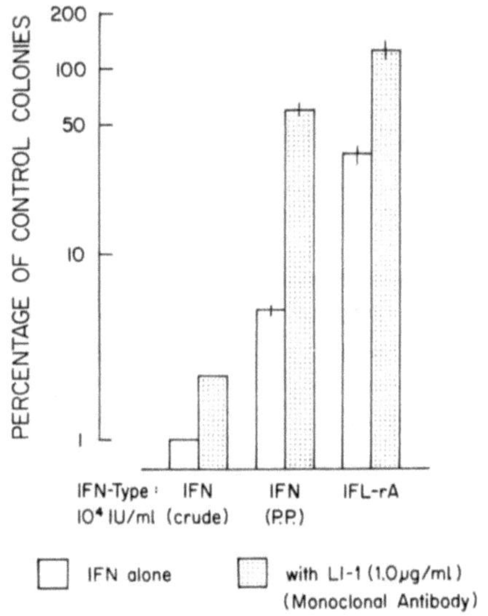

Fig. 6. WiDr growth inhibition by continuous exposure to interferon; neutralization by monoclonal antibody LI-1 significant for IFN P.P. ($p < 0.01$) and IFL-rA ($p < 0.001$). Experiments in quadruplicate ± SD *P.P.*, partially purified

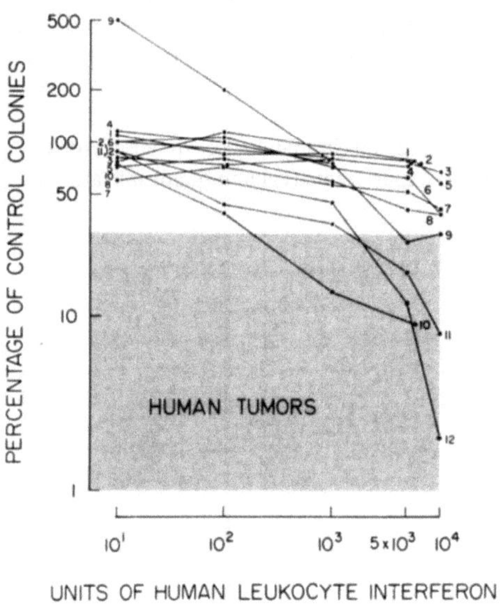

Fig. 7. Human leukocyte interferon (crude) tested against a panel of freshly explanted human tumors. Ovarian carcinomas nos. 1, 4, 5, 7, 11; breast carcinomas nos. 6, 8, 12; melanoma no. 9; adenocarcinoma (lung) no. 2; pancreatic carcinoma no 3; gastric carcinoma no. 10. *Each point* represents the mean of quadruplicate dishes; *shaded area,* 70% growth inhibition range

in four tumor samples: a gastric cancer (10^3 U/ml, no. 10), a melanoma (no. 9), an ovarian carcinoma (no. 11), and a breast carcinoma (10^4 U/ml, no. 12).

Discussion

We have shown that leukocyte interferon preparations inhibit the in vitro growth of human granulocyte/macrophage precursor cells (GM-CFU-c), human tumor cell lines (MDF-7 and WiDr), and freshly explanted human tumor cells (HCTC). In view of the substantial pharmacokinetic variability of the various types of interferon (Hilfenhaus et al. 1981;

Sarkar 1982) as demonstrated in experimental animal models, and the wide range of interferon plasma peak levels in man (Table 1), it seemed reasonable to test the activity of interferon over a wide range of concentrations. Continuous exposure of human bone marrow cells to leukocyte interferon resulted in a dose-dependent inhibition of colony growth with 50% of GM-CFU-c surviving a treatment with 100 U crude IFN per milliliter. Using different LeIFN-α preparations with various specific activities, dose-related colony growth suppression has been described by several other authors for granulocyte precursor cells (CFU-c), erythroid progenitor cells (BFU-e), and also for noncommitted precursor cells (CFU-GEMM) (Table 2). CFU-c has been shown to be more sensitive to interferon than CFU-e (Van t'Hull et al. 1978), which (together with the findings of other authors) suggests that this substance possesses tissue specificity (Borden et al. 1982a; Einhorn and Strander 1977; Ho and Enders 1959; Kataoka et al. 1982; Ludwig and Swetly 1980; Van t'Hull et al. 1978). BFU-e, in contrast, has been found to be highly sensitive to IFN, with 40% suppression at 40 U/ml (Ortego et al. 1979). Crude IFN consistently displayed higher colony growth suppression than IFN p.p. and IFL-rA, expecially at IFN concentrations of 10^4 U/ml. Greenberg and Mosny (1977) have also found increased growth inhibition with less pure interferon preparations, although their results were not statistically significant. One explanation is that the lower specific activity (6×10^4 U/ml vs 5×10^5 U/mg protein) of our crude IFN involves an increased chance of endotoxin contamination. Concentrations of ≥ 0.11 μg endotoxin per liter have been reported to suppress bone marrow CFU-c (Greenberg and Mosny 1977); since, in addition, endotoxin is a potent interferon inducer (Stewart 1979), both these direct and indirect effects may have contributed to the higher growth-inhibitory effect that we have found with crude IFN at 10^4 U/ml. Some bone marrows have displayed increasing cluster formation at IFN concentration ≤ 100 U/ml, resulting in a constant plating efficiency (colonies and clusters). Interferon concentration of 1,000 U/ml generally caused reduction in the number of clusters which is in contrast to the findings reported by Verma et al. (1979). No colony growth was observed when the conditioned medium was replaced by low-dose interferon (IFN crude; 10 U/ml), which confirms the results reported by Greenberg and Mosny (1977). Our data have demonstrated that only prolonged exposure to crude IFN results in a dose-related inhibition of GM-CFU-c. Similar results were reported by others (Balkwill and Oliver

Table 1. Plasma peak levels of human leukocyte interferon (IFN-α) and bacterial cloned leukocyte interferon (IFL-rA) in man

	Application	Dose of drug ($\times 10^6$ IU/ml)	Plasma peak level (IU/ml)	Reference
IFN-α	IV	30	100– 200	Emödi et al. (1975)
	IM	1	100– 200	Emödi et al. (1975)
	SC	1	100	Emödi et al. (1975)
	IM	9	230	Borden et al. (1982b)
	IM	9	140	Gutterman et al. (1980)
	IM	0.5–0.7/kg	3,000	Arvin et al. (1982)
IFL-rA	IM	198	1,000	Gutterman et al. (1982)
	IM	108	1,228 ± 330	Horning et al. (1982)
	IM	144	3,402 ± 2,028	Horning et al. (1982)
	IM	198	1,244 ± 1,396	Horning et al. (1982)
	IM	50	100– 200	Sherwin et al. (1982)

Table 2. Effect of human leukocyte interferon-α on erythroid and granulocyte precursor cells

Interferon (IFN): spec. activity (IU/mg protein)	Source of cells	Conditioned medium	Test system	Results	Reference
5×10^5 $2\text{–}3 \times 10^7$	Human bone	Leukocyte	Soft agar	1- and 4-h exposure: no suppression; cont. exp. (7 days): dose-dependent colony growth suppression 45% surviving colonies at 10^2 IU/ml	Greenberg and Mosny (1977)
5×10^5 2×10^7 1.7×10^5	Human bone marrow	Leukocyte	^{3}H-TdR uptake	At 5×10^3 IU/ml there was a 60% reduction in ^{3}H-TdR uptake	Balkwill and Oliver (1977)
8.6×10^5	Human bone marrow	Leukocyte feeder layer	Soft agar	Dose-dependent suppression of CFU-c (40% at 10 IU/ml) and CFU-e (40% at 1,000 IU/ml)	Van t'Hull et al. (1978)
7.5×10^5	Human bone marrow	Human placenta	Soft agar	Dose-dependent suppression of CFU-c with increasing cluster formation – differentiation block. Antagonistic effect of IFN (= CIA) and CSA	Verma et al. (1979)
8.8×10^5	Human bone marrow	Leukocyte	Soft agar	Dose-dependent suppression of non-committed precursor cells (CFU-GEMM). 30–40% inhibition at 100 IU/ml. Drug exposure delayed by 72 h still resulted in reduction of colonies	Neumann and Fauser (1982)
5×10^5	Human peripheral blood	Erythropoietin	Methyl-cellulose	Dose-related growth inhibition of human BFU-e. 68% suppression at 40 IU/ml	Ortega et al. (1979)

^{3}*H-TdR*, ^{3}*H*-thymidine; *CIA*, colony-inhibiting activity; *CSA*, colony-stimulating activity

1977; Gidali et al. 1981; Verma et al. 1981), who have found that inhibition of CFU-c after short exposure times (1 h or 4 h) does not affect GM-CFU-c formation. Paucker et al. (1962a) have described an increasing growth delay of L cells when exposed to a increasing interferon concentration. After interferon was removed a growth curve was observed identical to that in nonexposed cells. This has been interpreted as a transient block of proliferation rather than a cytotoxic effect of interferon. Similar findings supporting this interpretation have been found for the pluripotent bone marrow stem cells of the mouse (Gidali et al. 1981) and for human bone marrow CFU-c and acute leukemic colony-forming cells (Balkwill and Oliver 1977; Verma et al. 1981). Interferon-α has also been tested against a range of various human tumor cell lines and has been shown to be effective (≥ 50% inhibition) against monocytic leukemia (J 111), leukemia KG-1, and various Burkitt's lymphoma cell lines, as well as osteosarcoma and HEC-50 endometrial carcinoma (Table 3). Most other tumor cell lines have been resistant to IFN-α.

In contrast to Einhorn and Strander (1978), we did not find MCF-7 mammary carcinoma to be sensitive to IFN-α in this test system unless high concentrations of IFN (10^4 U/ml) were used with continuous exposure. At lower concentrations, we found an increase of smaller colonies, while the total number of colonies did not significantly differ from the control plates. Taylor-Papadimitriou (1980) has observed a significant inhibition of MCF-7 proliferation at IFN concentration of 10^3 U/ml in liquid culture and demonstrated the non-cycle-specific action of interferon. Short-term exposure (1 h) did not result in decreased colony growth even at high concentrations. Similar data have also been shown for a melanoma cell line (Ratner et al. 1980). Borden et al. (1982a) found no difference in the antiproliferative effect of IFN-α on various human tumor cell lines whether the exposure time was 72 h or 120 h. IFN-β, however, was more potent following a 120-h treatment than after 72 h. Bradley and Ruscetti (1981) found identical colony growth inhibition after 1-h and continuous exposure to IFN-β when given to the myeloma cell line ATCC 8226 and ovarian carcinomas from fresh ascites. The WiDr colon carcinoma cell line was poorly inhibited (< 35%) by all three interferons (crude, p.p., recombinant) at 10^3 U/ml. IFN at higher concentrations (10^4 U/ml), however, significantly suppressed colony growth and crude IFN demonstrated a higher suppressive effect than IFN p.p. and IFL-rA, a phenomenon we have already described for GM-CFU-c. LI-1, a monoclonal antibody against IFL-rA, neutralized the antiproliferative effect of IFL-rA on GM-CFU-c and WiDr cell lines, which supports the assumption that most of the inhibitory effects of less pure interferon result from interferon rather than from other contaminants (Greenberg and Mosny 1977; Stewart et al. 1976). The growth-inhibitory effect of IFN p.p. was also neutralized (not completely) by LI-1, which suggests that some IFN subspecies can be bound to LI-1.

Overall, human tumor cells from solid tumor tissues, malignant pleural effusions, or ascites grew colonies in 56% of samples, which is in good agreement with data reported by others (Pavelic et al. 1980; Schlag et al. 1982; Von Hoff et al. 1981). However, because of poor plating efficiencies only 25% of the specimens were adequate for drug testing. Although crude IFN caused a significant reduction in colony growth in 50% of the specimens, a greater than 70% suppression was only observed in 8% of the samples. Von Hoff et al. (1982) reported over 70% inhibition in 8% of the specimens at 10^3 U IFN-α per milliliter and Salmon et al. (1983) found greater than 70% inhibition in 18.7% of the cases when exposed to 800 U per milliliter IFL-rA. Higher inhibition rates were found in some malignant hematological tumors (Table 4). For 17% of the samples we have even noticed an increased tumor colony formation in vitro at IFN concentrations ≤ 100 U/ml. Growth-stimulatory effects were also reported for CFU-c and human tumors (Bradley and

Table 3. Effect of interferon (IFN) on the growth of human tumor cell lines

Spec. activity of IFN (IU/mg protein)	Source of cells	Test system	Results	Reference
–	Burkitt's lymphoma, non-Burkitt's lymphoma, leukemia, osteosarcoma, breast carcinoma	Liquid culture	Burkitt's lymphoma cell lines; 2 groups, 1 sensitive to low-dose IFN (2–300 IU/ml) for 50% inhibition; 2nd group only inhibited by very high dose IFN (10^4 IU/ml) Osteosarcoma sensitive (= 50% inhib.) at 10–300 IU/ml Mammary carcinoma (BT-20 and MCF-7) sensitive to 10 and 100 IU/ml respectively	Einhorn and Strander (1978)
2–1 × 10^8	Breast carcinoma	Liquid culture and semisolid agar culture	IFN showed cytostatic (colony growth inhibition) and cytotoxic effect (increased cell shedding from plastic surface), e.g., 45% inhibition of colony growth (≥ 200 μm in diameter) of T47D in agar at 1,000 IU/ml	Shibata and Taylor-Papadimitriou (1981)
× 10^6	25 human cell lines or strains	Liquid culture	Dose- and time-dependent antiproliferative effect of IFN; at 100 IU/ml over 5 days, 36% of the cell lines showed 20% growth inhibition; growth inhibition predominantly due to cytostasis	Borden et al. (1982a)
× 10^6	Endometrial adenocarcinoma (HEC-1D, HEC-50, HEC-1C)	Liquid culture	HEC-1C resistant up to 1,000 IU/ml; HEC-1D significant suppression at 1,000 IU/ml; HEC-50 highly sensitive with 47% suppression at 10 IU/ml	Morinaga et al. (1983)
	Osteogenic sarcoma, myeloma, leukemia	Semisolid agar	Osteosarcoma 292 suppressed by 68%, type MG-63 stimulated by factor 6.3, both at 1,000 IU/ml and cont. exp.; myeloma RPMI 8326 suppressed by 45% at 1,000 IU/ml; leukemia KG-1 suppressed by 95%; no difference in short- vs long-term exposure for ATCC 8226 myeloma cell line	Bradley and Ruscetti (1981)
× 10^6 × 10^7	Burkitt's lymphoma, laryngeal tumor (H.Ep = 2) Monocytic leukemia (J-111) ALL (CCRF-SB)	Liquid culture	Daudi cells highly sensitive to IFN. High doses (4.8×10^3 IU/ml) necessary for H.Ep = 2. J-111 showed 50% inhibition at 1.2×10^3 IU/ml, CCRF-SB only 40% inhibition with 10×10^3 IU/ml	Kataoka et al. (1982)

ALL, acute lymphocytic leukemia

Table 4. Effect of human leukocyte interferon-α (IFL-α) on human clonogenic tumor cells in vitro

Spec. activity of IFN (IU/mg protein)	Source of cells	Test system	Results	Reference
5×10^5; 2×10^7; 1.7×10^5 (lymphoblast)	Peripheral blood (AML)	^{3}H-TdR uptake (16 h)	Closely similar dose-response curves of ^{3}H-TdR uptake independent of specific activity; 10^5 IU/ml almost complete suppression (89%)	Balkwill and Oliver (1977)
1.8×10^6 (lymphoblast)	Bone marrow (myeloma)	Plasma clot and soft agar	50% of 14 samples showed 45% colony growth inhibition	Ludwig and Swetly (1980)
NR	Ascites, pleural effusion (stomach 1, ovary 3, breast 1, rhabdomyosa. 1, testis 1)	Soft agar	Colony inhibition was 74% at 1,000 IU/ml cont. exposure for stomach, 47% for ovary, 58%, 51%, and 38% for breast, rhabdomyosa., and testis respectively	Bradley and Ruscetti (1981)
2.3×10^5	Ovarian carcinoma (solid + ascites)	Soft agar	In 48% of the samples a 25% colony inhibition was obtained with 300 IU/ml	Epstein and Mareis (1981) Epstein et al. (1980)
NR	Breast 23, melanoma 6, lung 9, ovarian 5, N.H. lymphoma 6, neuroblastoma 3, and various other primaries 10	Soft agar	With 1,000 IU/ml a 70% decrease in TCFU was found in breast (4%), lung (22%), pancreas (1 of 2), and leukemia (1 of 2)	Von Hoff et al. (1982)
* 2×10^8 ** 5×10^8	273 tumors tested 71 tumors tested	Soft agar	At 800 IU/ml (= 4 ng/ml), 38% and 18% of the tumors showed ≥ 50% and ≥ 70% suppression of TCFU by IFL-α A respectively; 16% and 8% of the tumors showed ≥ 50% and ≥ 70% suppression of TCFU by IFL-α D respectively	Salmon et al. (1983)

* Recombinant IFL-α A; ** recombinant IFL-α D
NR, not reported; *AML*, acute myelocytic leukemia; *N.H.*, non-Hodgkin's; ^{3}*H-TdR*, ^{3}*H*-thymidine; *TCFU*, tumor-colony-forming units

Ruscetti 1981; Salmon et al. 1983; Von Hoff et al. 1982). Overall, interferon exhibited definite growth-inhibitory activity on normal as well as on malignant proliferating cells in vitro. Enough data have now accumulated to show this growth inhibition to be dependent on the test system used, the type of interferon, the concentration, and the exposure time. Studies are now necessary better to define tumor cell sensitivity to different types and concentrations of IFN, however, with minimal inhibitory effects on normal proliferating cells such as bone marrow cells.

Acknowledgements. We are grateful to Prof. Th. Staehelin, Hoffman La Roche Company, Basel, Switzerland, for making the interferon (IFL-rA) and monoclonal antibody (LI-1) available. Dr. S. Arrenbrecht, Division of Oncology, Department of Medicine, University of Zurich kindly performed the assay evaluating the antiviral activity of IFN (crude). We thank Maya Muller for excellent technical assistance and Irène E. Schuster for help in preparation of the manuscript. We thank Dr. Victor Hofmann and Dr. David Zava for their helpful comments on and criticisms of this work.

References

Anfinsen CB (1981) Human interferon. Interdisc Sci Rev 6: 110–118

Arvin AM, Schmidt NJ, Cantell K, Merigan TC (1982) Alpha interferon administration to infants with congenital rubella. Antimicrob Agents Chemother 21: 259–261

Balkwill FR, Oliver RTD (1977) Growth inhibitory effects of interferon on normal and malignant human haemopoietic cells. Int J Cancer 20: 500–505

Borden EC, Hogan TF, Voelkel JG (1982a) Comparative antiproliferative activity in vitro of natural interferon alpha and beta for diploid and transformed human cells. Cancer Res 42: 4948–4953

Borden EC, Holland JF, Dao TL et al. (1982b) Leukocyte-derived interferon (alpha) in human breast carcinoma. Ann Intern Med 97: 1–6

Bradley EC, Ruscetti FW (1981) Effect of fibroblast, lymphoid, and myeloid interferons on human tumor colony formation in vitro. Cancer Res 41: 244–249

Bridgen PJ, Anfinsen CB, Corley L et al. (1977) Human lymphoblastoid interferon. Large scale production and partial purification. J Biol Chem 252: 6585–6587

De Clercq E, Zhang ZX, Huygen K, Leyten R (1982) Inhibitory effect of interferon on the growth of spontaneous mammary tumors in mice. J Natl Cancer Inst 69: 653–657

Einhorn S, Strander H (1977) Is interferon tissue-specific? Effect of human leukocyte and fibroblast interferons on the growth of lymphoblastoid and osteosarcoma cell lines. J Gen Virol 35: 573–577

Einhorn S, Strander H (1978) Interferon therapy for neoplastic diseases in man: in vitro and in vivo studies. Adv Exp Med Biol 110: 159–174

Emödi G, Just M, Hernandez R, Hirt HR (1975) Circulating interferon in man after administration of exogenous human leukocyte interferon. J Natl Cancer Inst 54: 1045–1049

Epstein LB, Marcis SG (1981) Review of experience with interferon and drug sensitivity testing of ovarian carcinoma in semisolid agar culture. Cancer Chemother Pharmacol 6: 273–277

Epstein LB, Shen JT, Abele JS, Reese CC (1980) Further experience in testing the sensitivity of human ovarian carcinoma cells to interferon in an in vitro semisolid agar culture system: comparison of solid and ascitic forms of the tumor. Proc Clin Biol Res 48: 277–290

Gewert DR, Shah S, Clemens MJ (1981) Inhibition of cell division by interferons. Eur J Biochem 116: 487–492

Gidali J, Fehér I, Talas M (1981) Proliferation inhibition of murine pluripotent haemopoietic stem cells by interferon or poly I:C. Cell Tissue Kinet 14: 1–7

Glasgow LA, Crane JL, Kern ER, Younger JS (1978) Antitumor activity of interferon against osteogenic sarcoma in vitro and in vivo. Cancer Treat Rep 62: 1881–1888
Greenberg PL, Mosny SA (1977) Cytotoxic effects of interferon in vitro on granulocytic progenitor cells. Cancer Res 37: 1794–1799
Gresser I, Mauri C, Brouty-Boyé D (1972) On the mechanism of the antitumor effect of interferon in mice. Nature 239: 167–168
Gutterman JU, Blumenschein GR, Alexanian R et al. (1980) Leukocyte interferon-induced tumor regression in human metastatic breast cancer, multiple myeloma, and malignant lymphoma. Ann Intern Med 93: 399–406
Gutterman JU, Fine S, Quesada J et al. (1982) Recombinant leukocyte A interferon: pharmacokinetics, single-dose tolerance, and biological effects in cancer patients. Ann Intern Med 96: 549–556
Hamburger AW, Salmon SE (1977) Primary bioassay of human tumor stem cells. Science 197: 461–463
Hamburger AW, Salmon SE, Kim MB et al. (1978) Direct cloning of human ovarian carcinoma cells in agar. Cancer Res 38: 3438–3444
Hilfenhaus J, Karges HE (1974) Growth inhibition of human lymphoblastoid cells by interferon preparations, obtained from human leukocytes. Z Naturforsch [C] 29: 618–622
Hilfenhaus J, Damm H, Hofstaetter T, Mauler R, Ronneberger H, Weinmann E (1981) Pharmacokinetics of human interferon-beta in monkeys. J Interferon Res 1: 427–436
Ho M, Enders JF (1959) Further studies on an inhibitor of viral activity appearing in infected cell cultures and its role in chronic viral infections. Virology 9: 446–477
Holdener EE, Park CH, Belt RJ, Stephens RL, Hoogstraten B (1983) Effect of mannitol and plasma on the cytotoxicity of cisplatin. Eur J Cancer Clin Oncol 19: 515–518
Horning SJ, Levine JF, Miller RA, Rosenberg SA, Merigan TC (1982) Clinical and immunologic effects of recombinant leukocyte A interferon in eight patients with advanced cancer. JAMA 247: 1718–1722
Horoszewicz JS, Leong SS, Ito M, Buffett RF, Karakousis C, Holyoke E et al. (1978) Human fibroblast interferon in human neoplasia: clinical and laboratory study. Cancer Treat Rep 62: 1899–1906
Isaacs A, Lindenmann J (1957) Virus interference. I. The interferon. Proc R Soc Lond 147: 258–267
Isaacs A, Lindenmann J, Valentine RC (1957) Virus interference. II. Some properties of interferon. Proc R Soc Lond [Biol] 147: 268–273
Ito H, Murakami K, Yanagawa T et al. (1980) Effect of human leukocyte interferon on the metastatic lung tumor of osteosarcoma. Cancer 46: 1562–1565
Kataoka T, Oh-hashi F, Sakurai Y, Ida N (1982) Characteristics of in vitro antiproliferation activity of human interferon-beta. Cancer Chemother Pharmacol 9: 75–80
Ludwig H, Swetly P (1980) In vitro inhibitory effect of interferon on colony formation of myeloma stem cells. Cancer Immunol Immunother 9: 139–143
McNeill TA, Gresser I (1973) Inhibition of haemopoietic colony growth by interferon preparations from different sources. Nature [New Biol] 244: 173–174
Morinaga N, Yonehara S, Tomita Y, Kuwata T (1983) Insensitivity to interferon of two subclones of human endometrial carcinoma cell line, HEC-1. Int J Cancer 31: 21–28
Neumann HA, Fauser AA (1982) Effect of interferon on pluripotent hemopoietic progenitors (CFU-GEMM) derived from human bone marrow. Exp Hematol 10: 587–590
Ortega JA, Ma A, Shore NA, Dukes PP, Merigan TC (1979) Suppressive effect of interferon on erythroid cell proliferation. Exp Hematol 7: 145–150
Paucker K, Cantell K, Henle W (1962a) Quantitative studies on viral interference in suspended L cells. III. Effect of interferon viruses and interferon on the growth rate of cells. Virology 17: 324–334
Paucker K, Skurska Z, Henle W (1962b) Quantitative studies on viral interference in suspended L cells. I. Growth characteristics and interfering activities of vesicular stomatitis, Newcastle disease, and influenza A viruses. Virology 17: 301–311

Pavelic ZP, Slocum HK, Rustum YM et al. (1980) Growth of cell colonies in soft agar from biopsies of different human solid tumors. Cancer Res 40: 4151–4158

Pestka S, Maeda S, Staehelin T (1981) The human interferons. Annu Rep Med Chem 16: 229–241

Pike BL, Robinson WA (1970) Human bone marrow colony growth in agar-gel. J Cell Physiol 76: 77–84

Ratner L, Nordlund JJ, Lengyel P (1980) Interferon as an inhibitor of cell growth: studies with mouse melanoma cells (40760). Proc Soc Exp Biol Med 163: 267–272

Rohatiner AZS, Balkwill FR, Griffin DB, Malpas JS, Lister TA (1982) A phase I study of human lymphoblastoid interferon administered by continuous intravenous infusion. Cancer Chemother Pharmacol 9: 97–102

Salmon SE, Durie BGM, Young L, Liu RM, Trown PW, Stebbing N (1983) Effects of cloned human leukocyte interferons in the human tumor stem cell assay. J Clin Oncol 1: 217–225

Sarkar FH (1982) Pharmacokinetic comparison of leukocyte and *Escherichia coli*-derived human interferon type alpha. Antiviral Res 2: 103–106

Schlag P, Wolfrum J, Schremel W, Herfarth C (1982) Effect of fibroblast interferon on colony tumor growth of miscellaneous solid human tumors. Klin Wochenschr 60: 1455–1459

Scott GM, Secher DS, Flowers D, Bate J, Cantell K, Tyrell DAJ (1981) Toxicity of interferon. Br Med J 282: 1345–1348

Scott GM, Wallace J, Tyrrell DAJ, Cantell K, Secher DS, Stewart WE II (1982) Interim report on studies on "toxic" effects of human leucocyte-derived interferon-alpha (HuIFN-Alpha). J Interferon Res 2: 127–130

Sherwin SA, Knost JA, Fein S et al. (1982) A multiple-dose phase I trial of recombinant leukocyte A interferon in cancer patients. JAMA 248: 2461–2466

Shibata H, Taylor-Papadimitriou J (1981) Effects of human lymphoblastoid interferon on cultured breast cancer cells. Int J Cancer 28: 447–453

Soule HD, Vazquez J, Long A, Albert S, Brennan M (1973) A human cell line from a pleural effusion derived from a breast carcinoma. J Natl Cancer Inst 51: 1409–1416

Staehelin T, Hobbs DS, Kung H, Lai CY, Pestka S (1981) Purification and characterization of recombinant human leukocyte interferon (IFLrA) with monoclonal antibodies. J Biol Chem 256: 9750–9754

Stewart WE II (1979) The interferon system. Springer, Berlin Heidelberg New York

Stewart WE, Gresser I, Tovey MG, Bandu M, LeGoff S (1976) Identification of the cell mulitplication inhibitory factors in interferon preparations as interferons. Nature 262: 300–302

Strander H, Cantell K, Carlström G, Jakobsson PA (1973) Clinical and laboratory investigations on man: systemic administration of potent interferon to man. J Natl Cancer Inst 51: 733–742

Taylor-Papadimitriou J (1980) Inhibition of cell growth by interferons. In: Collier LH, Oxford J (eds) Developments in antiviral therapy. Academic Press, London, pp 189–200

Van t'Hull E, Schellenkens H, Löwenberg B, de Vries MJ (1978) Influence of interferon preparations on the proliferative capacity of human and mouse bone marrow cells in vitro. Cancer Res 38: 911–914

Verma DS, Spitzer G, Gutterman JU, Zander AR, McCredie KB, Dicke KA (1979) Human leukocyte interferon preparation blocks granulopoietic differentiation. Blood 54: 1423–1427

Verma DS, Spitzer G, Dicke KA (1980) Human leukocyte interferon: a possible regulator of myelopoietic differentiation. In: Khan A, Hill NO, Dorn GL (eds) Interferon: properties and clinical uses. Leland Fikes, Dallas, pp 543–559

Verma DS, Spitzer G, Zander AR et al. (1981) Human leukocyte interferon preparation-mediated block of granulopoietic differentiation in vitro. Exp Hematol 9: 63–76

Von Hoff DD, Casper J, Bradley E, Sandbach J, Jones D, Makuch R (1981) Association between human tumor colony-forming assay results and response of an individual patient's tumor to chemotherapy. Am J Med 70: 1027–1032

Von Hoff DD, Guttermann J, Portnoy B, Coltman CA Jr (1982) Activity of human leukocyte interferon in a human tumor cloning system. Cancer Chemother Pharmacol 8:99–103
WHO Scientific Group (1982) Interferon therapy. WHO Tech Rep Ser 676:B569R

Effect of Leukocyte Interferons on Cell Proliferation of Human Tumors In Vitro

C.U. Ludwig, B.G.M. Durie, S.E. Salmon, and T.E. Moon

Kantonsspital Basel, Departement für Innere Medizin,
Onkologische Abteilung, Petersgraben 4, 4031 Basel, Switzerland

Introduction

Interferons (IFNs) were originally defined as antiviral substances but were subsequently found also to modulate the immune response and to inhibit the growth of normal and transformed cells. When the antiproliferative effect of interferon α-A in vitro was analyzed using the human tumor clonogenic assay (HTCA), tumor inhibition of greater than 50% of colony formation was found in about 36% of the human tumors tested. However, some of the fresh tumor specimens exhibited growth stimulation on exposure to interferon, manifested by a significant increase in the number of tumor colonies compared to the controls. This is in parallel to IFN's immune modulatory effect: interferon has a stimulatory as well as an inhibitory effect on many different immunological reactions. In this paper we present the potential growth-stimulatory effect of IFNs on human tumors.

Material and Methods

We studied the effect of leukocyte IFN prepared by the method of Cantell (referred to as IFNα) and also the highly purified leukocyte IFNs prepared by the recombinant technique (referred to as IFNα-A and IFNα-D) from Hoffmann-La Roche.
The growth-regulatory effect of these IFNs was analyzed in clonogenic assays and by thymidine uptake studies. For solid tumors, the HTCA as described by Hamburger and Salmon (1977) was used, for leukemias, the method using methylcellulose as described by Buick et al. (1977) was applied. The IFNs were included in the upper layer at a concentration of 0.4–4.0 ng/ml, corresponding to 80–800 units/ml for the IFNα-A and IFNα-D. IFNα was tested at 50, 100, 250, and 500 units/ml.
In fresh human myeloma samples, tritiated thymidine uptake was measured: cells were incubated for 3 h at 37° with and without IFNs at the same concentrations as used in the HTCA. Then 5 μCi tritiated thymidine per milliliter was added and the cells incubated for another hour at 37° C. Finally, slides for high-speed scintillation autoradiography as described by Durie (Durie and Salmon 1975) were prepared.
As an operational definition, growth stimulation in the HTCA was defined as tumor colony formation more than 2 standard errors above the control. For the thymidine uptake, a twofold increase or an increase from 0 to $\geq 1\%$ labeling index was regarded as stimulation of proliferation.

Recent Results in Cancer Research. Vol. 94

Results

A total of 500 HTCAs were carried out with samples from 225 patients. In 30 (13.3%) of these samples, growth stimulation was seen in the HTCA with one or more of the different types of IFN. The frequency of growth stimulation varied with tumor cell type: most frequently growth stimulation was seen with acute myelocytic leukemia (AML), renal cancer, breast cancer, and ovarian cancer, but rarely with melanoma samples (Table 1).

Table 1. Overall growth stimulation in fresh human tumor samples by interferons (IFNs)

Tumor type	Number of samples tested[a]	Samples with growth stimulation no. (percent)
Acute myelocytic leukemia	22	6 (27.3)
Renal carcinom	10	2 (20.0)
Breast carcinom	21	4 (19.0)
Ovarian carcinom	43	5 (11.6)
Lung carcinom	18	2 (11.0)
Unknown primary	12	1 (8.3)
Melanoma	34	2 (5.9)
Myeloma	23	1 (4.3)
Others	42	7 (16.7)
Total	225	30 (13.3)

[a] A total of 500 human tumor clonogenic assays were performed with the various leukocyte IFNs at different concentrations

Table 2. Growth stimulation by interferons (IFNs) in the human tumor clonogenic assay (HTCA): dose-response relationships

Type of interferon	Concentration of interferon	Number of HTCAs performed[a]	No. of tests with percentage survival > 2 SEs above control (100%)	Percentage of tests exhibiting stimulation
IFNα	50 u/ml	140	8	15.7
	100 u/ml	19	5	26.3
	250 u/ml	38	0	0.0
	500 u/ml	18	0	0.0
IFNαA	4×10^{-4} μg/ml	90	6	6.6
	8×10^{-4} μg/ml	9	2	22.2[b]
	4×10^{-3} μg/ml	93	6	6.5
IFNαD	4×10^{-4} μg/ml	45	4	8.9
	4×10^{-3} μg/ml	48	3	6.3
Total		500	34	(mean) 6.8

[a] A total of 225 tumors were tested against three different types of IFNs in a total of 500 assays

[b] Only acute leukemias were tested at 8×10^{-4} μg/ml

Table 3. Tritiated thymidine uptake [labeling index (*LI*) %] induced by interferon (*IFN*) in multiple myeloma (41 assays)

Type of IFN	Number of myeloma marrows tested	Number of samples with increased LI (200% of control)	Percentage of marrows exhibiting stimulation
Leuk IFN[a]	21 (16)[b]	3 (3)[b]	14.2 (18.8)[b]
IFNα-A[c]	20	3	15
Total	41	6	14.6

[a] Tested at 100, 500, and 1,000 U/ml
[b] Only samples tested at 100 U/ml
[c] Tested at 4×10^{-4}, 8×10^{-4}, and 4×10^{-3} µg/ml

Analyzing different dose levels of IFNs, a dose dependency for growth stimulation could be demonstrated: at 50 units IFNα per milliliter, 5.7% of samples showed evidence of growth stimulation, but at 100 units/ml this was true for 26.3% of the samples. On the other hand, at the highest concentration of IFNα tested, 250 and 500 units/ml, no growth stimulation was found (Table 2).

A similar pattern of dose-response relationship was demonstrated for the IFNα-A: there was little growth stimulation at the lowest and highest concentrations tested (6.6% at 4×10^{-4} µg/ml and 6.5% at 4×10^{-3} µg/ml), but 22.2% of the samples showed growth stimulation at 8×10^{-4} µg/ml (Table 2).

Using IFNα or IFNα-A, an increased thymidine uptake by fresh human myeloma cells could be demonstrated in about 15% of all samples. With the IFNα, an increased thymidine uptake was again only seen at the lowest dose (100 units/ml) tested (Table 3).

Discussion

Interferons have been shown to have several properties, including antiviral activity, inhibitory effects upon cell division, and the power to enhance certain specialized cell functions, including cell surface effects. On the basis of the antiproliferative effect, IFNα has been evaluated as treatment for tumors in animals and in humans. Our data indicate that IFNα, even in its pure form as recombinant IFN, can in a percentage of cases stimulate rather than inhibit the growth of some human tumors in vitro. This stimulation could be demonstrated by means of two different in vitro systems: clonal growth in semisolid medium (agar or methylcellulose) and thymidine incorporation after short-term culture. Our findings are in agreement with data reported by Bradley and Ruscetti (1981), showing a stimulatory effect on human tumor proliferation in vitro using fibroblast interferon. The growth-stimulatory effect of IFNs observed in the HTCA could be mediated directly via a proliferative action on tumor cells or indirectly through an effect on host immunoreactive cells present in the HTCA. The fact that we could demonstrate a growth-stimulatory effect also in the myeloma cell line 8226 (data not shown), which lacks any contaminating immunoreactive cells, strongly suggests that IFNs can directly stimulate proliferation of tumor cells.

How can we explain our data? Regulation of cell growth by IFNs appears to be a function of both the level of the (2′-5′)-oligoisoadenyl synthetase E, referred to as 2′-5′-A-synthetase, and the amount of 2′-phosphodiesterase, which are both induced by IFNs (Revel et al. 1980). The 2′-5′-A-synthetase does inhibit protein synthesis and finally cell proliferation. In contrast, the 2′-phosphodiesterase is responsible for the degradation of the 2′-5′-pppApAp into adenosine monophosphate, leading to cell proliferation. In L-cells, Revel found the ratio of the 2′-5′-A-synthetase to the 2′-phosphodiesterase to decrease at lower dosages of IFNs, and only to increase over the basal level at higher concentrations of IFNs. Chapekar and Glazer recently stressed again the importance of the ratio of these two enzymes. An increase in the ratio seems to be a prerequisite for the antiproliferative action of IFNs (Chapekar and Glazer 1983). One may speculate that growth stimulation may occur in samples where at lower IFN concentrations the ratio of these two enzymes does decrease.

An in vivo immunoregulatory effect may counterbalance the growth-stimulatory effect on tumor cells which we have observed in vitro. However, in animals tumor growth stimulation has been described using IFN inducers (Gazdar 1972).

Our observation of growth stimulation mainly at lower IFN concentrations and growth inhibition preferentially at higher concentrations resembles the dose-response pattern observed for the immunoregulatory effect of IFN: very low concentration of IFN can cause enhancement of antibody response, while high concentrations result in inhibition of the antibody response.

The growth-stimulatory effect of IFNs in vitro was mainly observed at relatively low levels of 50–160 units/ml. However, serum peak concentrations achieved in patients after intramuscular applications of IFNs are in a similar range. Therefore, awareness of potential tumor growth stimulation by IFNs may be important for the clinician.

References

Bradley EC, Ruscetti FW (1981) Effect of fibroblast lymphoid and myeloid interferons on human tumor colony formation in vitro. Cancer Res 41: 244

Buick RN, Till JE, McCulloch EA (1977) Colony assay for proliferative blast cells circulating in myeloblastic leukemia. Lancet 1: 862–863

Chapekar MS, Glazer RI (1983) Effects of fibroblast and recombinant leukocyte interferons and double-stranded RNA on ppp (2′-5′)A_n-synthesis and cell proliferation in human colon carcinoma cells in vitro. Cancer Res 43: 2683–2687

Durie BGM, Salmon SE (1975) High speed scintillation autoradiography. Science 190: 1093–1095

Gazdar AF (1972) Enhancement of tumor growth rate by interferon inducers. J Natl Cancer Inst 49: 1435

Hamburger AW, Salmon SE (1977) Primary bioassay of human tumor stem cells. Science 197: 461–463

Revel M, Kimchi A, Shulman L et al. (1980) Role of interferon induced enzymes in the antiviral and antimitogenic effects of interferon. Ann NY Acad Sci 350: 459

Drug Combination Testing with In Vitro Clonal Cultures

M.S. Aapro

Hôpital Cantonal Universitaire, Division für d'Onco-Hématologie, 1211 Genève 4, Switzerland

Introduction

Development of clinically useful antitumor drug combinations has been largely empirical and there is a pressing need to evaluate experimental systems that can provide predictive data useful in the rational design of combination regimens (Goldin and Carter 1982). In vitro colony formation by single cells, i.e., clonal cultures, seems to provide the best method of assessing cell lethality induced by chemotherapeutic agents; doubling time, labeling index, dye exclusion, ^{51}Cr release, and rate of tritiated thymidine uptake have not been found to be satisfactory means of determining reproductive death (Roper and Drewinko 1976). Recent developments of some of these techniques might, however, allow their use for in vitro single-drug and possibly drug combination screening. Nucleic acid precursor uptake assays have been shown to predict in vivo tumor response (Sanfilippo et al. 1981; KSST 1981). The possible complications associated with variable isotope uptake into nonmalignant cells within the tumors render these results surprising (Editorial 1982) and use of short-term isotope uptake assays to test sensitivity of cell lines has been reported either always to predict for resistance (Roobol et al. 1983) or to need considerably higher drug concentrations than clonogenic assays to achieve the same apparent activity (Zirvi et al. 1983). A novel dye exclusion method for testing in vitro chemosensitivity of human tumors has been compared with a clonogenic assay and found to be in excellent qualitative agreement with it (Weisenthal et al. 1983). The same authors conclude that, if the purpose of in vitro assays is to serve as a model of in vivo response, then the true test of their validity will be to have a direct comparative study of their predictive ability. Therefore, most of the above objections should be regarded as theoretical. Updated information on these and other tests appears elsewhere in this volume. The potential clinical usefulness of a "human tumor stem cell assay" was reported by Salmon et al. (1978). Since then several retrospective studies have shown that a similar test has clinical validity and can identify chemoresistant patients (Johnson and Rossof 1983). In spite of several limitations (Selby et al. 1983) the human tumor cloning system has recently been reported to have a 60% true positive and a 85% true negative rate for predicting response or lack of response of an individual patient's tumor to a single agent (Von Hoff et al. 1983). The present review will address several aspects of the use of clonal cultures for in vitro evaluation of drug combinations.

Recent Results in Cancer Research. Vol. 94

Definitions Used for In Vitro Evaluation of Drug Combinations

The use of various terms to describe the increased or decreased cytotoxic effect of antitumor drugs studied in vitro can lead to some confusion. Valeriote and Lin (1975) have proposed several definitions which have since then been discussed by Momparler (1980). The cornerstone of their definitions is the one for additivity, which is the product of the individual survival fractions of a cell population treated by agents A and B. It is assumed that each agent acts independently of the other and the definition is the consequence of probability theory. Considering that the survival fraction (SF_A) for drug A is greater than or equal to the survival fraction (SF_B) for drug B, several possibilities exist. The additive effect will be $(SF_A) \times (SF_B)$. If the observed SF_{A+B} of the combination is greater than SF_A, then the term "antagonistic" is used. If SF_{A+B} has a value between SF_A and SF_B, one should speak of "interference". For practical purposes, however, this is frequently called antagonism, as one drug alone would have been more effective. Any value of SF_{A+B} lying between SF_B and the theoretical value for additivity $(SF_A) \times (SF_B)$ will be considered to be subadditive. Any value of SF_{A+B} smaller than (SF_A) 0 (SF_B) will indicate synergism (Fig. 1 and Table 1). The particular case of SF_A being 100%, i.e., of drug A having no effect, can still be considered with the above definitions. As $(SF_A) \times (SF_B) = SF_B$ in this case, any value of SF_{A+B} beyond SF_B will be synergism. The terms "enhancement", "potentiation", and "sensitization" are not used. When several data points are available, dose-response curves for drug A alone, for the theoretical additive effect of drug B, and for the observed

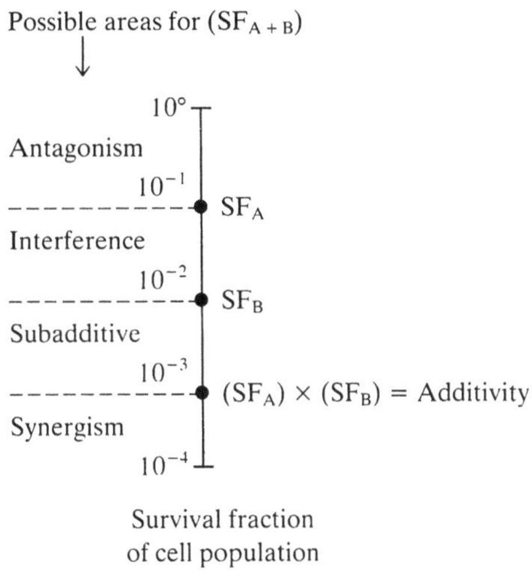

Fig. 1. Definitions according to Valeriote and Lin (1975) of the observed effect of a combination of two drugs A and B (SF_{A+B}) on the survival fraction of a cell population. Its value can lie in different areas limited by the survival fraction induced by A (SF_A), B (SF_B), or the theoretical additive effect $(SF_A) \times (SF_B)$

Table 1. Definitions of drug combination effects [after Valeriote and Lin (1975) and Momparler (1980)]

Synergistic	$SF_{A+B} < (SF_A) \times (SF_B)$
Additive	$SF_{A+B} = (SF_A) \times (SF_B)$
Subadditive	$SF_{A+B} > (SF_A) \times (SF_B)$ and $< SF_B$ when $SF_A > SF_B$
Interference	$SF_{A+B} > SF_B$ and $< SF_A$
Antagonistic	$SF_{A+B} > SF_A$ when $SF_A > SF_B$

SF, surviving fraction; *A, B,* individual drugs; *A+B,* drugs used in combination

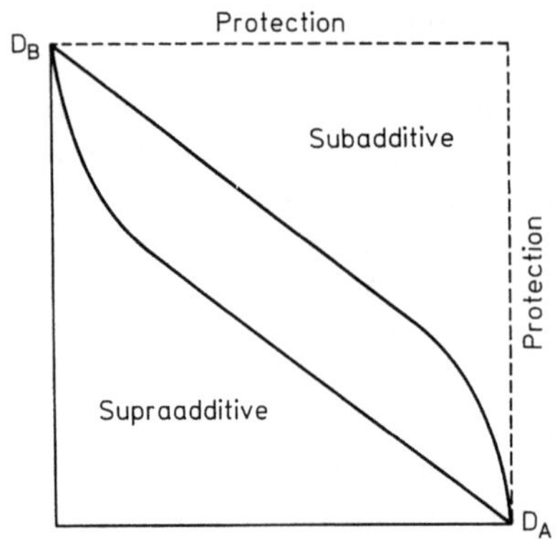

Fig. 2. Theoretical isobologram (isoeffect plot) representation of a drug combination. D_A and D_B are the doses of drug A and of drug B which have the same cytotoxic effect. The *area between the two curves* is the "envelope" of additivity, obtained according to Steel and Peckham (1979)

results are drawn. Single points or the overall curve can be assessed for deviation from additivity using a statistical test described by Drewinko et al. (1976).

Another approach to the description of interactions in the response of tumors to combined modality therapy can be adopted from Steel's work (Steel 1979). The suggested terminology is particularly relevant for drug-radiation interactions. Several situations are considered. When one agent is inactive by itself, it is said either to sensitize cells to or to protect cells from the effects of another agent. When fixed doses are used, the combination is said to have a cooperative or an antagonistic effect when compared to the addition of the single-agent responses. When the dose-effect curve of one agent is studied with and without the addition of a fixed dose of another agent, a positive interaction is called "enhancement", the expected curve for additivity "noninteractive," the area of the experimental curve between additivity and the single-agent curve "inhibition", and the combination dose-effect curve that shows less effect than the single-agent curve is said to show "protection".

The most sophisticated approach to the study of antitumor agent interactions is to analyze the response to the combination through an isobologram approach (Fig. 2) (Steel and Peckham 1979), which necessitates the production of a full dose-reponse curve for both of the single agents. An area called the "envelope of additivity", wherein enhancement or inhibition can be determined, is calculated and is a straight line if the dose-response curves of both single agents are linear. Any observed value which lies to the left of this area (line) will be called "supraadditive" and any value to the right of this area (line) will be called "subadditive". These terms are probably an unfortunate choice. Supraadditive might correspond to synergism according to Valeriote's definitions, but Steel's subadditive can indicate a negative interaction, when for Valeriote it indicates a weakly positive interaction.

Studies of Drug Interactions Using Clonal Cultures

Clonal cultures have been widely used to determine whether combinations given in different doses and sequences would modify the cytotoxic effects of either agent when given alone. The terms used to describe the strength of positive or negative effects are often confusing and do not correspond either to Valeriote's or to Steel's definitions. One recent abstract can serve as an example. Cohen et al. (1983) have reported that A-10, an

Table 2. Combination of interferon-αA (continuous exposure) and cytotoxic agents

	VBL		VCR		VDS		VZL		PLAT		DOXO
Time of exposure	1 h	CT	1 h	CT	1 h	CT	1 h	CT	1 h	CT	1 h
Experiments	19	13	3	9	4	4	3	4	7	6	5
Synergistic	7	8	1	1	1	1	0	0	0	3	0
Additive	10	5	1	5	2	1	1	4	3	2	4
Subadditive	1	0	1	3	1	2	1	0	3	1	1
Antagonistic	1	0	0	0	0	0	1	0	1	0	0

Pooled results from three human cell lines (colon WiDR, myeloma 8226, breast MCF-7)
Synergism or antagonism if $P < 0.05$ between observed and expected (Drewinko 1976)
VBL, vinblastine; *VCR,* vincristine; *VDS,* vindesine; *VZL,* vinzolidine; *PLAT,* *Cis*-platinum; *DOXO,* doxorubicin; *CT,* continuous

antineoplastic agent derived from bovine aorta, shows synergism with doxorubicin hydrochloride (Adriamycin) when tested with a clonogenic assay of murine TA3Ha mammary carcinoma. Growth was assayed using a growth index calculated by counting the number of "colonies" across one diameter of a 35-mm culture dish and giving several coefficients of value to "colonies" of different sizes (groups of less than 20 cells being called colonies) to obtain the growth index. Analysis of variance showed that the growth index of dishes treated with the combination was significantly different from the growth index of dishes treated with either single drug. Let us now assume that the growth index represents the actual number of colonies observed. The survival fraction of cells treated with doxorubicin would then be 41% or 10% depending on the run of the assay that is studied. The survival fraction of A-10 treated cells would be 76% or 41%, and the expected value for additivity would be 41% × 76% = 31% or 10% × 41% = 4.1%. The observed survival fractions are 28% and 5.5%. Provided that this way of calculating the presented data is valid, doxorubicin and A-10 show an additive and not a synergistic interaction.

We have recently studied the interactions of purified human recombinant leukocyte interferon (Hoffmann-La Roche) with vinblastine, vincristine, vindesine, vinzolidine, *cis*-platinum, and doxorubicin (Aapro et al. 1983a). The activity of these agents alone or in combination was tested against myeloma RPMI 8226, breast MCF-7, and colon WiDR cell lines. Statistically significant synergistic activity against in vitro colony formation was observed with the combination of vinblastine and interferon-αA, and possibly *cis*-platinum and interferon-αA (only one cell line studied) (Table 2). Such experiments do not allow an explanation for the mechanism of action of this possible synergism. Welander et al. (1983) have studied the effects of recombinant interferon-α2 (Schering) in combination with doxorubicin, *cis*-platin, vinblastine, bleomycin, 5-fluorouracil, methotrexate, and triethylenethiophosphoramide. Single drugs and their combinations were tested with various schedules and exposure times against two ovarian carcinoma cell lines (BG-1 and BG-2). The most active synergistic combination was felt to be with doxorubicin. The sequencing of interferon followed by doxorubicin seemed to be important to obtain the best positive interaction. It is interesting to observe that these authors used short exposure times to interferon, which have been reported to have no effect on colony survival (Holdener et al. 1983; Aapro, unpublished results). We have, however, preliminary data that tend to confirm that even when there is no apparent effect of interferon on colony survival, its combination with another agent can lead to an increased effect of this agent.

Table 3. Combination chemotherapy against fresh tumor cells

	Antagonistic	Subadditive	Additive	Synergistic	Reference
Doxorubicin–*cis*-platinum	4[a]		14	1	Alberts (1981)
Bleomycin–vinblastine	6[a]		8	3	Alberts (1981)
Cis-platinum–vinblastine	1[a]		8	1	Alberts (1981)
Actinomycin D–vindesine	3[a]		5	2	Alberts (1981)
Interferon–doxorubicin	2	3	1	5	Welander (1983)

Experiments using a double-layer soft agar clonogenic assay
[a] Called "inhibitory" by the author

Several preliminary reports on studies of drug combinations against fresh human tumor cells are available (Table 3). Using different drug combinations, both Alberts et al. (1981) and Welander et al. (1983) have reported additive and sometimes synergistic effects. There are no clinical correlations available. These data are interesting but, in 98 assays comparing the in vitro sensitivities of tumors to single drugs and their combination, Bertelsen et al. (1982) report that the combination was active in vitro in only two out of 44 cases when one of its component drugs was inactive and that the combination was inactive in vitro in eight of 46 cases where all single drugs of the combination were active. When all component drugs were active, three combination assays were synergistic, 25 additive, and 18 antagonistic. The authors suggest that testing combinations is probably an ineffective means of finding active antitumor drug treatment. The same group has subsequently reported (Sondak et al. 1983) that in vitro durg combinations have a degree of activity that is comparable to the activity of the most active single drug component. Nevertheless, 21 out of 289 patients were found to be sensitive to the combination when no single drug was felt to be active.

Problems with Drug Combination Testing

Prospective clinically valid in vitro prediction of human tumor chemosensitivity using clonal assays is limited by several factors that also affect testing of combinations. Inadequate material and low plating efficiencies have been reported to allow drug testing in vitro in only 31% of more than 8,000 tumors cultured in a single laboratory (Von Hoff 1983). Several commonly used drugs are inactive in vitro and therefore cannot be tested unless an active metabolite is used, e.g., 4-hydroperaxycyclo phosphamide (4-HC) for cyclophosphamide. Drug doses and exposure times need to be adapted to be representative of in vivo conditions [but this is often difficult (e.g., Alberts and van Daalen Wetters 1976)], also taking into account their cycle specificity characteristics (e.g., cytosine arabinoside; Preisler and Epstein 1981). Numerous other problems relative to the use of clonal cultures are reviewed by Selby et al. (1983) and elsewhere in this volume.
Testing drug combinations in vitro requires that consideration be paid to possible chemical interactions that can activate or inactivate one of the compounds. Relevant to our studies on the effect of dexamethasone on the cytotoxicity of antitumor agents (Aapro et al. 1983b) is the fact that dexamethasone sodium phosphate causes gross precipitation of

daunorubicin and doxorubicin (Dorr 1979). Examples of similar interactions are numerous and some may be relevant to the observations of in vitro antagonistic interactions. A "hidden" combination is the presence in the medium used when the compounds are tested of substances that can by themselves modify the cytotoxic effect of a drug. Antimetabolites can be greatly influenced by an excess of some factor in vitro (e.g., thymidine for methotrexate or 5-fluorouracil), but other drugs can also have their cytotoxic activity modified by the presence of known or unknown substances, such as leucine, which inhibits melphalan cytotoxicity (Vistica et al. 1978). Testing of several antitumor agents might be facilitated by the development of defined serum-free media (see contribution by Eliason et al. in this volume), but the clinical relevance of results observed in those conditions will need further study. More or less complex metabolic interactions can occur during in vitro testing of drug combination activity. The very precise schedule dependency of the methotrexate-5-fluorouracil combination is one example. Synergistic results are obtained in some cell culture studies when methotrexate precedes 5-fluorouracil, because methotrexate increases the phosphoribosyl pyrophosphate pool and enhances nucleotide formation (Cadman et al. 1979); the reverse sequence (5-fluorouracil preceding methotrexate) will decrease the cytotoxic effect of methotrexate because of inhibition of thymidylate synthetase (Bowen et al. 1978). The in vivo relevance of similar interactions has been discussed by Tattersall et al. (1973), who state that "pure empiricism in the concoction of multiple drug regimens can have an uncertain outcome". These data help us to understand a recent abstract reporting that 11 of 13 tests of fresh human tumor colony survival after exposure to a combination of chlorambucil, methotrexate, and 5-fluorouracil showed drug antagonism (Osborne et al. 1982). This observation can thus be interpreted as showing either an in vitro artifact of drug combination testing or a clinically relevant phenomenon. The drug schedule has to be reproducible in vivo; doses need to be at clinically achievable levels and any possible antitumor synergism should not be paralleled by synergism in host toxicity before in vivo transposition of in vitro results is attempted. The use of clonal cultures should nevertheless allow for rapid exploration of drug combinations, which will then need to be explored in animal models before clinical trials are begun.

Conclusions

Clonal cultures have been used for several years to study in vitro chemotherapy and there is good evidence that they can predict for in vivo tumor response. Reports are often confusing because the same words to describe drug interactions will have different meanings for different authors. In vitro testing of combinations has several problems and its in vivo relevance remains to be proved. Clonal cultures of cell lines are, however, a relatively inexpensive technique for the study of drug interactions and might allow the discovery of useful antitumor combinations. It is possible that the use of fresh human tumor cloning to find individualized combinations is still an ineffective technique, but better definition of the conditions in which to test drug combinations (and which ones) and better cloning techniques should allow for predictive combination drug testing in the near future.

References

Aapro MS, Alberts DS, Salmon SE (1983a) Interactions of human leukocyte interferon with vinca alkaloids and other chemotherapeutic agents against human tumors in clonogenic assay. Cancer Chemother Pharmacol 10: 161–166

Aapro MS, Alberts DS, Serokman R (1983b) Lack of dexamethasone effect on the antitumor activity of cisplatin. Cancer Treat Rep 67: 1013–1017

Alberts DS, van Daalen Wetters T (1976) The effect of phenobarbital on cyclophosphamide antitumor activity. Cancer Res 36: 2785–2789

Alberts DS, Salmon SE, Chen HSG, Moon TE, Young L, Surwit EA (1981) Pharmacologic studies of anticancer drugs with the human tumor stem cell assay. Cancer Chemother Pharmacol 6: 253–264

Bertelsen CA, Kern DA, Mann BD, Campbell MA, Storm FK, Morton DL (1982) Selection of combination chemotherapy using the clonogenic assay. Proc Am Soc Clin Oncol 1: 41

Bowen D, White JC, Goldman ID (1978) Basis for fluoropyrimidine-induced antagonism to methotrexate in Ehrlich ascites tumor cells in vitro. Cancer Res 38: 219–222

Cadman E, Heimer R, Davis L (1979) Enhanced 5-fluorouracil nucleotide formation after methotrexate administration: explanation for drug synergism. Science 205: 1135–1137

Cohen J, Haid M, Eisenstein R, Murthy S, Berlin NI (1983) In vitro synergism between an antineoplastic agent derived from bovine aorta (A-10) and Adriamycin in murine TA3Ha mammary adenocarcinoma. Proc Am Soc Clin Oncol 2: 45

Dorr RT (1979) Incompatibilities with parenteral anticancer drugs. Am J Intraven Ther 6: 42, 45–46, 52

Drewinko B, Loo TL, Brown B, Gottlieb JA, Freireich EJ (1976) Combination chemotherapy in vitro with Adriamycin. Observations of additive, antagonistic, and synergistic effects when used in two-drug combinations on cultured human lymphoma cells. Cancer Biochem Biophys 1: 187–195

Editorial (1982) Clonogenic assays for the chemotherapeutic sensitivity of human tumors. Lancet 1: 779–781

Goldin A, Carter SK (1982) Screening and evaluation of antitumor agents. In: Holland JF, Frei E (eds) Cancer medicine. Lea Febiger, Philadelphia, pp 633–663

Holdener EE, Schnell P, Spieler P, Senn HJ (1983) Antiproliferative activity of interferon: in vitro studies of human granulocyte precursor cells (CFU-C) and human tumors. Proc Am Assoc Cancer Res 24: 312

Johnson PA, Rossof AH (1983) The role of the human tumor stem cell assay in medical oncology. Arch Intern Med 143: 111–114

KSST (1981) In vitro short-term test to determine the resistance of human tumors to chemotherapy. Cancer 48: 2127–2135

Momparler RL (1980) In vitro systems for evaluation of combination chemotherapy. Pharmacol Ther 8: 21–35

Osborne CK, von Hoff DD, Clark GM, Sandbach J, O'Brien M, and the South Central Texas Human Tumor Cloning Group (1982) The clonogenic assay identifies antagonism in combination chemotherapy of breast cancer. Proc Am Assoc Cancer Res 23: 185

Preisler HD, Epstein J (1981) A Comparison of two methods for determining the sensitivity of human myeloid colony-forming units to cytosine arabinoside. Br J Haematol 47: 519–527

Roobol K, Sips H, Theunissen J, Atassi G, Bernheim J (1983) Possibilities and improvements of in vitro drug testing assays. Proc Am Assoc Cancer Res 24: 311

Roper PR, Drewinko B (1976) Comparison of in vitro methods to determine drug-induced cell lethality. Cancer Res 36: 2182–2188

Salmon SE, Hamburger AW, Soehnlen B, Durie BGM, Alberts DS, Moon TE (1978) Quantitation of differential sensitivity of human-tumor stem cells to anticancer drugs. N Engl J Med 298: 1321–1327

Sanfilippo O, Daidone MG, Costa A, Canetta R, Silvestrini R (1981) Estimation of differential in vitro sensitivity of non-Hodgkin lymphomas to anticancer drugs. Eur J Cancer 17: 217–226

Selby P, Buick RN, Tannock I (1983) A critical appraisal of the "human tumor stem-cell assay". N Engl J Med 308: 129–134

Sondak VK, Korn EL, Morton DL, Kern DH (1983) Absence of in vitro synergy for chemotherapeutic combinations tested in the clonogenic assay. Proc Am Assoc Cancer Res 24: 316

Steel GG (1979) Terminology in the description of drug-radiation interactions. Int J Rad Oncol Biol Phys 5: 1145–1150

Steel GG, Peckham MJ (1979) Exploitable mechanisms in combined radiotherapy-chemotherapy: the concept of additivity. Int J Rad Oncol Biol Phys 3: 85–91

Tattersall MHN, Jackson RC, Connors TA, Harrap KR (1973) Combination chemotherapy: the interaction of methotrexate and 5-fluorouracil. Eur J Cancer 9: 733–739

Valeriote F, Lin H (1975) Synergistic interaction of anticancer agents: a cellular perspective. Cancer Chemother Rep 59: 895–900

Vistica DT, Toal JN, Rabinowitz M (1978) Amino acid conferred protection against melphalan. Characterization of melphalan transport and correlation of uptake with cytotoxicity in cultured L1210 murine leukemia cells. Biochem Pharmacol 27: 2865–2871

Von Hoff DD (1983) Send this patient's tumor for culture and sensitivity. Editorial. N Engl J Med 308: 154–155

Von Hoff DD, Clark GM, Stogdill BJ, Sarosdy MF, O'Brien MT, Casper JT, Mattox DE, Page CP, Cruz AB, Sandbach JF (1983) Prospective clinical trial of a human tumor cloning system. Cancer Res 43: 1926–1931

Weisenthal LM, Dill PL, Kurnick NB, Lippman ME (1983) Comparison of dye exclusion assays with a clonogenic assay in the determination of drug-induced cytotoxicity. Cancer Res 43: 258–264

Welander C, Gaines J, Homesley H, Rudnick S (1983) In vitro synergistic effects of recombinant human interferon $alpha_2$ (rIFN-$\alpha 2$) and doxorubicin on human tumor cell lines. Proc Am Soc Clin Oncol 2: 42

Zirvi KA, Hill ZH, Hill GJ (1983) Comparative studies of chemotherapy of human tumor cells in vitro by thymidine uptake inhibition and clonogenic assay. Proc Am Assoc Cancer Res 24: 306

*Neutralization of cis-Dichlorodiammineplatinum II and Nitrogen Mustard by Thiols**

C. Sauter, M. Cogoli, S. Arrenbrecht, and C. Marti

Universitätsspital Zürich, Departement für Innere Medizin, Abteilung für Onkologie, Rämistrasse 100, 8091 Zürich, Switzerland

Introduction

Alkylating substances were the first and are still among the most important agents used in the treatment of human cancer. The mechanism of action of alkylating substances in vivo is still a matter of debate (Wheeler 1982). DNA has been considered to be the primary target for the cytotoxic action of these drugs on susceptible cells. The DNA alkylation hypothesis is attractive on structural grounds because alkylating agents bind to the nitrogen-7 position of guanine, cross-linking DNA strands, and thus inhibiting DNA replication and cell division (Wheeler 1962). There are reasons, however, to question the importance of this mechanism in vivo, since binding sites of other molecules important for cell survival are readily available, as for instance sulfhydryl-containing molecules.

After an intravenous injection of the highly reactive agent nitrogen mustard (HN2) it is unlikely that the drug arrives unaltered at the nuclear DNA of dividing cells to exert its alkylating action only then. En route from the injection site to the DNA of, for instance, the bone marrow cells, the drug encounters many molecules essential for cell metabolism that might also be inactivated by alkylation. One such group of molecules is the sulfhydryl-containing molecules, which are known to interact with alkylating substances (Connors 1966).

In order to determine the capacity of different molecules to interact with alkylating agents, an assay was developed to measure the loss of cytotoxic activity of alkylating substances when exposed to various compounds. As representative and clinically important alkylating agents, HN2 and *cis*-dichlorodiammineplatinum II (*cis*-DDP-II) were studied.

Materials and Methods

BT 20 human mammary carcinoma cells (Sauter et al. 1975) or human hypernephroma cells (Groscurth and Kistler 1977) were seeded in sterile flat-bottomed plastic plates (24 wells, diameter 16 mm, Costar, Cambridge, Maryland) in 0.5 ml RPMI 1640 medium supplemented with 8% fetal bovine serum and grown in a 5% CO_2 atmosphere at 37° C. To the cell monolayers which formed after 1 day of incubation, 0.5 ml HN2 (Mustargen;

* This work was supported by the Krebsliga des Kantons Zürich, and the Heinz Kaiser Fonds. We thank Rosmarie Ammann for perfect secretarial assistance

Recent Results in Cancer Research. Vol. 94

Merck, Sharp & Dohme, West Point, Pennsylvania) or *cis*-DDP-II (Platinol, Bristol-Myers, Syracuse, New York) diluted in medium was added at different concentrations and incubated for an additional 18–24 h. Subsequent to examination for a cytopathogenic effect (CPE) by phase constrast microscopy, the monolayers were stained by methylene blue-parafuchsin (Kistler and Bischoff 1962). Dead cells were removed from the wells by the staining procedure. The definitive evaluation of the CPE was done on the stained plates. This evaluation always corresponded to the phase contrast results. BT 20 and hypernephroma cells were equally sensitive to the cytopathogenic action of the two alkylating agents.

The procedure of the neutralization assay was the same except that a constant amount of HN2 final concentration 0.32 μmol/ml) or *cis*-DDP-II (final concentration 0.07 μmol/ml) was mixed with different amounts of thiols, purines, pyrimidine, or DNA 5–10 min prior to addition to the monolayers. Final volume per well was 1.0 ml.

The substances used were obtained from the following sources: glutathione oxidized or reduced, crystallized; coenzyme A grade I lyophilized (Böhringer, Mannheim, FRG); α-thioglycerol; 2,6,8-trihydroxypurine; guanosine 5′-triphosphate (Fluka, Buchs, Switzerland); *N*-acetyl-L-cysteine (Impharzam, Cadempino, Switzerland); sodium-2-mercapto-ethane sulfonate (Asta, Bielefeld, FRG); 6-mercaptopurine; 6-thioguanine (Wellcome, London, England); guanylyl (3′, 5′) guanosine; DNA (degraded free acid type IV from herring sperm) (Sigma, St. Louis, Missouri); cytosine arabinoside (Upjohn, Kalamazoo, Michigan).

Results

In Table 1 the concentrations of HN2 and *cis*-DDP-II are shown which produced either complete or no CPE in our assay. Intermediate concentrations showed appropriate dose-response characteristics.

The neutralizing capacity of different molecules (thiols, purines, and pyrimidines) on HN2 and *cis*-DDP-II was investigated. The concentrations of the different substances at which the CPE of HN2 and *cis*-DDP-II were completely or just detectable neutralized are given in Table 2. Figure 1 illustrates such a neutralization assay. The highest concentrations of the substances studied where no neutralization could be observed are shown in Table 3.

In further experiments hypernephroma cell monolayers were incubated with 0.32 μmol/ml HN2. After various periods of incubation with the drug, the medium was aspirated and the cell sheets were rinsed twice with 2.0 ml fresh medium; 1.0 ml fresh medium was added and the incubation was continued. Evaluation of these plates after 24 h showed that a minimum contact time of 30 min with HN2 was necessary for the production of CPE in hypernephroma cells. The addition of thiols during these 30 min could neutralize the HN2.

Table 1. Cytopathogenic effect (CPE) of nitrogen mustard (HN2) and *cis*-dichlorodiammineplatinum II (*cis*-DDP-II)

	Complete CPE	No CPE
HN2 (μmol/ml)	0.13	0.03
cis-DDP-II (μmol/ml)	0.07	0.02

Table 2. Neutralization of HN2 and *cis*-DDP-II by different thiols

By	Neutralization of			
	HN2[a]		*cis*-DDP-II[b]	
	Complete	Initial	Complete	Initial
Glutathione (reduced)	1.63	0.32	3.25	0.32
Coenzyme A	1.3	0.32	Not done	
α-Thioglycerol	1.5	0.25	3.0	0.1
N-acetyl-L-cysteine	3.06	0.61	3.06	0.31
Sodium-2-mercaptoethane sulfonate	3.02	0.3	1.5	0.3
6-Mercaptopurine	1.6	0.32	1.6	0.32
6-Thioguanine	0.6	0.06	0.6	0.06

The figures represent μmol/ml which completely or just detectably (see Fig. 1a 6) neutralize HN2 or *cis*-DDP-II

[a] 0.32 μmol/ml

[b] 0.07 μmol/ml

Table 3. Failure of neutralization by oxidized glutathione, purines without SH groups, pyrimidine, or DNA

By	Failure of neutralization of	
	HN2[a]	*cis*-DDP-II[b]
Glutathione (oxidized)	3.3	3.3
2,6,8-trihydrooxypurine	11.9	11.9
Guanosine 5′-triphosphate	1.65	1.65
Guanylyl (3′ → 5′) guanosine	0.77	0.77
Cytosine arabinoside	4.1	4.1
DNA	(2,500 μg/ml)	(2,500 μg/ml)

Highest concentration (μmol/ml) tested

[a] 0.32 μmol/ml

[b] 0.07 μmol/ml

Under the conditions of our assay, among the range of molecules tested, only thiols were capable of neutralizing the two representative alkylating agents used, i.e., HN2 and *cis*-DDP-II. Guanine and cytosine without SH groups, or whole DNA molecules, had no-such neutralizing properties.

Discussion

The results of our neutralization assays were somewhat surprising, since the generally accepted mechanism of antitumor action of HN2 and *cis*-DDP-II postulates a binding to the nitrogen-7 of guanine and, to a lesser extent, to the nitrogen-3 of cytosine (Wheeler 1982). Although interactions of alkylating agents with thiols have been described (Connors 1966)

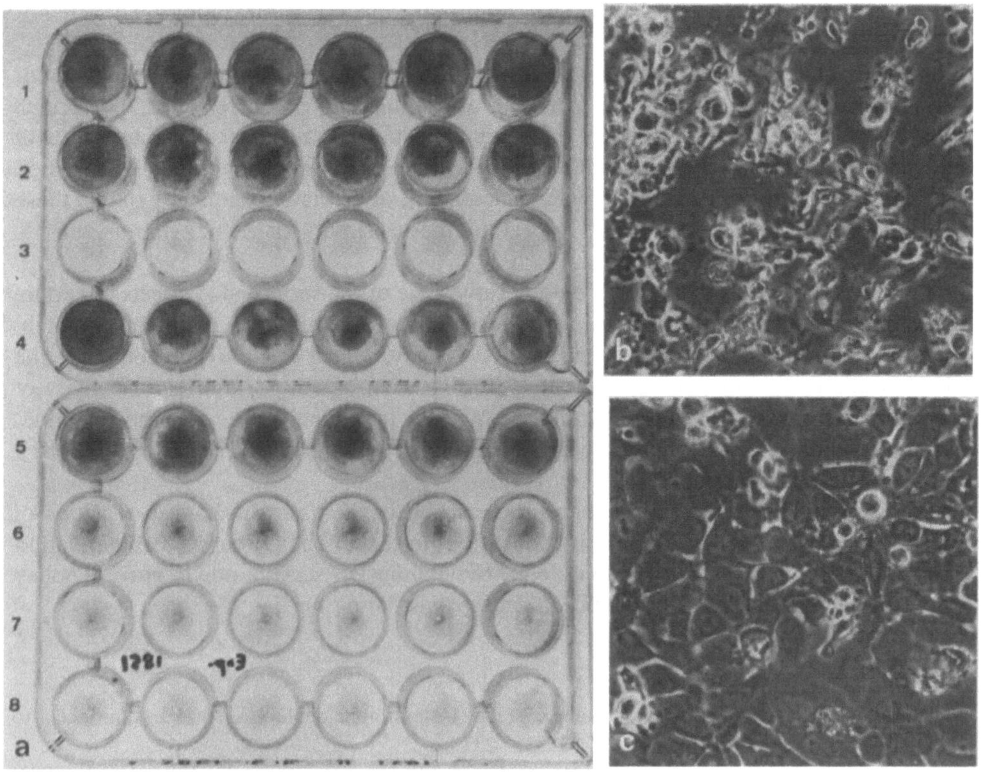

Fig. 1a–c. Neutralization assay of HN2 or *cis*-DDP-II by thiols. **a** BT 20 monolayers. Stained 24 h after addition of HN2 (0.32 μmol/ml) with or without reduced glutathione. *1,* Medium control; *2,* reduced glutathione (GSH) 3.25 μmol/ml; *3,* HN_2; *4,* HN_2 + GSH 3.25 μmol/ml *5,* HN_2 + GSH 1.63 μmol/ml; *6,* HN_2 + GSH 0.32 μmol/ml; *7,* HN_2 + GSH 0.16 μmol/ml; *8,* HN_2 + GSH 0.03 μmol/ml. Note the increasing cytopathogenic effect with decreasing reduced glutathione concentrations. **b** BT 20 cells, phase contrast, × 300; 22 h after addition of 0.07 μmol/ml *cis*-DDP-II. Almost complete destruction of the monolayer. **c** BT 20 cells, phase contrast, × 300; 22 h after addition of 0.07 μmol/ml *cis*-DDP-II plus 6 μmol/ml sodium-2-mercaptoethane sulfonate. Intact BT 20 monolayer

they have mainly been regarded as a detoxification mechanism (Wheeler 1982). In this report we show by neutralization experiments that the avidity of HN2 and *cis*-DDP-II for thiols is more pronounced than their avidity for guanine or cytosine. In general, initial neutralization occurred with 1 mol thiol for 1 mol HN2 and with 4–5 mol thiol for 1 mol *cis*-DDP-II (Table 2). The higher activities of α-thioglycerol and 6-thioguanine are difficult to explain, but might be related to their liposolubility.

In the light of these results three issues warrant discussion:

1. Mechanism of action of alkylating substances. Our experiments suggest that thiols, not nucleotides, are the primary site of alkylation. By alkylation of essential biothiols (e.g., reduced glutathione) the tumor cell replication might be hampered, since tumor cells are often sulfhydryl dependent (Toohey 1975). The antioxidant defenses are reduced or abolished since the oxidation – reduction cycle of glutathione serves as one of the major defense mechanisms of tumor cells against oxidant injury (Nathan et al. 1981). Inhibition of the glutathione redox cycle enhances macrophage – mediated cytolysis by enzymatically

generated H_2O_2 (Arrick et al. 1982). This could also explain why alkylating agents are not cell cycle specific.

2. Predictive tests for cancer chemotherapy. Quantitation of sensitivity of human tumor cells in vitro to anticancer drugs, including alkylating agents, is frequently performed (Salmon et al. 1978). In these assays thiols are almost always necessary for tumor growth and they are therefore routinely added to the culture medium (Hamburger and Salmon 1977). In the light of our results showing inactivation of alkylating agents by thiols, these assays should be reevaluated with respect to the sensitivity of human tumor cells to alkylating agents in vitro.

3. Application of thiols in cancer chemotherapy. One of the serious side effects of certain alkylating agents (isophosphamide, cyclophosphamide is urotoxicity. Acrolein, a metabolite of these drugs, is responsible for this toxicity, which can be prevented by the administration of thiols (Berrigan et al. 1982). To what extent the antitumor activity of alkylating agents is compromised by thiols, expecially in regimens where isophosphamide, *cis*-DDP-II, and sodium-2-mercaptoethane sulfonate are combined (Berdel et al. 1982), must be carefully evaluated.

The present findings may thus be relevant to aspects of the basic mechanism of action of alkylating agents as well as to clinical applications in cancer chemotherapy. Manipulation of essential biothiols, as can be done, e.g., by vitamin B_{12} compounds in vitro (Toohey 1975), may increase the therapeutic efficacy of alkylating agents.

References

Arrick BA, Nathan CF, Griffith OW, Cohn ZA (1982) Glutathione depletion sensitizes tumor cells to oxidative cytolysis. J Biol Chem 257: 1231–1237

Berdel WE, Fink U, Emmerich B, Maubach PA, Busch U, Remy W, Rastetter J (1982) Chemotherapie maligner Melanome mit *cis*-Diaminodichloroplatinum und Ifosfamid. Dtsch Med Wochenschr 104: 26–28

Berrigan MJ, Marinello AJ, Pavelic Z, Williams CJ, Struck RF, Gurtoo HL (1982) Protective role of thiols in cyclophosphamide-induced urotoxicity and depression of hepatic drug metabolism. Cancer Res 42: 3688–3695

Connors TA (1966) Protection against the toxicity of alkylating agents by thiols: the mechanism of protection and its relevance to cancer chemotherapy. Eur J Cancer 2: 293–305

Groscurth P, Kistler GS (1977) Human renal cell carcinoma in the nude mouse: long-term observations. Beitr Path 160: 337–360

Hamburger AW, Salmon SE (1977) Primary bioassay of human tumor stem cells. Science 197: 461–463

Kistler GS, Bischoff A (1962) Zur exfoliativen Zytologie kleiner Flüssigkeitsmengen. Schweiz Med Wochenschr 92: 863–865

Nahan CF, Arrick BA, Murray HW, DeSantis NM, Cohn ZA (1981) Tumor cell anti-oxidant defenses. J Exp Med 153: 766–782

Salmon SE, Hamburger AW, Soehnlen B, Durie BGM, Alberts DS, Moon TE (1978) Quantitation of differential sensitivity of human-tumor stem cells to anticancer drugs. N Engl J Med 298: 1321–1327

Sauter Chr, Bächi T, Lindenmann J (1975) Huamn mammary carcinoma cell line: infection by an avian myxovirus as a prerequisite for immunopotentiation. Eur J Cancer 11: 59–63

Toohey JI (1975) Sulfhydryl dependence in primary explant hematopoietic cells. Inhibition of growth in vitro with vitamin B 12 compounds. Proc Natl Acad Sci USA 72: 73–77

Wheeler GP (1962) Studies related to the mechanism of action of cytotoxic alkylating agents: a review. Cancer Res 22: 651–688

Wheeler GP (1982) Alkylating substances. In Holland JF, Frei E III (eds) Cancer medicine. Lea and Febiger, Philadelphia, pp 824–843

Usefulness of the Human Tumor Colony Forming Assay for New Drug Development

B.F. Issell, J.J. Catino, and E.C. Bradley*

Cetus Corporation, 1400 Fifty-Third Street, Emeryville, CA 94608, USA

The human tumor colony forming, or "clonogenic', assay is an in vitro primary culture system of human tumors. It was initially developed for predicting which chemotherapeutic drug would give the best response for individual cancer patients in a manner analogous to in vitro antibiotic testing, and encouraging in vitro-in vivo correlations have been derived from retrospective trials. Overall approximate correlation rates in a broad spectrum of solid tumors and myeloma of 70% true positive and 90% true negative have been obtained in over 1,600 in vitro-in vivo trials reported in 40 published summaries from separate laboratories.

After it was shown that in vitro response correlated very well with in vivo results for a large number of tumor types and chemotherapeutic agents, the obvious question to be answered next was whether the system could be used to screen for novel anticancer drugs. Historically, the identification of clinically useful anticancer drugs has been an extremely inefficient process. Between the years 1965 and 1975, only 30–40 clinically useful anticancer drugs were identified from the 325,000 synthetic and naturally occurring compounds which were initially tested (Von Hoff et al. 1977). A further concern is that our present systems may allow compounds with useful activity in man to be discarded because of poor activity in animal tumors.

This paper will examine the potential usefulness of the colony-forming assay in new drug development. Limitations will be addressed and methods for reducing these limitations will also be outlined. A study aimed at validating the assay in analog development will be presented, and interim study results available to date will be discussed.

Limitations of In Vitro Systems

Any in vitro system suffers limitation when considered as a "stand-alone" test in new drug development. These limitations and methods aimed at reducing them are listed in Table 1.

* The authors acknowledge the excellent assistance of Mrs. Judi Schurig in manuscript preparation

Table 1. Limitations of in vitro assays in drug development

Limitation	Possible solution
Differentiation of nonspecific cytotoxicity from specific antitumor effect	Integrate with in vivo animal system Integrate with human CFU_c assay for myelosuppressive drugs
Absence of relevant in vivo metabolic effects	Incubate with liver microsomes Test known and putative metabolites where available Compare with in vivo animal system
Difficulty of determining appropriate drug concentration and exposure	Test range of concentration Determine relative tolerable potencies by using animal in vivo and CFU_c systems Assess different exposure times using characterized cell line bioassays
Possible drug instability under storage and assay incubation	Determine stability using chemical or bioassays and modify conditions accordingly

Nonspecific Cytotoxicity

A cytotoxic effect observed in tumor cell cultures may be a manifestation of toxicity toward both tumor and normal cells and not reflect specific antitumor activity. For example, ethanol at concentrations of greater than 1% will inhibit colony formation in vitro of most tumor cell strains, but it is obvious that this does not represent biologically relevant antitumor activity. To determine the potential therapeutic index for a new drug requires in vivo testing or in vitro methods of comparing the relative toxicities of a compound against both normal and tumor tissues. In our program, the clonogenic assay is used alongside tumor-bearing and non-tumor-bearing in vivo models. Furthermore, we are also integrating the normal human bone marrow colony-forming unit (CFU_c) assay into the assessment of compounds presumed to have myelosuppression as their dose-limiting toxicity. We were encouraged to pursue this in vitro assay as an indicator of myelosuppression in man following an impressive correlation between th ranking of anthracyclines by the CFU_c assay and their relative myelosuppressive effects in phase I clinical trials (Issell et al. 1982).

In Vivo Metabolism

The administered form of a drug may be pharmacologically quite different from its metabolized products in vivo. Some compounds, such as cyclophosphamide, require bioactivation, while other compounds may be rapidly changed to inactive metabolites in vivo. For this reason, it is useful to incubate new uncharacterized compounds with and without liver microsomes (S-9 fraction) for in vitro testing. Comparisons with animal systems, such as the subrenal capsular assay developed by Bogden et al. (1979), would also identify compounds in which in vivo metabolism affected antitumor response.

Clinical pharmacologic studies showed that the anthracycline carminomycin was rapidly converted to carminomycinol to such an extent that the parent drug species was probably acting as a prodrug (Comis et al. 1982) In our previous experience with anthracycline development, we had found that the in vitro antitumor and CFU_c activity of carminomycinol were quite different from those of the parent drug, carminomycin (Issell et al. 1982; Tihon and Issell et al. 1982). It is thus important to determine the pharmacokinetic and metabolic profiles of new drugs as soon as possible, and animal work may be helpful in this respect. We now routinely test known and putative metabolites along with the parent drug where this is possible in vitro.

Drug Concentration and Exposure

Tolerable tissue concentrations in man are impossible to define for a compound in early development. For this reason, initially test all new compounds at a wide concentration range of 10^{-4}, 10^{-5}, 10^{-6}, and 10^{-7} *M*. Once animal tolerance data and achievable tissue concentration information are available, then the concentration range can be narrowed. Furthermore, it is important to run cytotoxicity tests at more than one concentration in order to demonstrate a dose-response relationship. We have been able to examine dose-response data by regression analysis to compare relative potencies of two compounds and to identify outliers and influential values in the assay (White et al. 1983). The response at a single concentration may be artifactual and, if seen, would require further verification.

The duration of drug exposure may be a critical determinant of cytoxicity for compounds which are highly phase specific, and a 1-h drug incubation time may not be sufficient to fully demonstrate antitumor activity for some compounds. Experiments comparing 1-h to longer drug exposure times have been run for a number of drugs and some show marked enhancement of colony inhibition with longer exposure times. However, the inherent variability of the primary culture assay and the inability to repeat each individual experiment make it difficult to interpret comparisons. It is our practice to test various incubation times for each new drug against established cell lines. If, in moderately sensitive cell lines, we are unable to show a prolonged incubation advantage in excess of that anticipated on a concentration x time basis, then the new drug is tested by 1 h incubation.

Drug Assay Stability

The stability of a compound under storage and assay incubation conditions is of critical importance in determining the cytotoxicity of any new drug. It is often impossible to have a chemical assay available for a drug early in its development, and bioassays using sensitive microorganisms or cell lines are useful in assuring biological stability. There are difficulties in determining the stability profile of a compound after it has been added to the complex soft agar mixture. We have, therefore, elected to incubate tumor cells with drug for a defined period prior to plating rather than adding drug directly to the soft agar-tumor cell plate.

Our experience with cisplatin in the colony-forming assay reinforces the need to evaluate storage conditions. Our initial practice was to solubilize and store cisplatin in amino-acid-containing medium which was ready for tumor cell addition. However, the binding of cisplatin to sulfhydryl-containing amino acids such as cysteine and methionine resulted in inactivity in the assay. This was corrected by preparing cisplatin in normal saline solution.

Limitations of Primary Culture Systems

In addition to the limitations inherent to any in vitro system, the colony-forming assay has further important limitations by virtue of its primary culture nature. These limitations and methods proposed for their reduction are outlined in Table 2.

Inability to Reproduce Individual Assay

The technical problems associated with isolating colonies which have grown in the soft agar matrix and successfully repassaging them make this a one-time assay. Because of the inherent variability of this biological system, multiple control and drug-treated plates are required to assure the validity of each assay end point.

In order to have confidence in the validity of each primary culture assay, we run established cell lines with known chemosensitivity profiles in parallel with the primary culture. In addition, we compare the new drugs to standard reference compounds with well-characterized activity profiles. Analogs are always tested along with their parent compound.

Cryopreservation may also allow sample retesting under some conditions. Salmon and co-workers found encouraging growth and chemosensitivity correlations when they compared cryopreserved alliquots with the previous fresh tumor results in a preliminary analysis (Salmon et al. 1981)

Selective Cell Growth by Assay

A theoretical advantage of a primary culture system over established cell lines is that there should be less selection in primary culture, since only one passage from the original tumor cell disaggregate has occurred. However, considerable selection does occur, since the cloning efficiency is usually $< 1\%$. Furthermore, only a proportion (20%–50%) of all tumor specimens received give rise to sufficient numbers of colony-forming cells in enough

Table 2. Limitations of primary culture systems

Limitation	Possible solution
Inhability to reproduce individual assay	Integrate established cell line bioassays as quality controls
	Cryopreservation may allow sample retesting under some conditions
Cell growth selection by assay	Increase plating efficiency by technical improvements and specific tumor growth factors and media

plates to allow the meaningful evaluation of an assay. Carney et al. (1981) and Mackillop et al. (1983) have provided evidence that the colony-forming cells do, in fact, represent an important component of the cells responsible for tumor growth in vivo and that many are truly stem cells. However, many examples of clonal heterogeneity of tumors have been found. This raises the question whether the cells growing in the assay always represent the cells causing uncontrolled tumor proliferation in vivo. Clonal selection may explain why there appears to be an apparent chemosensitivity overprediction for malignant melanoma, a tumor in which demonstrable clonal heterogeneity to drug response is often found (Salmon et al. 1980).

Methods to improve the cloning efficiency and harvest a greater proportion of clonogenic cells from each tumor specimen are being actively pursued in our laboratory. These include the addition of specific tumor growth factors and other technical improvements (Bradley et al. 1983; Tihon et al. 1981).

Study Design for Testing the Colony-Forming Assay in Drug Development

There are several methods of assessing the usefulness of the colony-forming assay in drug development. However, the true worth of a specific method will not be known until drugs which have been selected according to assay results have been fully profiled in terms of clinical activity. With present technology it is logistically impossible to use this system for initial identification of antitumor activity in crude fermentation broths and chemical mixtures. Validation is therefore confined mainly to the assay being used at a later preclinical stage of drug development. The National Cancer Institute has contracted several institutions to investigate the role of this assay in the identification of new drug types. They have also evaluated its ability to identify compounds with known clinical usefulness, and the results so far presented have been largely encouraging (Shoemaker et al. 1981, 1983) Because of the complexities and time involved with this approach, we have elected to focus our efforts on analog development. Our present study is aimed at validating the colony-forming assay as a method of selecting a mitomycin C analog with better clinical activity.

The experimental design we are using closely corresponds to a phase II clinical study where drug activity is assessed against numerous specimens of each tumor type so that response rates can be derived. The mitomycin analog which has the best profile in the animal tumor models will be undergoing clinical study regardless of the results of the clonogenic assay, so validation will be feasible. In addition, several back-up compounds will also be tested in vitro, as well as an analog which has toxicity but no demonstrable antitumor activity (P388 leukemia and B-16 melanoma) in mice.

Assay End Points

End points of cytotoxicity for each assay are arbitrarily defined according to whether IC_{50} and IC_{70} effects are or are not achieved at each drug concentration. A 70% colony number reduction due to drug treatment in the assay (IC_{70}) has been most commonly compared to a 50% or greater measured tumor regression in patients with encouraging correlations (Von Hoff et al. 1981). However, a 50% colony number reduction (IC_{50}) may also correlate with a beneficial antitumor effect in patients, and we are therefore assessing both IC_{70} and IC_{50} and points in vitro.

Determining Appropriate Drug Concentrations

Clinical pharmacologic studies often allow an assessment of achievable tissue concentrations in vivo for drugs in clinical development. When using the assay to predict which clinically established drug may have the best effect in an individual patient, assay drug concentrations approximating 10% of in vivo peak plasma or concentrations derived from the areas under concentration by time curves have been used. However, for drugs in preclinical development, there is no knowledge of achievable human plasma or tissue concentrations. Comparison with clinically effective drugs is achieved by estimating tolerable potency differences at concentration ranges which seem clinically appropriate.

If a new mitomycin analog appeared to be ten times more potent than mitomycin C, for example, then the number of mitomycin C assays achieving IC_{70} at 10^{-6} *M* would be compared to the new analog at 10^{-7} *M*. Clinical pharmacologic data suggest that 10^{-6} *M* is probably near an appropriate concentration range for mitomycin C. In addition, the results of CFU_c human bone marrow testing and animal tissue pharmacologic data will also be used to determine relative achievable tissue concentrations.

Criteria for Decision Making in Drug Development

In a way analogous to the assessment of phase II clinical study results, response rates using both IC_{50} and IC_{70} criteria will be made after 20 comparable assays have accrued in each tumor type. A comparison is considered valid only when there are data for both analog and parent compound within the same assay at a given drug concentration. A superiority for one compound over another at a specific drug concentration will be assumed if statistically derived confidence limits are nonoverlapping (e.g., 15/20 compared to 5/20 positive assays for colorectal cancer at 10^{-6} *M*). If the analog is 10 times more toxic in mice, for example, then a superiority would be considered if there were 15/20 positive assays at 10^{-7} *M* for the analog compared to 5/20 positive assays at 10^{-6} *M* for the parent compound.

When the statistically derived confidence limits are close but still overlapping, then the tumor denominator will be increased to 30 in the hope of achieving nonoverlapping limits. If there is still no statistically significant difference after 30 tumor tests, then it will be considered that no significant advantage of one compound over another can be demonstrated for that specific tumor type.

Interim Study Results

An example of analog analysis as described above is presented in Table 3. The IC_{50} and IC_{70} end points are summarized for mitomycin C and an analog (BL-6782) in colorectal specimens. Both the parent drug and the analog were required to have evaluable data for a given concentration in any sample to be tallied in the end point analysis. Altough a decision-making end point (20 evaluable assays) as not been reached, there is a significant difference in the response of colorectal specimens to the two compounds at 1×10^{-5} *M* in vitro. This does not take into consideration any potency difference of the two compounds which may exist, and 1×10^{-5} *M* is possibly an inappropriately high tissue concentration of mitomycin C to compare to a clinically relevant dose and schedule in man.

Table 3. Mitomycin C (MMC) and BL-6782: direct comparison in colorectal cancer

	$1 \times 10^{-7}\,M$	$1 \times 10^{-6}\,M$	$1 \times 10^{-5}\,M$	$1 \times 10^{-4}\,M$
IC_{50}				
MMC	2/10	5/15	4/14[a]	7/9
BL-6782	3/10	7/15	13/14[a]	9/9
IC_{70}				
MMC	0/10	0/15	1/14[a]	5/9
BL-6782	0/10	1/15	8/14[a]	8/9

IC_{50}, 50% colony number reduction; *IC_{70}*, 70% colony number reduction
[a] Statistically significant difference by McNemar's test

Conclusions

The in vitro colony-forming assay may be a valuable method for more accurately and efficiently developing new anticancer therapies. However, its acceptance in this role has not yet been proven. Much work has been reported without many firm conclusions being reached. The full potential of this system can only be assessed after well-controlled studies aimed at clinical validation and addressing the above limitations have been completed.

References

Bogden AE, Haskell PM, LePage DJ, Kelton DE, Cobb WR, Esber HJ (1979) Growth of human tumor xenografts implanted under the renal capsule of normal immunocompetent mice. Exp Cell Biol 47: 281–293

Bradley EC, Catino JJ, Dalton TC, Tihon C, Issell BF (1983) A comparison of tumor specimen disaggregation methods, ascites, and agar v. agarose on plating efficiency in the human tumor chemosensitivity assay. Proc Am Assoc Cancer Res 24: 314

Carney DN, Gazdar AF, Bunn PA Jr (1981) Demonstration of the stem cell nature of clonogenic tumor cells from lung cancer patients. Stem Cells 1: 149–164

Comis R, Issell B, Ginsberg S, Pittman K, Rudolph A, Aust J, Difino S, Tinsley R, Poiesz B, Crooke ST (1982) A phase 1 clinical pharmacology study of intravenously administered carminomycin from the United States. Cancer Res 42: 2944–2948

Issell BF, Ginsberg SJ, Tihon C, Rudolph AR, Comis RL (1982) Combining the in vitro human tumor and bone marrow clonogenic assays in cancer therapy development. Proc Am Soc Clin Oncol 1: 25

Mackillop WJ, Ciampi A, Till JE, Buick RN (1983) A stem cell model of human tumor growth: Implications for tumor cell clonogenic assays. J Natl Cancer Inst 70: 9–16

Salmon SE, Alberts DS, Meyskens FL Jr, Durie BGM, Jones SE, Soehnlen B, Young L, Chen HSG, Moon TE (1980) Clinical correlations of in vitro drug senstivity. In: Salmon SE (ed) Cloning of human tumor stem cells. Liss, New York, pp 223–245

Salmon SE, Liu RM, Casazza AM (1981) Evaluation of new anthracycline analogs with the human tumor stem cell assay. Cancer Chemother Pharmacol 6: 103–110

Shoemaker RH, Wolpert-DeFilippes MK, Makuch RW, Venditti JM (1981) Application of the human tumor clonogenic assay to drug screening. Stem Cells 1: 308

Shoemaker RH, Wolpert-DeFilippes MK, Makuch RW, Venditti JM (1981) Use of the human tumor clonogenic assay for new drug screening. Proc Am Assoc Cancer Res 24: 311

Tihon C, Issell BF (1982) Activity of anthracyclines on human tumors in vitro. In: Muggia FM, Young CW, Carter SK (eds) Anthracycline antibiotics in cancer therapy. Nijhoff, The Hague, pp 247–253

Tihon C, Catino JJ, Issell BF (1981) Methodology development for human tumor cloning assay. Stem Cells 1: 314–315

Von Hoff DD, Rozencweig M, Soper WM, Helman LJ, Penta JS, Davis HL, Muggia FM (1977) Whatever happened to NSC – An analysis of clinical results of discontinued anticancer agents. Cancer Treat Rep 61: 759–768

Von Hoff DD, Casper J, Bradley E, Sandbach J, Jones D, Makuch R (1981) Association between human tumor colony-forming assay results and response of an individual patient's tumor to chemotherapy. Am J Med 70: 1027–1032

White JM, Catino JJ, Issell BF (1983) Statistical analysis for a drug screening program utilizing the human tumor cloning assay (HTCA). Proc Am Assoc Can Res 24: 1244

In Vitro Characterization of New Antiestrogens in Human Mammary Tumor Cells

U. Eppenberger, W. Küng, R. Löser, and W. Roos

Kantonsspital Basel, Laboratorien Frauenklinik, Schanzenstrasse 46,
4031 Basel, Switzerland

Introduction

Nonsteroidal antiestrogens, whose chemical structures are based on or closely related to triphenylethylene, appear to have considerable potential for the treatment of hormone-dependent breast cancer. Especially Tamoxifen [*trans*-1-(4-β-dimethylaminoethoxyphenyl)-1.2-diphenyl but-1-ene] is now used routinely for the treatment of advanced breast cancer in women (Smith et al. 1981). Therapeutic response to Tamoxifen is correlated with the presence of estrogen receptors in mammary tumors (McGuire et al. 1978) In addition, these nonsteroidal compounds have successfully been used to suppress the growth of human breast cancer cell lines containing estrogen receptors (Lippman et al. 1976) and to elicit the regression of hormone-dependent mammary tumors in experimental animals (Rorke and Katzenellenbogen 1981). Although it is known that antiestrogens bind to the estrogen receptor in the respective target cells (Horwitz and McGuire 1978), the precise mechanism of antiestrogen action is not fully understood.

This study describes the antiestrogenic properties of the following new synthetic antiestrogens which could possibly contribute to an improved treatment of hormone-dependent tumors:

K060, *trans*-1-[4′-(2-di-methylaminoethoxy)-phenyl]-1-1-(3′-hydroxyphenyl)-2-phenyl-1-buten

K089, *trans*-1-[4′-(2-diethylaminoethoxy)-phenyl]-1-(3′-hydroxyphenyl)-2-phenyl-1-buten

K106, *trans*-1-[4′-(2-methylaminoethoxy)-phenyl]-1-(3′-hydroxyphenyl)-2-phenyl-1-buten

K122, *trans*-1-[4′-(2-dimethylaminoethoxy)-phenyl]-1-(3′-hydroxyphenyl)-2-(*p*-tolyl)-1-buten

K135, *trans*-1-[4′-(2-dimethylaminoethoxy)-phenyl]-1-(3′-hydroxyphenyl)-2-(4′-methoxyphenyl)-1-buten.

The experiments were performed in vitro using MCF-7 cells, a human breast cancer cell line originally isolated from metastatic breast cancer by Soule et al. (1973). This cell line is of particular interest, since it contains receptors for several hormones, including estrogen and progesterone (Horwitz et al. 1975). Estrogen treatment leads to an increased proliferation rate and to the induction of the progesterone receptor (Roos et al. 1982) and specific proteins (Lippman and Bolan 1975). In the present study the competitive action of

* This work was supported by the Swiss National Science Foundation Grant Nr. 3.557.0.79 and in part by Klinge Pharma GmbH & Co., Federal Republic of Germany

several antiestrogens on the binding of 17β-estradiol to its receptor is measured in the cytosol of MCF-7 cells and human mammary carcinomas. The results obtained are correlated with the effect of these compounds on the proliferation of MCF-7 cells.

Materials and Methods

MCF-7 cells (passage 123) were obtained from the Mason Research Institute (Rockville, Maryland). The cells were cultured in improved minimum essential medium-zinc option (IMEM-ZO) (Richter et al. 1972) containing 50 IU penicillin per milliliter, 50 μg streptomycin per milliliter, 60 μg tyclocin per milliliter, 50 μg gentamycin per milliliter, 0.2 μg insulin per milliliter, and 10% (vol: vol) of heat-inactivated fetal calf serum (FCS). Medium change was carried out every 2 days. No mycoplasma contamination was found. Dextran-charcoal-treated (DC-) FCS was prepared according to Armelin et al. (1974 as modified by Roos et al. 1983). Efficient removal of 17β-estradiol was confirmed by radioimmunoassay.

Binding Studies

Cells were grown in 75-cm^2 Falcon tissue culture flasks. At confluence ($\approx 2 \times 10^7$ cells per flask), medium was replaced by IMEM-ZO containing 5% DC-FCS to withdraw endogenous hormones. Cells were harvested 24 h later by trypsinization and washed twice at 4° C in 10 ml Ca^{2+}- and Mg^{2+}-free phosphate-buffered saline (CMF-PBS) per flask by repeated resuspension and centrifugation. Finally, the cells were suspended in TEMG buffer and homogenized with a Dounce homogenizer (40 strokes). The homogenate was centrifuged at 100,000 *g* for 1 h. Human mammary carcinoma cytosol was prepared from tissue biopsy specimens showing estrogen receptor concentrations of more than 100 fmol per milligram protein. The biopsy specimens had been stored frozen at −70° C for up to 6 months. The tissue was pulverized in liquid nitrogen, homogenized in three volumes to TEMG buffer by three 10-s bursts of an Ultra-Turrax, and centrifuged for 1 h at 100,000 *g* as described (Zava et al. 1982). The protein concentration of the supernatants was measured by means of the methods of Lowry et al. (1951) or Kalb and Bernlohr (1977), giving identical results.

Receptor ligands were dissolved in ethanol; the concentration of the stock solutions was controlled by UV spectrophotometry. Dilutions were prepared in TEMG containing 0.5 mg bovine serum albumin per milliliter to reduce adsorption artifacts. Cytosol (100 μl) was added to 20 μl hormone solution containing 1 n*M* 17β-^3H-estradiol (137 Ci/mmol; New England Nuclear Corp.) and increasing amounts of unlabeled competing ligand. Incubation was carried out for 16 h at 4° C. Bound 17β-^3H-estradiol was detected by means of the dextran-charcoal method according to Korenman and Dukes (1970). All experiments were done in triplicate. The relative binding affinity is defined as the ratio of the concentration of radioinert 17β-estradiol to the competitor which is necessary to achieve a 50% inhibition of the specific 17β-^3H-estradiol binding. Bound radioactivity at the highest concentration of radioinert 17β-estradiol (111 n*M*) was taken as nonspecific binding and subtracted from all values. Fifty percent inhibition was calculated from linear regression lines of the logit-log plot.

Growth Experiments

MCF-7 cells were plated at a density of 5×10^4 cells per well into 16-mm multidishes with 24 wells (Nunc) containing 0.5 ml medium. Two to four days later, the cells were incubated twice for 2 h in serum-free medium (1 ml per well) to reduce endogenous steroid levels, then incubated in media containing antiestrogens $\pm 10^{-10}$ *M* 17β-estradiol, which were changed every 2 days. Media containing final concentrations of 10^{-6} to 10^{-9} *M* of the ligands were prepared from 1 m*M* stock solutions in ethanol. Ethanol concentration did not exceed 0.1%, an amount that does not affect the growth of MCF-7 cells. After 5 days, cells were trypsinized and enumerated in a Sysmex CC 108 cell counter. Thymidine incorporation was measured by the addition of 0.1 μCi ^3H-thymidine (20 Ci/mmol, New England Nuclear Corp.) to the medium. After 2 h of incubation (during which time incorporation was linear), the cells were washed twice in cold CMF-PBS, three times in 5% trichloroacetic acid, and once in ethanol. The dry trichloroacetic acid insoluble material was then dissolved in 1 ml 1 *N* NaOH, neutralized with 1 ml 1 *N* HCl, and counted in 8 ml Instagel (Packard Instrument Co.).

Progesterone Receptor Assay

Cell cultures were grown to a density of 2×10^7 cells per 75-cm² flask. After two washes in serum-free medium, as described above for the growth experiments, the medium was changed to 5% DC-FCS with the respective concentrations of estradiol. Cytosol was prepared as described above, except that 10 m*M* sodium molybdate was included in the buffer to stabilize the receptor (Nin et al. 1981). A dextran-charcoal single-point assay was performed with 5 n*M* ^3H-R5020[8] (87 Ci/mmol; New England Nuclear Corp.) and 500 n*M* radioinert progesterone to measure nonspecific binding. All other conditions were the same as in the estrogen receptor binding studies described before.

52 KD Protein

This estrogen-induced protein was assayed as previously described by Westley and Rochefort (1980).

Results

The dependency of the MCF-7 cells on 17β-estradiol for proliferation makes it a useful model for the study of the action of antiestrogens. After withdrawal of endogenous steroids by appropriate medium conditions (DC-FCS), the growth of these cells can be stimulated more than threefold by physiological doses of 17β-estradiol (10^{-10}*M*) as demonstrated in Fig. 1B. This can be confirmed by ^3H-thymidine incorporation experiments (Fig. 1A). A similar dose-response relationship can be shown for the induction of 52 KD protein by 17β-estradiol (Fig. 1C), originally described by Westley and Rochefort (1980). This protein, recently also detected in primary cultures taken from human breast cancer biopsy material (Veith et al. 1983), is secreted into the medium and can be measured by ^{35}S-methionine-labeling followed by SDS-electrophoresis and autoradiography. Another parameter demonstrating the estrogen dependency of the MCF-7 cells is the progesterone

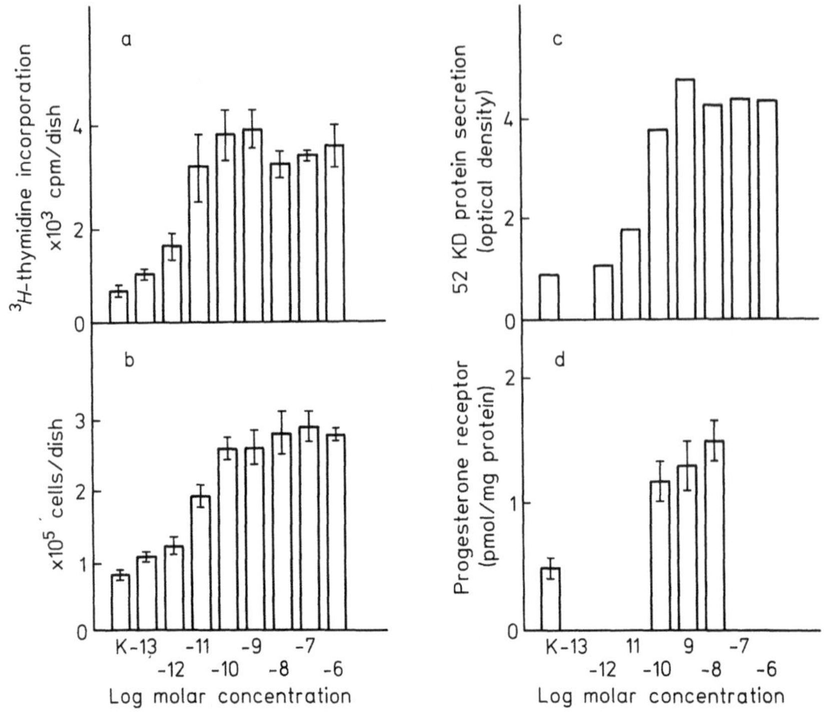

Fig. 1a–d. Dose-response curve of the influence of 17β-estradiol on MCF-7 cells. **a** ^3H-thymidine incorporation after 2 days incubation and **b** cell counts after 4 days incubation in medium containing 5% DC-FCS and the indicated hormone concentrations. Means ± standard deviations of triplicate determinations are shown in both cases. **c** Release of 52 KD estradiol-induced protein: Cells were incubated for 2 days in 5 DC-FCS ± estradiol and labeled with ^{35}S-methionine for 6 h in serum-free medium. The supernatant was analyzed by SDS-electrophoresis, autoradiography, and densitometry. The *ordinate* shows the peak height of the 52 KD protein band in the densitogram. **d** Induction of progesterone receptor, measured after 5 days of incubation. Ninety-five per cent confidential limits of specific binding are shown in the plot

receptor concentration (Fig. 1D), which is increased by 17β-estradiol treatment, a phenomenon well known from in vivo conditions.

In competitive binding experiments MCF-7 cytosol was incubated at 4° C with $10^{-9}M$ 17β-^3H-estradiol and increasing amounts of radioinert antiestrogens (K060, K089, K106, K122, and K135) and Tamoxifen (Fig. 2A). Specific binding of 17β-estradiol was completely inhibited by all compounds tested. The best competitor found was K135 and the weakest Tamoxifen, with intermediate relative binding affinities of K060, K106, K089, and K122 for the estrogen receptor, as summarized in Table 1. The relative binding affinities of K135 was found to be 7.6% of the value obtained for 17β-estradiol. According to these results, K135 is a more than 40-fold better ligand for the estrogen receptor than Tamoxifen, with a relative binding affinity of 0.18% in MCF-7 cytosol (Table 1). The same experiments were repeated with the cytosol of human mammary carcinomas containing estrogen receptor, resulting in a very similar pattern (Fig. 2B). The relative binding affinities of the antiestrogens K060, K106, and K122 as shown in Table 1 were significantly lower than that of the antiestrogen K135, but still much higher than that of Tamoxifen.

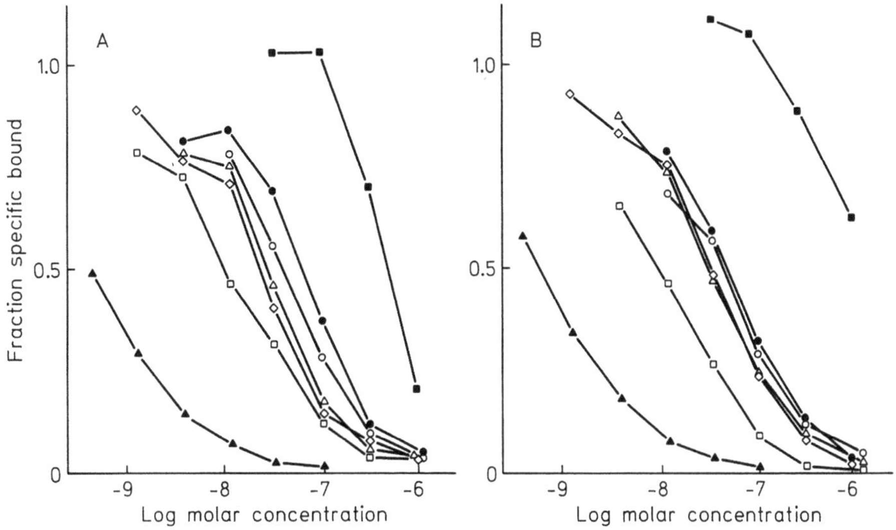

Fig. 2. Competitive binding of antiestrogens to cytosolic estrogen receptor of **A** MCF-7 cells (protein concentration 2.8 mg/l) and **B** human mammary carcinomas (protein concentration 3.7 mg/ml). ▲, 17β-estradiol; ■, Tamoxifen; △, K060; ◇, K089; ○, K106; ●, K122; □, K135. The concentration of ^3H-17β-estradiol was $10^{-9} M$; the *points* represent means of triplicate determinations the standard deviations were generally less than 10%

Table 1. Competitive binding assay of Tamoxifen and the respective antiestrogens (K060, K089, K106, K122, and K135) to the cytosolic estrogen receptors of MCF-7 cells and human mammary carcinomas. MCF-7 cytosol and human mammary carcinoma tissue cytosol were incubated with $1 \times 10^{-9} M$ ^3H-17β-estradiol and the respective radioinert competitor ligand, as shown in Fig. 2. The log concentrations at 50% inhibition were used to calculate the relative binding affinity, which is defined as the ratio of the concentrations of radioinert estradiol to the competitor which are necessary to achieve a 50% inhibition of the specific ^3H-estradiol binding

Antiestrogen	MCF-7 cells			Human mammary carcinomas		
	n	Log concentration at 50% inhibition	Relative binding affinity (%)	*n*	Log concentration at 50% inhibition	Relative binding affinity (%)
Tamoxifen	3	− 6.25 ± 0.26	0.18	4	− 5.87 ± 0.41	0.07
K060	4	− 7.45 ± 0.36	2.82	6	− 7.27 ± 0.19	1.86
K089	2	− 7.54 ± 0.46	3.47	2	− 7.55 ± 0.17	3.54
K106	4	− 7.29 ± 0.29	1.95	6	− 6.98 ± 0.29	0.95
K122	4	− 7.29 ± 0.34	1.95	6	− 7.14 ± 0.28	1.38
K135	4	− 7.88 ± 0.23	7.59	6	− 7.51 ± 0.36	3.23

n, represents the number of experiments carried out

Since MCF-7 cells can be stimulated by physiological concentrations of 17β-estradiol, our approach to test antiestrogenic effects on the proliferation makes use of cell cultures which are stimulated by exogenously added 17β-estradiol ($10^{-10} M$). The ability of potential drugs to reverse such stimulation characterizes their antiestrogenic capacity. The effects

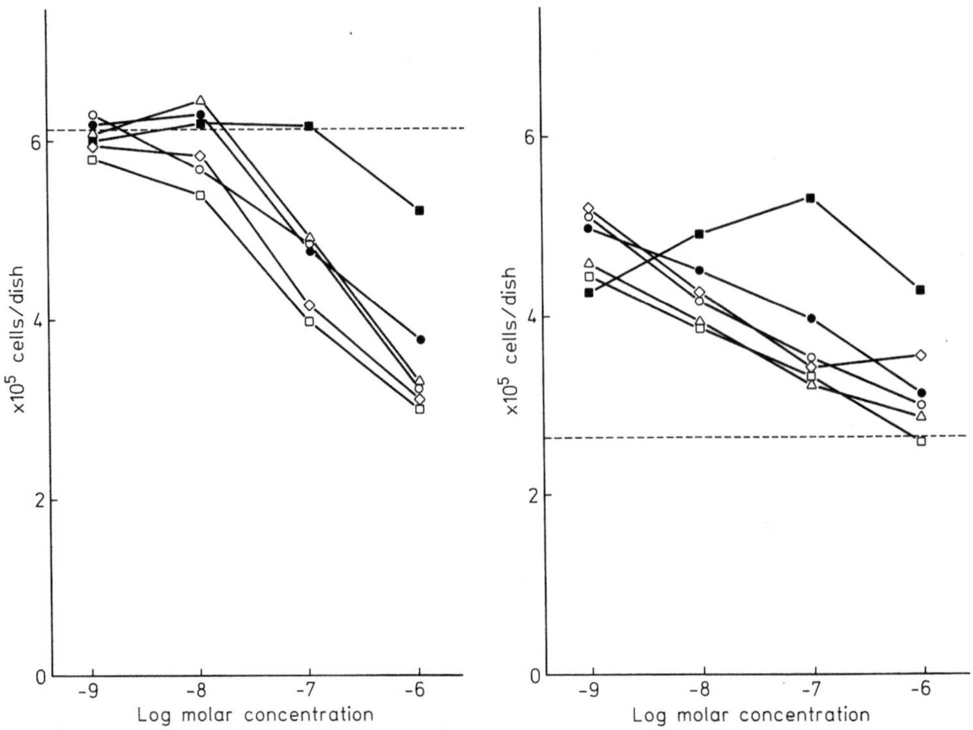

Fig. 3. Dose-dependent effect of antiestrogens on the proliferation of MCF-7 cells **(left)** in the presence of 10^{-10} *M* 17β-estradiol and **(right)** in the absence of 17β-estradiol. The incubation time was 5 days. Control levels measured in the absence of antiestrogens are indicated by *dotted lines*. The *points* are means of triplicate determinations. ▲, 17β-estradiol; ■, Tamoxifen; △, K060; ◇, K089 ○, K106; ●, K122; □, K135

measured in this system are specifically antiestrogenic and are obviously mediated via the estrogen receptor and not by specific antiestrogen-binding sites, which have been described recently by different authors (Sutherland et al. 1980). A comparison of the effects of these newly developed antiestrogens to those of Tamoxifen is presented in Fig. 3A. The growth stimulation elicited by 10^{-10} *M* 17β-estradiol cannot be suppressed by 10^{-7} *M* Tamoxifen. In contrast, 10^{-7} *M* K135 is sufficient to block the growth stimulation by approximately 30%, whereas more than 10^{-6} *M* Tamoxifen is necessary to achieve the same effect. The antiestrogenic compounds K060, K089, K106, and K122 exhibit at 10^{-7} *M* concentration a less pronounced antiproliferative effect than K135 but are still far better than Tamoxifen.

In the absence of exogenously added 17β-estradiol the inhibitory action of these antiestrogens (K060, K089, K106, K122, and K135) and Tamoxifen on the proliferation of MCF-7 cells is completely absent (Fig. 3B), a result which demonstrates that under our experimental conditions cells are on a true basal level with respect to estrogen stimulation of growth. Under these circumstances the partially estrogenic nature of the triphenylethylene derivatives becomes evident. All compounds tested are able to stimulate the growth of the MCF-7 cells. However, the concentrations which are effective are about one order of magnitude lower than those inhibiting the proliferation of 17β-estradiol-stimulated cells (Fig. 3A). Such an agonistic behavior is well known from various animal systems (Cowan and Leake 1979). In MCF-7 cells it has also been shown for the induction of progesterone receptor and 52 KD-induced protein (data not shown).

Discussion

There are strong indications that the development of new and better antiestrogens than Tamoxifen is possible. Frequently, the recurrence of mammary carcinomas after successful therapy with Tamoxifen can be stopped again by means of an additional method of endocrine therapy, such as application of medroxyprogesterone acetate (MPA) or aminoglutethimide. In these cases the hormone dependency of the tumor cells persists and the degree of growth suppression by Tamoxifen seems to be insufficient.

Different authors have demonstrated that the estrogen-dependent MCF-7 cell line can be used to characterize new antiestrogens in vitro (Eckert and Katzenellenbogen 1982; Roos et al., 1983). In the present study, we used this system to investigate the antiestrogenic capacity of the newly developed triphenylethylene derivatives K060, K089, K106, K122, and K135 by their ability to compete for the estrogen receptor and by their inhibitory action on the proliferation of MCF-7 cells. In competitive binding studies all compounds tested proved to be better ligands for the cytosolic estrogen receptor of MCF-7 cells and human breast carcinoma biopsies than Tamoxifen. The inhibitory effect of the compounds, measured as ability to reverse the growth stimulation exerted by 10^{-10} *M* 17β-estradiol, closely reflects their affinity for the estrogen receptor. This is in accordance with a hypothesis introduced by Rochefort and Borgna (1981), postulating a direct correlation between the affinity of an antiestrogen for the estrogen receptor and its efficiency.

In the absence of 17β-estradiol all antiestrogens are able to stimulate the growth of MCF-7 cells at concentrations which are not sufficient to inhibit 17β-estradiol-induced proliferation. Therefore, a complete characterization of such compounds must cover both growth inhibition and agonistic activity. The ratio of the two effects elicited by such compounds at in vivo concentrations seems best to describe the antiestrogenic capacity. Under these aspects the antiestrogens investigated in this study appear to be more effective than Tamoxifen. Further research dealing with toxicological and pharmacokinetic studies of these compounds should reveal whether clinical applications are possible.

References

Armelin HA, Wishikawa K, Sao GH (1974) Control of mammalian cell growth in culture: The action of protein and steroid hormones as effector substances. In: Clarkson B, Baserga R (eds) Control of proliferation in animal cells. Cold Spring Harbor Conferences on Cell Proliferation. vol 1. Cold Spring Harbor Laboratory, New York, pp 97–104

Cowan S, Leake R (1979) The influence of the antiestrogen tamoxifen on DNA synthesis in the rat uterus. In: Agarwal M (ed) Antihormones. Elsevier, Amsterdam, pp 283–292

Eckert LE, Katzenellenbogen BS (1982) Effects of estrogens and antiestrogens on estrogen receptor dynamics and the induction of progesterone receptor in MCF-7 human breast cancer cells. Cancer 42: 139–144

Horwitz KB, McGuire WL (1978) Antiestrogens: Mechanism of action and effects in breast cancer. In: McGuire WL Jr (ed) Breast cancer: advances in research and treatment, vol 2. Plenum, New York, pp 155–204

Horwitz KB, Costlow ME, McGuire WL (1975) A human breast cancer cell line with estrogen, androgen, progesterone and glucocorticoid receptors. Steroids 26: 785–795

Kalb VF, Bernlohr RW (1977) A new spectrophotometric assay for protein in cell extracts. Anal Biochem 82: 362–371

Korenman SG, Dukes BA (1970) Specific estrogen binding by the cytoplasm of human breast carcinoma. J Clin Endocrinol Metab 30: 639–645

Lippman ME, Bolan G (1975) Oestrogen-responsive human breast cancer in long-term tissue culture. Nature 256: 592–593

Lippman ME, Bolan G, Hiff K (1976) The effect of estrogens and antiestrogens on hormone-responsive human breast cancer in long-term tissue culture. Cancer Res 36: 4595–4601

Lowry OH, Rosebrough NJ, Farr AL, Randall RJ (1951) Protein measurement with the Folin phenol reagent. J Biol Chem 193: 265–275

McGuire WL, Zava DT, Horwitz KB, Chamness G (1978) Steroid receptors in breast tumours: current status. Curr Top Exp Endocrinol 3: 93–129

Nin EM, Neal RM, Pierce VK, Sherman MR (1981) Structural similarity of molybdate-stabilized steroid receptors in human breast tumors, uteri and leukolytes. J Steroid Biochem 15: 1–10

Richter A, Sanford KK, Evans VJ (1972) Influence of oxygen and culture media on plating efficiency of some mammalian tissue cells. J Natl Cancer Inst 49: 1705–1712

Rochefort H, Borgna J (1981) Differences between oestrogen receptor activation by oestrogen and anti-oestrogen. Nature 292: 257–259

Roos W, Huber P, Oeze L, Eppenberger U (1982) Hormone dependency and the action of Tamoxifen in human mammary carcinoma cells. Anticancer Res 2: 157–162

Roos W, Oeze L, Löser R, Eppenberger U (1983) Antiestrogenic action of 3-hydroxytamoxifen in the human cancer cell line MCF-7. J Natl Cancer Inst 71: 55–59

Rorke EA, Katzenellenbogen BS (1981) Antitumor activities and estrogen receptor interactions of the metabolites of the antiestrogens C1628 and U23,469 in the 7.12-dimethylbenz(a)anthracene-induced rat mammary tumor system. Cancer Res 41: 1257–1262

Smith IE, Harris AL, Morgan M, Ford HT, Gazel JC, Harmer CL, White H, Parsons CA, Villardo A, Walsh G, McKinna JA (1981) Tamoxifen versus aminoglutethimide in advanced breast carcinoma: a randomized cross-over trial. Br Med J 283: 1432–1434

Soule HD, Vazquez J, Long A, Albert S, Brennan M (1973) A human cell line from a pleural effusion derived from a breast carcinoma. J Natl Cancer Inst 51: 1409–1416

Sutherland RL, Murphy LC, San Foo M, Green MD, Whybourne AM, Krozowski ZS (1980) High affinity antioestrogen-binding site distinct from the oestrogen receptor. Nature 288: 273–275

Veith FO, Capony F, Garcia M, Chantelard J, Pujol H, Veith F, Zajdela A, Rochefort H (1983) Release of estrogen-induced glycoprotein with a molecular weight of 52.000 by breast cancer cells in primary culture. Cancer Res 43: 1861–1868

Westley B, Rochefort H (1980) A secreted glycoprotein induced by estrogen in human breast cancer cell lines. Cell 20: 353–362

Zava DT, Wyler-von Ballmoos A, Goldhirsch A, Roos W, Takahashi A, Eppenberger U, Arrenbrecht S, Martz G, Losa G, Gomez F, Guelpa C (1982) A quality control study to assess the interlaboratory variability of routine estrogen and progesterone receptor assays. Eur J Cancer Clin Oncol 18: 713–721

Modulation of Tumor Growth by Non-Chemotherapeutic Intervention

*Growth Factor Enhancement of the In Vitro Stem Cell Assay**

G. Spitzer**, F. Baker, G. Umbach***, V. Hug, B. Tomasovic, J. Ajani, M. Haynes, and S.K. Sahu

Department of Hematology, The University of Texas System Cancer Center, M.D. Anderson Hospital and Tumor Institute, 6723 Bertner Avenue, Houston, TX 77030, USA

Introduction

In vitro cultures in which clonal growth of cells was sufficient to form visible colonies in semisolid matrix were first developed for human cells several decades ago (Bradley and Metcalf 1966). These cultures were initially used to grow granulocyte macrophage precursors (GM-CFC and under presently defined conditions colonies of several hundred cells are realized after 7–14 days of culture. An absolute requirement for colony development of GM-CFC growth factors (GFs) are CSAs (colony-stimulating activities) (Metcalf 1977). With the identification of other hematopoietic GFs, it has become possible to clone cells of multilineage potentiality and colonies approximating 1,000 cells are formed in culture, and within these colonies are cells with the ability to form secondary colonies (self-renewal) (Johnson and Metcalf 1977). In these systems GFs are an essential requirement unless cells are cultured in high numbers, in which case colony growth can occur secondary to release of GFs from cells plated (Moore et al. 1973). Furthermore, using lower cell numbers in cultures supplemented with hematopoietic GFs, true clonogeneic growth has been demonstrated by the subsequent reculture in agar of single cells appearing early in cultures (Metcalf 1980). These initial observations stimulated investigations into the identity of GF requirements for leukemic and normal bone marrow cells. Studies showed that in almost all instances colony growth of leukemia cells was dependent on GFs, although in some instances growth was initiated at a lower concentration of these factors (Spitzer et al. 1978; Metcalf et al. 1974).
There remain some major technical problems with the supposedly clonogenic assays for stem cell assays of human nonhematopoietic neoplasms. At present, there are no convincing experiments to show that colonies in this system originate from a single cell. Occasional experiments have been graphically presented to show a linear cell dose response, but such responses could also be generated if growth was related to the number of clumps plated. Clonal growth has usually been generated when high numbers of cells

* Supported in part by the NIH grants CA 28153 and CA 14528, LifeTrac Research and Development, Inc. grant JMV:bg 11783, and 174208 Allotment for the Solid Tumor Cloning Laboratory
** Recipient of a scholarship from the Leukemia Society of America
*** Recipient of a postdoctoral fellowship research grant from the Max Kade Foundation, New York

Recent Results in Cancer Research. Vol. 94

have been plated, usually 5×10^5 cells, suggesting that growth may be secondary to multiple interacting cell populations, not just autonomous growth of a tumor clonogenic cell. It has been both our experience and the experience of a number of investigators that growth is substantially compromised when something close to a single-cell suspension is cultured in soft agar after prolonged enzymatic digestion. This failure of colony growth is not associated with a simultaneous decrease in viabilities as revealed by staining with trypan blue (Morgan et al. 1983). Finally, there is the well-known problem of the lack of consistent success with certain tumors, such as head and neck tumors, and a significant percentage of failures with common tumors such as breast and lung cancer.

For this assay to be successful and reproducible and to be accepted by the scientific community, a number of future criteria have to be met. These include colony growth from cell numbers lower than 5×10^5 per dish, growth from real single-cell suspensions, demonstration of clonogenicity, consistent demonstration of self-renewal properties, and reproduction of basic biological principles, such as exponential drug and irradiation survival curves. This, we believe, will require a combination of factors necessary for tumor growth. Such possibilities include: better tissue culture medium with hormonal supplementation nonspecific polypeptide GFs, such as epidermal GF (EGF), fibroblast GF (FGF), and platelet-derived GFs (PDGFs); and tissue-specific polypeptide GFs. Other GFs ill understood at this moment, which may or may not resemble the transforming growth factors (TGFs) described by Todaro and Moses, or components of extracellular matrix, may also be required. Without such potential improvements we will continue to select for only subpopulations of tumor clonogenic cells Hepner and Miller 1983). This manuscript will review the literature on GFs and also the small amount of work which has been done on supplementing primary human tumor clonal cultures in agar. We shall describe our short experience of an attempted systematic approach to medium supplementation with hormones, a search for tumor GFs, and the developement of a tumor extracellular matrix.

Tumor Growth Factors

Todaro (Sporn and Todaro 1980) initially hypothesized that tumors may have a selective growth advantage because the tumors themselves may either secrete locally into the surrounding extracellular space (paracrine secretion) or produce GFs which predominantly remain localized intracellularly (autocrine secretion). Todaro first described the so-called TGFs. These factors were called "transforming" because (a) they induced morphological changes in both rat and mouse fibroblast lines; (b) they allowed cellular growth to higher cell density; (c) they lowered serum requirements in monolayer cultures and (d) they induced anchorage-independent growth of these fibroblasts in agar culture. These factors were acid extractable from a variety of cells that were derived from either viral or chemically transformed cells. Since this initial description, TGFs have been described in platelet-rich serum, embryonic tissue, benign tumors, and normal tissues (Childs et al. 1982; Assoian et al. 1983; Tucker et al. 1983). It is also suggested that the occasional stimulation of anchorage-independent growth by high serum concentrations is really due to platelet-derived factors. Recently, these TGFs have also been described in a number of human tumor cell lines; rhabdomoyosarcoma, another human sarcoma, and a melanoma (Sherwin et al. 1981). These TGFs were subdivided into two types; the first type, which interacts with the EGF receptor, is usually responsible for 90% of the total activity; the second type is independent of the EGF receptor and enhances the growth-promoting activity of EGFs. Those derived from the mouse embryo, like TGFs derived from chemically transformed cell lines, do not bind to the EGF receptor. The effect of these TGFs is reversible, and the number of cycle cell divisions induced is many more cell divisions than that induced by EGFs or FGFs, which usually induce only one or two cell divisions. This ability to induce anchorage-independent growth of these fibroblast lines that are questionably normal is the main criterion for diffentiating TGFs from other growth-inducing hormones, such as EGFs and FGFs. Colony formation is a much more critical evaluation of growth-promoting activity than most systems, which measure only the ability to induce a round of cell division or tritiated thymidine uptake in density – inhibited or serum – restricted heterologous cells. By themselves EGFs, insulin, the somatomedins, melanocyte-stimulating hormone, and highly cationic PDGFs do not stimulate this anchorage-dependent growth of the fibroblast lines AKR-2B and NRK.

Recently, TGFs have been subdivided even further (Nickell et al. 1983). The terms "TGFa", for the type which stimulates ARK-2B cells, and "TGFn" for the type which stimulates NRK cells, have been introduced. The TGFn's are further subdivided into two subtypes: TGF ns, with a molecular weight of less than 6,000 daltons and a larger TGFn called TGF nl, with a molecular weight of 12,000–20,000 daltons. TGFa and TGFnl are derived from both malignant and nonmalignant tissues and the TGFns from benign tumors. All these TGFs are acid extractable, heat stable, and trypsin and dithiothiectol sensitive. It is difficult to evaluate the significance of these factors in human tumor progression and regulation. The significance of this family of acid-extractable molecules with disulfide bonds which stimulate anchorage-independent growth of both murine and rat fibroblastic cell lines is uncertain. Their isolation from such numerous sources as embryonic and normal tissue, benign and malignant tumors, and platelets and the multiplicity of molecules suggest at the least a normal physiological role in mesenchymal growth but an uncertain role in tumor growth. Specimens prepared as described by Todaro and Roberts have not been formally or adequately tested on primary human tumor cultures.

Other Tumor Cell Products

Work on other tumor cell products is a very difficult area to review. Most of this work has been performed with crude conditioned medium from tumor cells. Well known is the phenomenon of better cell growth in either suspension or semisolid culture with higher cell numbers. Theories proposed to explain this effect include the secretion of GF by tumor cells, the secretion of some essential nutrient, and the absorption of toxic factors. One of the exciting possibilities is the paracrine secretion of GF.

A number of human cell lines have been reported to produce growth-promoting factors. Two human prostatic cancer lines which were passaged through nude mice grew better with the use of either mitomycin-treated feeder layers or conditioned medium which had been ultracentrifuged (not filtered) (Kaughn et al. 1981). This conditioned medium only improved tumor colony growth, not the growth of fibroblastic cell lines in agar, nor did it transform fibroblastic tissue morphology suggesting a factor other than TGFs. Extracts from an epidermoid lung cancer line, A431, contain TGFs. This cell line is not stimulated in soft agar by autologous extracts. However, clonal growth of this cell line can be increased by physiological concentrations of steroids and by feeder layers of several human carcinoma cell lines. Conditioned medium extracts from these human carcinoma cell lines were inactive (Halper and Moses 1983a). This does suggest that tumor colony growth is dependent on another set of factors than TGFs. Recently described are acid extracts derived from SW-13 cell lines derived from a human adrenal carcinoma (Halper and Moses 1983b). These extracts could induce anchorage-independent growth of SW-13. These factors were separable by molecular sieve and high-performance liquid chromatography from TGFs which induce anchorage-independent growth of supposedly nontransformed mouse AKR-2b and rat NRK fibroblast cells. The most interesting aspect of this activity is that when colonies of SW-13 cells stimulated by this acid extract were transferred from soft agar into monolayer culture for several passages and then back into soft agar, SW-13 cells started forming large colonies without further addition of stimulatory extracts. Colony formation now occurred with a high cloning efficiency. Similar extracts could be obtained from three other human carcinoma cells and a variety of freshly excised human epithelial solid neoplasms and neoplastic tissues, but not from nonepithelial cancers. A melanoma cell line did not secrete these activities. Overall 26 of 32 (80%) carcinoma extracts stimulated anchorage-independent growth of this cell line, whereas none of nine noncarcinomatous malignant neoplasms (mostly sarcomas and melanomas) stimulated the cell line. The greater success rate of establishing small-cell bronchogenic carcinoma (SCBC) cell lines with conditioned medium from SCBC cell lines was recently described (Gazdar et al. 1980). Using this approach six of 11 attempts were successful, compared to only two of 21 successful attempts when cell lines were originally established in only medium and serum. After a period of 1–3 months the cell lines were no longer dependent on conditioned medium. These same authors then examined the effects of these conditioned media on both the clonal growth of fresh tumor specimens and cell lines. In only two of the six reported primary specimens examined did the conditioned medium improve clonal growth, and this was by a factor of 1.8 and 2 respectively. The conditioned medium also improved the clonal growth of established lines in two of four experiments. Mention was made that the colony size was much increased. Smith and Hackett have reported a culture system which involves two steps (Smith et al. 1981). In the first step epithelial cell organoids from breast tissue are cultured in suspension and then after 1 week single-cell suspensions are cloned in a methylcellulose matrix. The secondary culture system consists of fibroblast feeder layers, hormonal supplementation, and conditioned medium from embryonic and breast cell lines. Both the feeder layer and the conditioned medium from these cell lines are essential. Cloning efficiency with such an approach ranges from 1% to 50%. Hug from our group has reported on the high cloning efficiency obtained from fresh biopsy specimens in breast cancer in an agar culture system which incorporates a number of modifications, such as hydrocortisone, estradiol, EGF, and conditioned medium pooled from three human breast carcinoma cell lines (Hug et al., 1983; Hug et al., to be published). Conditioned medium from these three cell lines appears to be essential to produce cloning efficiency high enough for drug assays to be performed in over 70% of the specimens. In an overall view there is little described on specific tumor GFs active on improving tumor growth in vitro to such a magnitude that this can be measured by increased colony size and number, and be detected when cultures are supplemented with serum. Unfortunately, our knowledge of the heterogeneity of human cancer, particularly in terms of the biology of response to potential GFs, is limited. How many GFs are there? What is their chemical nature? Are they specific for tumor cells, or specific for different stem cells of a particular tissue? Is metastatic disease responsive to GFs other than that of the primary tumor, and does primary or metastatic disease become independent of growth regulation?

From just the few publications reported in the literature on the effects of cell line conditioned medium on fresh tumor clonal culture, we anticipated that in our initial experiments we would have to look at many cell lines and many cultures to gain some insight into how these GFs may effect culture growth.

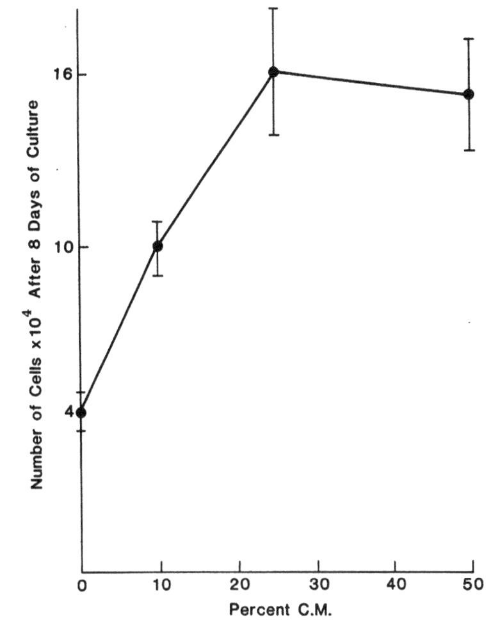

Fig. 1. The increased growth of low-density breast cell line cultures with increasing concentrations of autologous conditioned medium *(C.M.)*

The first cell line we looked at was a breast cell line, MDA435. Cells were originally seeded at approximately 500,000/ml and conditioned medium was harvested from exponentially growing cultures on days 2, 4, 6, and 8. This conditioned medium was then centrifuged and filtered and readded at various concentrations to low-density suspension cultures of the cell line. As can be seen from Fig. 1, growth was significantly enhanced in suspension cultures of this cell line by autologous conditioned medium. The effect was maximal with higher concentrations of conditioned medium and most efficient with conditioned medium harvested after 8 days (data not shown). We then tested this conditioned medium at 50% concentration in an underlayer (final concentration 25%) on a number of agar cultures of tumors of different origins. Initially, a total of 14 cultures were tested with this conditioned medium and four of these cultures responded significantly to the conditioned media (Table 1). It should be mentioned here that our culture system involves an overnight disaggregation of the tumor specimen in an enzyme cocktail of collagenase and DNase. We require prolonged enzyme incubation to obtain cultures approaching a single-cell suspension necessary to evaluate the role of culture manipulation. Table 2 shows the tumors not responding to this GF. From the small number of preliminary experiments it would appear that this particular conditioned medium may cause growth enhancement of only specific tumor types, such as breast cancer and oat cell, but more experiments are needed to prove this point. Simultaneously, we examined conditioned media from two ovarian carcinoma cell lines, a cervical carcinoma, a malignant melanoma, three colon carcinomas, two myelomas, and one vulval carcinoma. Of all these ten cell lines only one has been shown to have any activity for agar colony growth. This caused enhancement of colony growth in five of 11 experiments, interestingly in only one of four ovarian carcinoma specimens so far tested, but in a number of other epithelial carcinomas of different tissue origin. The conditions for preparing this conditioned medium are different from those for the previously described breast cell line; shorter periods of harvest are more effective, and interestingly, as distinct from TGFs, this factor is not trypsin sensitive. In only one

Table 1. Effect of breast cell conditioned medium (CM) on primary clonal growth tumors responding to CM

Tumor type	*p* value	Mean ± SD	
		−CM	+CM
Breast	< 0.02	350 ± 23	465 ± 53
	< 0.001	0	58 ± 6
Oat cell	< 0.01	61 ± 21	91 ± 6
	< 0.002	18 ± 3	42 ± 4

Table 2. Effect of breast cell conditioned medium (CM) on primary clonal growth tumors not responding to CM

Tumor type	Mean ± SD	
	−CM	+CM
Melanoma	71 ± 9	63 ± 7
	191 ± 10	180 ± 12
	165 ± 5	189 ± 20
	7 ± 4	8 ± 4
	244 ± 14	186 ± 22
Lung		
SCC	33 ± 7	37 ± 2
Adenocarcinoma	40 ± 11	33 ± 9
Adenocarcinoma	27 ± 4	27 ± 2
Colon	62 ± 7	64 ± 9
	18 ± 3	21 ± 2

experiment has EGF enhanced the colony growth induced by the factor. Experiments projected for the future include molecular sizing, the further screening of other cell lines for active factors, and an examination of whether factors derived from different cell lines are of different molecular size, are synergistic with or antagonistic to this conditioned medium, and are synergistic with or antagonistic to other polypeptide GFs, such as EGF. Our variable results are no surprise. The heterogeneity of human tumors would predict that we would have a variable response from a multitude of GFs theoretically present in humans acting at different levels in the stem cell hierarchy of any human tissue or tumor.

Fibroblast in Human Tumor Growth

Fibroblasts have been reported to enhance clonal growth in a number of experimental systems and in the agar colony growth of human tumors. One publication has shown how nonanchored fibroblast growth supports prostatic cell line clonal growth. Conditioned media derived from those fibroblasts also support clonal growth, but fibroblasts grown in an anchorage-dependent state inhibited growth (Kirk et al. 1981). Recently, normal adult bronchial epithelial cells could be routinely established from normal adult bronchus after an initial suspension culture and then a secondary clonal culture. Mitomycin-treated fibroblast feeder layers were essential and conditioned medium could not replace it (Lechner et al. 1981). Smith also required fibroblasts for clonal growth of breast tissue. Hamburger reported that fibroblasts also enhanced clonal growth of a number of human tumor specimens. This was with a small number of experiments, four of six experiments responding to a number of different fibroblast cell lines, and EGF sometimes synergized with this effect (Hamburger et al. 1981).

Other Growth Factors

Sato (Barnes and Sato 1980) first investigated serum replacement by a number of GFs, which include insulin, transferrin, somatomedins (insulin-like GFs), other hormones, such as thyroxin, prolactin, gonadotropins, estrogens, androgens, and other polypeptide GFs such as EGF, FGF, endothelial GF, and a number of other growth substitutions. It is difficult to evaluate how this work applies to the clonal growth in agar of human tumor specimens. At this moment it appears very unlikely that a serum will be eliminated from the present system. Little systematic work has been done on the necessity of this multitude of factors or hormones in serum-supplemented low-density clonal cultures of human tumors. We routinely substitute serum from our cultures with final concentrations of 2.5 μg insulin per milliliter and 5 μg transferrin per milliliter. In ten initial experiments we noted that two tumors absolutely required insulin and transferrin for growth. Again, as with other supplementations, we expect the response to be heterogeneous; we expect that serum will have adequate hormones and trace metals in the great majority of tumors but that some tumors will be exquisitely sensitive to these supplementations for growth. We have also investigated hydrocortisone supplementation final concentration 1.25 μg/ml and only in one of eight experiments (Table 3) have we seen any tissue culture response, this particular response being quite dramatic. Bombesin has been demonstrated to enhance both suspension culture and agar culture growth in a serum-free system (Carney et al. 1983). No data are available to show whether this would add to cultures supplemented with serum. Meyskens (Meyskens and Salmon 1980) examined the effects of follicle-stimulating hormone, nerve GF, and melatonin on the cloning efficiency of melanoma. As with all substances examined in these systems the response was variable. However, the response was usually a decrease in cloning efficiency with all three substances, and only occasionally

Table 3. Effect of hydrocortisone in vivo on tumor colony formation

Additive	Colonies
None	0
Hydrocortisone 1×10^{-6} *M*	264 ± 48
Insulin 5 μg/ml + hydrocortisone 1×10^{-6} *M*	257 ± 45.5
Insulin 5 μg/ml + transferrin 10 μg/ml + EGF 50 ng/ml + hydrocortisone 1×10^{-6} *M*	404 ± 18
Insulin 5 μg/ml + transferrin 10 μg/ml + EGF 50 ng/ml	0

EGF, epidermal growth factor

an increase. There was some modification of the nature of the colony growth. EGF has been examined by a number of investigators into tumor colony growth. Salmon (Pathak et al. 1982) reported that 16 of 40 specimens showed a 50% increase with EGF. In only eight of these samples was there a marked increase, that is a two- to eightfold increase. Four of these results were in cell lines and only four on fresh tumor specimens. Therefore, in only approximately 10% of cases examined did EGF significantly enhance tumor colony growth. The response was not related to the density of EGF receptors, a finding not unexpected considering the heterogeneity of the EGF receptor. EGF also inhibited growth significantly in two cases. Hamburger et al. (1981) reported that EGF increased the number of colonies in 36 of 58 (62%); this effect was most noticeable in those which had virtually no colony growth and the magnitude of the response varied from 1.64 to 46 times that of the control. The response occurred in all kinds of tumors and, as mentioned previously, in a number of experiments there was a further enhancement with fibroblastic cell lines. Other authors did not routinely document an EGF effect (Forseth et al. 1981). Our experience very much resembles that of Salmon when examining EGF at final concentrations of approximately 25 μg/l. At these concentrations we notice only that approximately 10% of tumors respond with a large increment in colony growth. Another GF which has received a lot of attention in experimental literature is PDGF. Tumor cells have even been described as releasing a PDGF (Graves et al. 1983), and as mentioned previously some platelet-derived products have TGF-like activity (Assoian et al. 1983). Platelet lysate has been reported to increase growth in 47% of growing tumors, with the increments ranging from 1.25 up to 6.69 (Cowan and Araham, to be published). Thirteen percent of specimens showed an actual decrement in response to lysate and 40% no change. A freeze-thaw preparation was better than collagen stimulation but thrombin stimulation released equivalent activity to the freeze-thaw preparations.

A number of logical additions to these systems have not been reported, such as a necessity for estradiol in the cloning of breast cancer, gonadotropins for ovarian cancer and prostatic carcinoma, or testosterone for prostatic carcinoma. Many other modifications of the tissue culture medium have not been systematically examined, such as the preferable origin of the serum (fetal calf, horse, or mixtures), the calcium content, the addition of certain trace metals, such as selenium, and many other modifications which have been used basically to prepare defined media (Tsao et al. 1982; Agy et al. 1981). It should be remembered that the "HITES" medium described as supporting the serum-free growth of small-cell bronchogenic carcinoma lines of humans (medium supplemented with hydrocortisone, insulin, transferrin, estradiol, and selenium) has not been reported to support the clonal growth of fresh tissue in agar. It has, however, been reported to enrich the clonogenic cells in liquid preculture before agar clonal growth. For reasons unknown at present, such media appear selective for small-cell bronchogenic carcinoma and could not be used routinely for all tumor types (Carney et al. 1981).

Initially, when the assay was described by Salmon, a conditioned medium derived from adherent spleen cells of mineral-oil-primed BALB/c mice was used (Salmon et al. 1978; Hamburger and Salmon 1977). This has not been shown by other investigators to be necessary for the growth of nonhematopoietic neoplasms (Von Hoff et al. 1981).

Extracellular Matrix Proteins

Tumor cells have been cloned in agar primarily for two reasons: (a) because anchorage-independent growth has been thought to be a general marker of the malignant

Table 4. Effect of ovarian cell conditioned medium (CM) on primary human tumor colony formation

Tumor type	Colony −CM	Number/5 × 10^5 cells +CM
Rectal	0	0
Colonic	2 ± 2	53 ± 21[a]
Colonic	21 ± 6	84 ± 17[a]
Cervical	0	0
Endometrial	0	14 ± 4[a]
Ovarian	0	0
Ovarian	0	16 ± 4[a]
Ovarian	4 ± 3	0
Ovarian	1 ± 1	0
Leiomyosarcoma	0	0
Adenocarcinoma lung	19 ± 2	67 ± 14[a]

[a] Significant changes

phenotype, and (b) because agar culture was felt to discriminate against fibroblast growth. However, in a number of experimental systems it has been determined that anchorage-independent growth does not necessarily select for malignant clones (Dodson et al. 1981), that anchorage-independent growth is selective and therefore does not represent the true heterogeneity (Heppner and Miller 1983) or total extent of all the potential tumor stem cell populations, and that monolayer cultures can be modified to support tumor growth clonally with better efficiency than that of agar or methylcellulose without fibroblast overgrowth (Tsao et al. 1982). The main modification of monolayer cultures is the placement of components of the extracellular matrix (ECM) on the bottom surface of plastic culture dishes. In general terms, ECM is composed of proteins such as collagen (one of at least five different types depending on the anatomical location of the matrix) and elastin; proteoglycans such as chondroitin sulfate, heparin sulfate, and hyaluronic acid; and glycoproteins such as fibronectin, laminin, and chondronectin. Also, in general terms, collagen and elastin form the structural components of ECM, the proteoglycans are involved in the normal assembly of collagen fibrils and are incorporated into the matrix, and the glycoproteins fibronectin and laminin are involved in the attachment of cells to the matrix. Fibroblasts synthesize collagen and fibronectin as well as do a variety of other cells. Tumor cells show reduced amounts of cell surface fibronectin and restoration of fibronectin to tumor cell cultures results in a normalization of the transformed phenotype (Alpin and Hughes 1982).

In view of the fact that clumping of cells is frequently necessary for tumor colony growth in the agar culture system, one cannot but ask the question whether tumor colony growth is actually secondary to an ECM being maintained because of an inadequate physical separation method prior to culture. We have added the collagenase digest, presumably containing material from the ECM, from our biopsy specimens to the agar culture and have occasionally induced colony growth (Table 5).

Gospodarowicz has examined ECM laid down on plastic dishes by confluent monolayers of bovine vascular endothelial cells and has demonstrated that the clonal variability in response to GFs is lost when the cells are grown over ECM. When individual components of ECM were examined, fibronectin was generally most active in enabling cell attachment

Table 5. Enzyme digest can occasionally promote tumor colonies from ovarian cancer

Control	With 25%–50% enzyme digest[a]
4 ± 0.57	63 ± 9
57 ± 24	95 ± 19
35 ± 4	0
3 ± 5	0
0	0

[a] Cultured medium harvested after overnight enzyme disaggregation with collagenase and DNase
Values are means ± SD

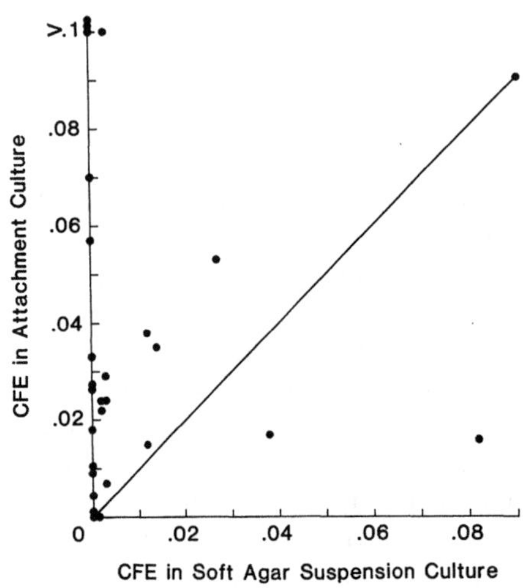

Fig. 2. The comparison of colony-forming efficiency *(CFE)* of tumor specimens cultured simultaneously in the attachment and agar culture assays

and spreading, although laminin appeared better in these respects for epithelial cells. Fibronectin-coated dishes are commonly used to support clonal growth in a variety of primary epithelial cell cultures (Longnecker et al. 1983; Lechner et al. 1982; Gospodarowicz et al. 1982).

The tumor microenvironment or, in broad terms, the tumor ECM differs from normal ECM in that it is predominantly composed of deposits of fibrin caused by the release of procoagulant activity by the tumor cells. Fibronectin has a high affinity for fibrin, with the result that the tumor ECM also contains fibronectin. Tumor cells also produce proteases which degrade the tumor ECM, thereby permitting expansion of the tumor mass within the fibrin/fibronectin environment and the concomitant production of fragments of these two entities, which have been shown to possess growth-promoting properties. Thus it has become apparent that tumor cells are capable of producing a local environment conducive to their own proliferation (Turner et al. 1983; Dvorak et al. 1983).

Cold insoluble material from ascitic fluid containing large amounts of fibronectin and either intact or fragmented fibrinogen as well as other components but void of laminin has been used to coat the bottom of plastic dishes. With this approach we have cultured approximately 100 human tumors, and in a number of these cultures simultaneously

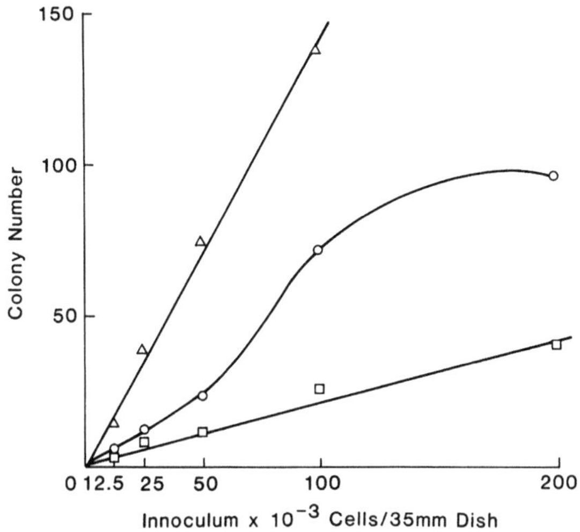

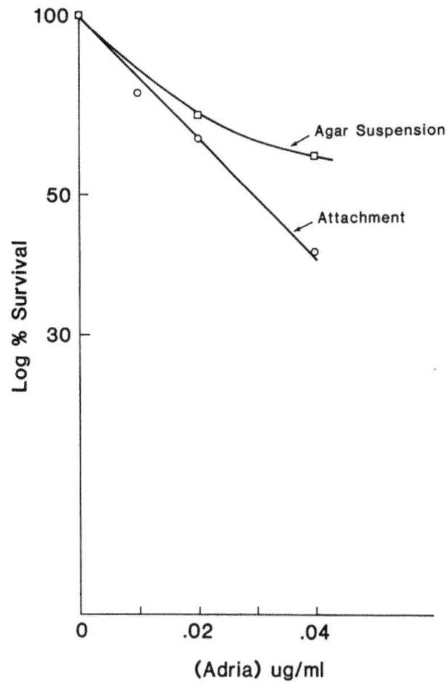

Fig. 3 (Left). Cell dose: colony response in the attachment assay. Note that 1×10^5 cells in the attachment assay may give a working assay

Fig. 4 (Right). Comparison of colony survival with and without continuous exposure to doxorubicin *(Adria)* in the attachment and agar colony assay

compared the growth of these tumors in the standard agar culture system. Figure 2 compares the plating efficiency in the two system, showing greatly enhanced growth in the monolayer culture system. Cell dose titrations (Fig. 3) suggest that small numbers of cells can be cultured in this system. Drug dose responses suggest linear relationships in the monolayer system and similar responsivity to the agar culture system (Fig. 4). This monolayer culture system is being tested for usefulness in chemosensitivity testing and promises to permit the evaluation of more drugs with greater efficiency than the agar culture method. Cytologic examination and evaluation of the malignant nucleolar antigen suggests malignant origin of the colonies.

Conclusion

There can be no argument about the fact that the agar culture system as presently defined by Salmon has clinical application. Numerous investigators have shown consistently that this assay predicts with a 50%–70% acuracy an in vivo tumor response. Because of the frequency of unsuccessful cultures, the low cloning efficiency, and the high numbers of cells needing to be cultured, a large number of specimens are excluded from a drug assay, and when successful the tumor cells can only be tested against a limited number of drugs. There are also many theoretical and technical problems with this assay, such as the tendency for more growing cultures to originate from clumped specimens, cytotoxic artifacts, and the major question of selection. No doubt in the future there will be major modifications in this method; this may involve a switch to monolayer cultures with the use of well-defined attachment factors to enhance clonal growth at low cell numbers. A switch to such culture systems should increase cloning efficiency and overcome some of the objections of selection. Both agar and monolayer culture systems may be improved by the addition of growth factors and other hormone-like substances. It is expected that there will be a large number of these factors. Individual factor and hormone requirements will have to be identified for each individual tumor type, both primary and metastatic. The identification of extracellular matrix tumor cell growth interaction and growth factor and hormone tumor cell growth interactions could have important therapeutic implications in the design of new therapeutic approaches.

References

Agy PC, Shiley CD, Aanr RG (1981) Protein-free medium for C-1300 mouse neuroblastoma cells. In Vitro 17: 671–680

Alpin JD, Hughes RD (1982) Complex carbohydrates of the extracellular matrix structures, interaction and biological roles. Biochem Biophysiol Acta 694: 375–418

Assoian RK, Komoriya A, Meyers CA, Sporn MB (1983) Platelet-derived transforming growth factor. Fed Proc 428: 1831

Barnes D, Sato G (1980) Methods for growth of cultured cells in serum-free medium. Biochemistry 102: 255–270

Bradley TR, Metcalf D (1966) The growth of mouse bone marrow in vitro. Aust J Exp Biol Med Sci 44: 287–299

Carney DN, Bunn PA, Gazdar AF, Pagan JA, Minna JD (1981) Selective growth in serum-free hormone-supplemented medium of tumor cells obtained by biopsy from patients with small cell carcinoma of the lung. Proc Natl Acad Sci USA 78: 3185–3189

Carney DN, Oie H, Moody T, Gazdar A, Cuttitta F, Minna J (1983) Bombesin, an autocrine growth factor for human small cell lung cancer cell lines. Clin Res 31: 2

Cowan DH, Araham JC (to be published) Stimulation of human tumor colony formation by platelet lysates. Exp Hematol

Childs CB, Proper JA, Tucker RF, Moses HL (1982) Serum contains a platelet-derived transforming growth factor. Proc Natl Acad Sci USA 79: 5312–5316

Dodson MG, Slota J, Lange C, Major E (1981) Distinction of phenotypes of in vitro anchorage-independent soft-agar growth and in vivo tumorigenicity in nude mouse. Cancer Res 41: 1441–1446

Dvorak HF, Senger DR, Dvoerak AM (1983) Fibrin as a component of the tumor stroma: origins and biological significance. Cancer Metastasis Rev 2: 41–73

Forseth B, Paque RE, Von Hoff DD (1981) Exploration of media changes and growth factors to enhance tumor growth in soft agar. Proc Am Assoc Cancer Res 22: 54

Gazdar AF, Carney DN, Russell EK, Sims HL, Baylin SB, Bunn PA, Guccion JG, Minna JD (1980) Establishment of continuous, clonable cultures of small-cell carcinoma of the lung which have amine precursor uptake and decarboxylation cell properties. Cancer Res 40: 3502–3507

Gospodarowicz D, Leri GM, Gonzalez R (1982) High-density lipoproteins and the proliferation of human tumor cells maintained on extracellular matrix-coated dishes and exposed to a defined medium. Cancer Res 42: 3704–3713

Graves DT, Owen AJ, Antoniades HN (1983) Evidence that a human osteosarcoma cell line which secretes a mitogen similar to platelet-derived growth factor requires growth factors present in platelet-poor plasma. Cancer Res 43: 83–87

Halper J, Moses HL (1983a) Epithelial tissue-derived growth factor-like polypeptides. Cancer Res 43: 1972–1979

Halper J, Moses HL (1983b) Stimulation of soft agar growth of A431 human carcinoma cells by glucocorticoids and diffusable factors. Fed Proc 525: 370

Hamburger AW, Salmon SE (1977) Primary bioassay of human tumor stem cells. Science 197: 461–463

Hamburger AW, White CP, Brown RW (1981) Effect of epidermal growth factor on proliferation of human tumor cells in soft agar. J Natl Cancer Inst 67: 825–830
Heppner G, Miller BJ (1983) Tumor heterogeneity: biological implications and therapeutic consequences. Cancer Metastasis Rev 2: 5–23
Hug V, Drewinko B, Spitzer G, Blumenschein GD (1983) Improved culture conditions for the in vitro growth of human breast tumors. Proc Am Assoc Cancer Res 24: 35
Hug V, Thames H, Spitzer G, Blumenschein GD (to be published) Normalization of in vitro sensitivity testing of human tumor cloning. Cancer Res
Johnson GR, Metcalf D (1977) Pure and mixed erythroid colony formation in vitro stimulated by spleen conditioned medium with no detectable erythropoietin. Proc Natl Acad Sci USA 74: 3879–3882
Kaughn EM, Kirk D, Szalay M, Lechner JF (1981) Growth control of prostatic carcinoma cells in serum-free media: interrelationship of hormone response, cell density, and nutrient media. Proc Natl Acad Sci USA 78: 5673–5676
Kirk D, Szalay MF, Kaign ME (1981) Modulation of growth of human prostatic cancer cell line (PC-3) in agar culture by normal human lung fibroblasts. Cancer Res 41: 1100–1103
Lechner JF, Haugen A, Autrup H, McClendon IA, Trump BF, Harris CC (1981) Clonal growth of epithelial cells from normal adult human bronchus. Cancer Res 41: 2294–2304
Lechner JF, Haugen A, McClendon IA, Pettis EW (1982) Clonal growth of normal adult human bronchial epithelial cells in a serum-free medium. In Vitro 18: 633–642
Longnecker JP, Kilty IA, Ridge JA, Miller DC, Johnson LK (1983) Proliferative variability of endothelial clones derived from adult bovine aorta: influence of fibroblast growth factor and smooth muscle cell extracellular matrix. J Cell Physiol 114: 7–15
Metcalf D (1977) Hemopoietic colonies. In vitro cloning of normal and leukemic cells. Springer, Berlin Heidelberg New York (Recent results in cancer research, vol 61)
Metcalf D (1980) Clonal analysis of proliferation and differentiation of paired daughter cells: action of granulocyte-macrophage colony-stimulating factor on granulocyte-macrophage precursors. Proc Natl Acad Sci USA 77: 5327–5330
Metcalf D, Moore M, Sheridan J, Spitzer G (1974) Responsiveness of human granulocytic leukemic cells to colony-stimulating factor. Blood 43: 847
Meyskens FL, Salmon SE (1980) Regulation of human melanoma clonogenic cell expression in soft agar by follicle stimulating hormone (FSH), nerve growth factor (NGF) and melantonin (MTN). Proc Am Assoc Cancer Res 21: 199
Moore MAS, Williams M, Metcalf D (1973) In vitro colony formation by normal and leukemic human hematopoietic cells: interaction between colony-forming and colony stimulating cells. J Natl Cancer Inst 50: 591–602
Morgan GR, Williams JA, Smallwood I, Taylor JMA (1983) Colony-forming cells from primary tumours: influence of dissociation method on colony growth in vitro. Proc Am Assoc Cancer Res 4: 1
Nickell KA, Halper J, Moses HL (1983) Transforming growth factors in solid human malignant neoplasms. Cancer Res 43: 1966–1971
Pathak MA, Matrisian LM, Magun BE, Salmon SE (1982) Effect of epidermal growth factor on clonogenic growth of primary human tumor cells. Int J Cancer 30: 745–750
Salmon SE, Hamburger AW, Soehnlen B, Durie BGM, Alberts DS, Moon TE (1978) Quantitation of differential sensitivity of human-tumor stem cells to anticancer drugs. N Engl J Med 298: 1321–1327
Sherwin SA, Minna JD, Gazdar AF, Todaro GJ (1981) Expression of epidermal and nerve growth factor receptors and soft agar growth factor production by human lung cancer cells. Cancer Res 41: 3538–3542
Smith HS, Lan S, Ceriani R, Hackett AJ, Stampfer MR (1981) Clonal proliferation of cultured nonmalignant and malignat human breast epithelia. Cancer Res 41: 4637–4643
Spitzer G, Verma DS, Dicke KA, McCredie KB (1978) Culture studies in vitro in human leukemia. Semin Hematol 15: 352–378

Sporn MB, Todaro GJ (1980) Autocrine secretion and malignant transformation of cells. N Engl J Med 303: 878–880

Tsao MC, Walthall BJ, Ham RG (1982) Clonal growth of normal human epidermal keratinocytes in a defined medium. J Cell Physiol 110: 219–229

Tucker RF, Valkenant ME, Branum EL, Moses HL (1983) Comparison of intra- and extracellular transforming growth factors from nontransformed and chemically transformed mouse embryo cells. Cancer Res 43: 1581–1586

Turner WA, Menter DG, Honn KV, Taylor JD (1983) Tumor cell induced platelet fibronectin release. Proc Am Assoc Cancer Res 97: 25

Von Hoff DD, Casper J, Bradley E, Sandbrach J, Jones P, Maruch R (1981) Association between human tumor colony-forming assay results and response of an individual patient's tumor to chemotherapy. Am J Med 70: 1027–1032

*Improving Techniques for Clonogenic Assays**

J.F. Eliason, A. Fekete, and N. Odartchenko

Schweizerisches Institut für Experimentelle Krebsforschung,
1066 Epalingess/Lausanne, Switzerland

Introduction

A number of technical problems prevent the widespread use of in vitro clonal assays for human tumor cells on a routine clinical basis (Selby et al. 1983). One area where improvements can be expected is in the culture conditions, since a major limitation of these assays is the low plating efficiencies (PEs) which are usually obtained. This, of course, means that anticancer drugs can be individually tested only on a minority of tumor samples. Furthermore, tumor colony growth in most assay cultures is dependent on the presence of animal sera, which contain variable amounts of various nutrients, hormones, and growth factors. Standardization of culture conditions between laboratories or even between serum batches is, therefore, impossible.

The approach we have taken to this problem has been to attempt to develop serum-free semisolid media for clonal growth of cells from human tumor cell lines. We have used cell lines for this work because their growth remains relatively constant with time, allowing experiments to be repeated and making step by step improvements feasible. Serum-free media have been described for a number of tumor cell lines in liquid culture (Barnes et al. 1981), but growth requirements for cells in semisolid media can be different (Eliason, to be published).

We have used the single-layer methylcellulose system described by Cillo et al. (1981; see also chapter by Cillo and Odartchenko in this volume) for these studies, methylcellulose being chemically defined and devoid of stimulatory activities which are present with agar. In order to obtain floating colonies in this system, it is important to plate the cells in bacterial petri dishes. This inhibits attachment of cells to the bottom (Eliason et al. 1979, 1982). Careful selection of fetal calf serum (FCS) batches is also helpful in this respect (Cillo et al. 1983).

We describe here a new serum-free medium (Table 1) which supports colony formation by cells from one colon adenocarcinoma and four melanoma cell lines. In the course of this work, the linearity of colony formation with numbers of cells plated was examined in detail. It was found that linearity is highly dependent on culture conditions in general. It thus should probably be monitored for every tumor sample.

---

* This work was supported by grants from the Swiss National Foundation for Scientific Research, the Swiss League Against Cancer, and the Fonds de Recherche sur les Lymphomes Malins

Table 1. Serum-free medium for tumor cell assay

Enriched Dulbecco's (EMED) Modified F-12 (FMED) (α-thioglycerol 7.5×10^{-5} *M*, HEPES 10 m*M*)	1 : 1 Mixture
Methylcellulose	0.9%
Bovine serum albumin	10 mg/ml
Transferrin	80 µg/ml
Insulin	3 µg/ml
Ethanolamine	1.2 µg/ml
Linoleic acid	2.8 µg/ml
Cholesterol	2.6 µg/ml
Nucleosides	10 µg/ml each
Trace elements: Se, Si, Mn, Mo, V, Ni, Sn	

Materials and Methods

Media

Iscove's modified Dulbecco's medium (IMDM, Gibco) was prepared as suggested by the manufacturer.

Our own enriched Dulbecco's formulation (EMED) was prepared from powdered Dulbecco's modified Eagle's medium (1,000 mg/l glucose; Gibco) by supplementing it with *N*-2-hydroxyethylpiperazine-*N'*-2-ethynesulfonic acid (HEPES; 2.38 g/l), $NaHCO_3$ (2.86 g/l), Na_2SeO_3 (0.0173 mg/l), L-alanine (25 mg/l), L-asparagine · H_2O (50 mg/l), L-aspartic acid (30 mg/l), L-cysteine (70 mg/l), L-glutamic acid (75 mg/l), L-proline (40 mg/l), sodium pyruvate (110 mg/l), vitamin B_{12} (0.025 mg/l), and biotin (0.03 mg/l).

Modified Ham's F-12 nutrient mixture (FMED) was prepared by supplementing the commercial product (Gibco) with HEPES (2.38 g/l), Na_2Se0_3 (0.0173 mg/l), and $NaHCO_3$ (0.336 g/l).

All media contained α-thioglycerol (7.5×10^{-5} *M*) and were supplemented immediately before use with L-glutamine (204 mg/l). Nucleosides (10 mg/l each of uridine, adenosine, cytidine, guanosine, thymidine, 2'-deoxyadenosine, 2'-deoxycytidine. HCl, and 2'-deoxyguanosine) were added to the 1 : 1 mixture of EMED and FMED. This latter supplement was prepared as a 100-fold concentrated stock solution in EMED. Ethanolamine was added at a final concentration of 2×10^{-5} *M*. The following trace elements were also added: $Na_2SiO_3 \cdot 9H_2O$ (5×10^{-7} *M*), $MnSO_4 \cdot H_2O$ (1×10^{-9} *M*), $(NH_4)_6Mo_7O_{24} \cdot 4H_2O$ (1×10^{-9} *M*), NH_4VO_3 (5×10^{-9} *M*), $NiCl_2 \cdot 6H_2O$ (5×10^{-10} *M*), and $SnCl_2 \cdot 2H_2O$ (5×10^{-10} *M*).

Fetal calf serum was purchased from KC Biological and was screened for growth-promoting activity for mouse hemopoietic colony formation. Purified human transferrin (TSF; Behringwerke) was fully iron saturated. Bovine serum albumin (BSA; Behringwerke) was stripped of lipids by dextran-charcoal treatment and deionized (Iscove et al. 1980). A mixture of linoleic acid (2.8 mg/ml) and cholesterol (2.6 mg/ml) was prepared in 95% ethanol (LIN-CHOL).

Cell Lines

The melanoma cell lines (Me43, MP6, Meluso, and Me85) were kindly provided by Dr. S. Carrel of the Ludwig Institute for Cancer Research, Lausanne Branch. The colon adenocarcinoma cell line (WiDr) was obtained from the American Type Culture Collection. Two lines, Me43 and WiDr, were maintained in our laboratory in a 1 : 1 mixture of EMED and FMED (hereafter referred to as EF) supplemented with 5% FCS. The cells were passaged weekly by treatment with a solution of 0.25% trypsin and 0.05% ethylenediaminetetraacetic acid for 10 min at 37° C. Cells were washed twice with serum-containing medium and the numbers of trypan-blue-excluding cells were counted on a hemocytometer. For subculture, cells were diluted to 5×10^4 cells per milliliter in fresh medium.

Methylcellulose Cultures

Cells obtained during passage of the lines were suspended at the desired concentration in various media (as described in the Results section) containing 0.9% methylcellulose. At least two replicate 1 ml aliquots were plated in bacterial petri dishes (Greiner, no. 627102). The cultures were incubated for 10–14 days at 37° C in a fully humidified atmosphere of 5% CO_2 in air. Colonies were counted using an inverted microscope. In cultures where there were more than 500 colonies, only a portion of the plate was counted and the results were multiplied by the appropriate factor to give a value for the whole plate.

Results

Preliminary experiments with different cell lines using IMDM as the nutrient medium indicated that colony formation was not linear with the numbers of cells plated. Therefore, we tested whether use of a different nutrient medium would improve the system. The medium tested, EF, was developed to support long-term hemopoiesis in liquid bone marrow cell cultures (Eliason, to be published). Table 2 shows the results of an experiment in which colony formation by Me43 cells was compared in the two media. Cells plated in EF plus 5% FCS gave rise to approximately the same numbers of colonies at the highest cell concentration and significantly higher numbers at lower concentrations than were seen in IMDM plus 15% FCS. In fact, growth was nearly linear in EF medium with PEs between 7.2% and 8.8%, whereas in IMDM, even with a higher FCS concentration, PEs were clearly dependent on the initial cell concentration.

By further supplementing EF with BSA, TSF, INS, and LIN-CHOL, we have been able to grow colonies from both cell lines in the complete absence of serum. This can be seen in Fig. 1 for WiDr cells, where serum-free growth was compared to that in cultures further supplemented with 5% FCS. Colony formation in both instances appeared to be linear with cell concentration, having correlation coefficient (r) for the regression lines equal to 0.998 for serum-free cultures and 0.993 for serum-containing cultures. The y-intercepts are not significantly different from zero (−78, without FCS; −52, with FCS). The PEs are thus given by the slopes of the lines and are 0.340 ± 0.070 colonies/cell (± 1 SD) in serum-free cultures and 0.260 ± 0.084 colonies/cell in cultures with FCS. These results show that the PE for WiDr in our serum-free medium is significantly higher ($p < 0.001$) than in the same

Table 2. Comparison of colony formation by Me43 cells in different media

Cells/ml	IMDM + 15% FCS		EF + 5% FCS	
	Colonies/ml	PE	Colonies/ml	PE
2×10^4	1,820 ± 80	0.091	1,740 ± 40	0.087
1×10^4	610 ± 40	0.061	880 ± 20	0.088
5×10^3	74 ± 10	0.015	360 ± 10	0.072

Results are expressed as mean ± 1 SD from two independent experiments
IMDM, Iscove's modified Dulbecco's medium; *FCS,* fetal calf serum; *EF,* 1 : 1 mixture of enriched Dulbecco's formulation and modified Ham's F-12 nutrient mixture; *PE,* plating efficiency

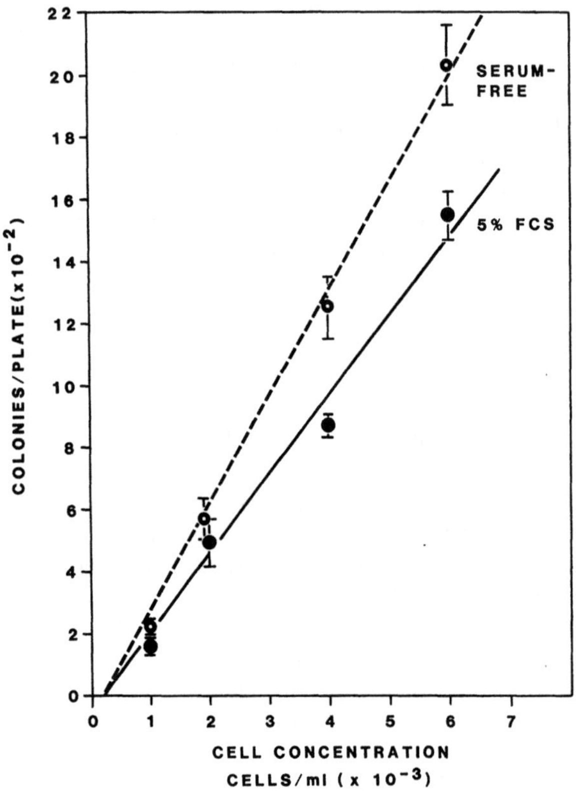

Fig. 1. Colony formation by WiDr cells in the presence and absence of 5% fetal calf serum *(FCS).* Various concentrations of WiDr cells were plated in EF supplemented with 0.9% methylcellulose, 1% bovine serum albumin, 80 μg/ml transferrin, 3 μg/ml insulin, 0.1% linoleic acid and cholesterol and no FCS (○) or 5% FCS (●). The results are expressed as mean colony counts ± 1 SD of eight replicates for serum-free cultures and two replicates for serum-containing cultures. *EF,* 1 : 1 mixture of enriched Dulbecco's formulation and modified Ham's F-12 nutrient mixture

medium containing FCS. Similar experiments with Me43 indicated that the PEs were the same in the presence and absence of serum.

Figure 2 shows the results of an experiment to determine the effect of BSA concentration on colony formation by WiDr cells. The PEs were determined by regression analysis of curves similar to those shown in Fig. 1. Colony formation was highly dependent on BSA concentration between 0.1%, where no colonies were seen, and 0.3%, where nearly

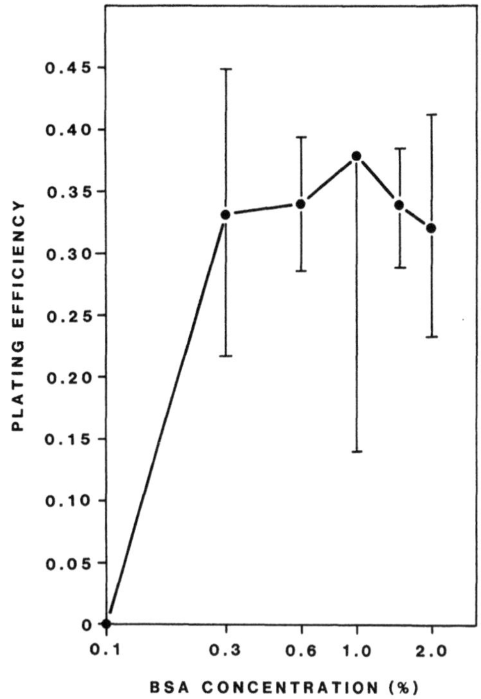

Fig. 2. Effect of bovine serum albumin (BSA) concentration on plating efficiency (PE) of WiDr cells in serum-free medium. WiDr cells were plated at 6×10^3, 4×10^3, 2×10^3, and 1×10^3 cells per milliliter in EF supplemented with 0.9% methylcellulose, 80 μg/ml transferrin, 3 μg/ml insulin, 0.1% linoleic acid and cholesterol and various concentrations of BSA. The PEs were derived from the slopes of least squares regression curves as described in the text. *Vertical bars* represent standard deviations of the slopes. *EF,* 1 : 1 mixture of enriched Dulbecco's formulation and modified Ham's F-12 nutrient mixture

maximum numbers of colonies were scored. Above this level, increasing BSA concentrations did not give higher numbers of colonies although colonies grown in the presence of higher concentrations of BSA were of a larger size than those grown with 0.3%. The shape of the BSA titration curve for Me43 cells was similar (data not shown), but these cells appeared to have a slightly higher requirement for BSA since colonies were seen only with concentrations above 0.3%. We have therefore chosen 1% BSA for all subsequent work.

The relationship between colony formation and numbers of Me43 cells plated in serum-free medium is shown in Fig. 3A. This relationship appears to be linear with a correlation coefficient for the regression line of 0.998. However, the y-intercept is equal to -560 colonies (with no cells plated), suggesting that the assay is, in fact, not linear. This conclusion is supported by plotting the results as a log-log relationship (Fig. 3B), where there is also good fit as determined by the correlation coefficient. A linear relationship will have a slope equal to 1 in such a plot. The slope of 1.46 for colony formation by Me43 cells in these cultures is significantly different from 1.00 ($p < 0.05$), indicating nonlinearity. The ability of our serum-free medium to support colony formation by other melanoma cell lines is demonstrated in Table 3. All three cell lines gave high numbers of colonies and the relationships with initial cell concentrations showed good fit to linear regression lines, with PEs determined from the slopes, ranging from 21% for MP6 to 70% for Meluso. Table 4 also gives the results of log-log regression analysis, which support the apparent linearity of the assay for MP6 and Meluso. However, colony formation by Me85 had a slope of 1.45, indicating that growth was nonlinear, similar to the situation with Me43.

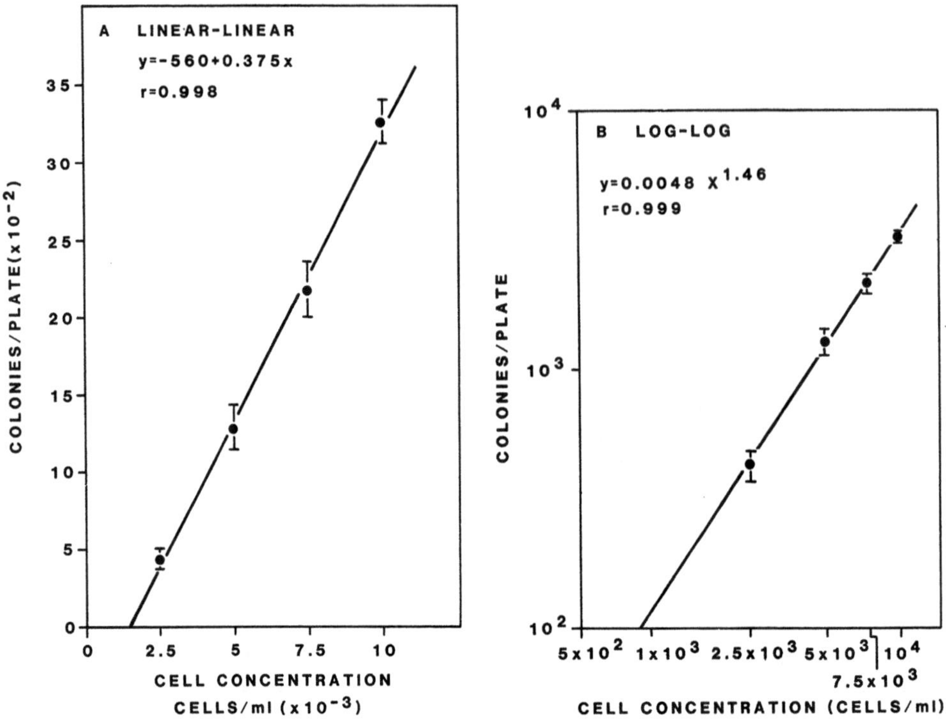

Fig. 3A, B. Relationship between numbers of Me43 cells plated and numbers of colonies formed in serum-free medium. Various concentrations of Me43 cells were plated in EF supplemented with 0.9% methylcellulose, 1% bovine serum albumin, 80 μg/ml transferrin, 3 μg/ml insulin and 0.1% linoleic acid and cholesterol. The results are expressed as mean colony counts ± 1 SD of 12 replicates. **A** Results plotted as a linear relationship; **B** results plotted as a log-log relationship. *EF*, 1 : 1 mixture of enriched Dulbecco's formulation and modified Ham's F-12 nutrient mixture

Table 3. Clonal growth of three melanoma cell lines in serum-free medium

Cells/ml	Colonies/plate ± 1 SD		
	MP6	Meluso	Me85
3×10^4	6,490 ± 380	TMTC	TMTC
1×10^4	2,010 ± 250	6,960 ± 420	3,090 ± 40
3×10^4	1,330 ± 130	2,380 ± 490	670 ± 30
1×10^3	190 ± 20	570 ± 100	110 ± 20

TMTC, too many to count

Table 4. Regression analysis

Linear			
r	0.992	0.998	0.996
Slope	0.21	0.70	0.32
Log-log			
r	0.963	0.994	0.998
Slope	0.96	1.08	1.45

Discussion

An essential requisite for an assay which is to be used as a routine clinical test is that it be standardized. That is, each laboratory must be able to obtain the same result with the same sample. This means that every variable capable of affecting the result must be controlled. In tissue culture systems containing animal sera or crude extracts of plant or animal origin, standardization is impossible, since the composition of these substances is variable, making some batches stimulatory and others inhibitory. In order to overcome this problem, we have been working to develop defined media for clonal assay of human tumor cells. The medium we have described in this paper is defined in the sense that the protein additives are highly purified. However, the possibility that they may still contain active contaminants cannot be ruled out.

Agar, for instance, has been shown to contain substances, probably highly sulfated agaropectins, which are mitogenic for β-lymphocytes (Kincade et al. 1976) and can induce the release of hemopoietic colony-stimulating activities by cells in normal human bone marrow (Hoang et al. 1981). Patient tumor samples may contain variable numbers of normal cells belonging to the immune system, especially lymphocytes and macrophages. Thus use of agar in clonogenic tumor cell assays introduces another level of complexity and may be, in part, responsible for observed growth modulation by normal cells seen with some samples (Buick et al. 1980; Hamburger and White 1981). Some laboratories add diethylaminoethyl-dextran (DEAE-dextran) to neutralize the negatively charged sulfate groups in agar. However, DEAE-dextran has recently been reported to inhibit tumor cell colony growth (Hug et al. 1983).

The use of methylcellulose avoids these problems, but also has some disadvantages for clonogenic assays. Since methylcellulose supports colony formation by increasing the viscosity of medium, cells tend to sediment to the bottom of these cultures. This can lead to attachment of cells to the dish. One way to eliminate cell attachment is to use an agar underlayer (Buick et al. 1979). Plating cells in dishes which have not been treated to make cells attach (i.e., nontissue culture dishes) and careful selection of serum batches work as well (Cillo et al. 1983). Sedimentation of cells in methylcellulose also means that the effective cell concentration found at the bottom of the culture is higher than in cultures with the same numbers of cells plated in more structured support substances such as agar, fibrin, or collagen gels. This can result in increased aggregation unless lower cell concentrations are plated initially.

Another potential benefit from the use of serum-free media for tumor cell assays is that the medium can, theoretically, be designed for selective growth of single cell types, since it appears that different cells have different growth requirements in vitro (Barnes and Sato 1980). This aspect of serum-free media has been exploited to develop a liquid culture system for selective growth of small-cell carcinoma of the lung (Carney et al. 1981). It can even be envisaged that serum-free systems could become a diagnostic tool, based on better growth in one medium than in another. It is interesting in this regard that we have obtained good growth of colon adenocarcinoma cells and melanoma cells in the same medium. However, differences between WiDr and Me43 were seen in the minimal concentrations of BSA required for colony formation. Furthermore, nonlinearity between cell numbers plated and colony formation for two of the melanoma cell lines suggests that this formulation is not yet optimal for melanoma cells.

Any clonal assay for progenitor cells should be linear with numbers of cells plated. Linearity of the human tumor colony assay has not been well documented, with only a few examples shown for any particular type of tumor. We have found that cell lines from the same type of tumor, namely melanoma, can reveal differences in their growth patterns. Two, MP6 and MeIuso, grew in a linear manner, whereas colony formation by two others, Me43 and Me85, appeared to be nonlinear. To complicate matters further, Me43 and Me85 gave good fit to linear relationships, with only the fact that the *y*-intercepts were not zero suggesting nonlinearity. However, we have shown that transforming the results of cell titration experiments to log-log relationships can provide a simple test for linearity based on the slope of the line.

Nonlinearity of the assay can have important implications for predictive drug testing. If drugs had been tested with Me43 in the assay shown in Fig. 3, having a log-log slope of 1.46, then a 90% reduction in colonies would, in fact, mean a 20% rather than 10% survival of colony-forming cells. On the basis of these considerations, it would clearly be advisable to include several control groups with lower cell concentrations in every drug test. These would provide an indication of linearity for each individual sample and an important internal standard if colony formation is not linear.

In conclusion, we have described a serum-free medium which supports clonal growth of cells from several human tumor cell lines. This medium should provide the basis for developing specific serum-free formulations for other tumor types and could, eventually, allow standardization of culture conditions for assays using primary tumor samples.

Summary

A serum-free medium has been developed which supports colony formation by cells from several human tumor cell lines, one colon adenocarcinoma (WiDr) and four melanoma (Me43, Me85, MP6, MeIuso). This medium consists of a 1 : 1 mixture of an enriched Dulbecco's modified Eagle's medium (EMED) and a modified Ham's F-12 nutrient mixture (FMED) supplemented with 0.9% methylcellulose, 1% bovine serum albumin, 80 μg/ml human transferrin, 3 μg/ml insulin, 2.8 μg/ml linoleic acid, 2.6 μg/ml cholesterol, 20 μ*M* ethanolamine, and trace elements. Colony formation by WiDr cells is linear with the numbers of cells plated, having a plating efficiency (PE) of 34%, as compared to 26% in serum-containing medium. Two of the melanoma cell lines, MP6 and MeIuso, exhibit linear relationships between colony numbers and cell concentration with PEs of 21% and 70% respectively. Colony formation by the other two melanoma cell lines appears to be nonlinear. This work represents a step toward standardizing culture conditions for human tumor clonogenic cell assays.

References

Barnes D, Sato G (1980) Serum-free cell culture: a unifying approach. Cell 22: 649–656

Barnes D, van der Bosch J, Masui H, Miyazaki K, Sato G (1981) The cultures of human tumor cells in serum-free medium. Methods Enzymol 79: 368–391

Buick RN, Stanisic TH, Fry SE, Salmon SE, Trent JM, Krasovich P (1979) Development of an agar-methylcellulose clonogenic assay for cells in transitional cell carcinoma of the human bladder. Cancer Res 39: 5051–5056

Buick RN, Fry SE, Salmon SE (1980) Effect of host-cell interactions on clonogenic carcinoma cells in human malignant effusions. Br J Cancer 41: 695–704

Carney DN, Bunn PA Jr, Gazdar AF, Pagan JA, Minna JD (1981) Selective growth in serum free hormone supplemented medium of tumor cells obtained by biopsy from patients with small cell carcinoma of the lung. Proc Natl Acad Sci USA 78: 3185–3189

Cillo C, Abele R, Alberto P, Odartchenko N (1981) Culture of colony-forming cells from human solid tumors. UICC Conference Clinical Oncology, Lausanne, p 114
Cillo C, Aapro MS, Eliason JF, Abele R, Fekete A, Odartchenko N, Alberto P (1983) Methylcellulose (MC) clonogenic assay for human solid tumors. AACR Anual meeting, San Diego, Abstract 1234
Eliason JF (to be published) Long-term production of hemopoietic progenitors in cultures containing low levels of serum
Eliason JF, Testa NG, Dexter TM (1979) Erythropoietin-stimulated erythropoiesis in long-term bone marrow cultures. Nature 281: 382–384
Eliason JF, Testa NG, Dexter TM (1982) Comparison of the burst enhancing activities of fetal calf serum, bone marrow cells and spleen cell conditioned medium. In: Baum SJ, Ledney GD, Thierfelder S (eds) Experimental hematology today 1982. Karger, Basel, pp 61–67
Hamburger AW, White CP (1981) Interaction between macrophages and human tumor clonogenic cells. Stem Cells 1: 209–223
Hoang T, Iscove NN, Odartchenko N (1981) Agar extract induces release of granulocyte colony-stimulating activity from human peripheral leukocytes. Exp Hematol 9: 499–505
Hug V, Spitzer G, Drewinko B, Blumenschein GR (1983) Effect of diethylaminoethyl-dextran on colony formation of human tumor cells in semi-solid suspension cultures. Cancer Res 43: 210–213
Iscove NN, Guilbert LJ, Weyman C (1980) Complete replacement of serum in primary cultures of eryhtropoietin-dependent red cell precursors (CFU-E) by albumin, transferrin, iron, unsaturated fatty acid, lecithin and cholesterol. Exp Cell Res 126: 121–126
Kincade PW, Ralph P, Moore MAS (1976) Growth of B lymphocyte clones in semi-solid culture is mitogen dependent. J Exp Med 143: 1265–1268
Selby P, Buick RN, Tannock I (1983) A critical appraisal of the "human tumor stem-cell assay". N Engl J Med 308: 129–134

*Relationship of Steroid Hormone Receptors to the Cloning of Fresh Breast Cancer Tissues**

G.A. Losa and G.J.M. Maestroni

Laboratory of Cellular Pathology, Istituto Cantonale di Patologia,
6604 Locarno, Switzerland

Introduction

Among the various laboratory chemosensitivity tests, clonogenic tumor cell survival in soft or semisolid media has recently been evaluated as one of the most promising assays for individualizing cancer chemotherapy. Several tumors form colonies in semisolid media (Hamburger and Salmon 1977; Von Hoff et al. 1980; Hamburger et al. 1978), but growing cultures are not always suitable for clinical or biological investigations, the plating efficiency (PE) being one of the main limiting factors. The proportion of cells which form colonies is usually below 1%, and it is not yet clear whether low PE reflects inadequate culture conditions or a low stem cell concentration in human tumors.

The definition of optimal culture conditions for frequent malignancies such as breast cancer is of obvious importance. Feeder layers, esoteric additives, and other often complicated procedures have been utilized for increasing the PE of solid tumors, including breast cancer, but no significant improvements have been yet reported (Pavelic et al. 1980; Sandbach et al. 1982; Touzet et al. 1982). Usually, the great majority of these studies have been performed using the agar culture technique as developed by the Tucson group (Hamburger and Salmon 1977; Hamburger et al. 1978; Sandbach et al. 1982; Touzet et al. 1982).

Methylcellulose (MC) viscous media have proven to be at least as good as agar and agarose semisolid media in supporting colony growth (Pavelic et al. 1980; Hoang et al. 1981). Furthermore, MC requires a simpler handling technique than agar and, most important, it allows an easier manipulation of the colonies grown. In this report we present results of human breast cancer cultures in MC media. In particular, the effect of autologous serum (AS) on the PE of breast cancers and the relation between the in vitro tumor growth and its original (biopsy) content of steroid hormone receptors are discussed.

Materials and Methods

Collection and Treatment of Cells

Biopsy samples of primary tumors diagnosed by histopathological examination were used for the study. Representative specimens of the tumor biopsies were processed for steroid

---

* Supported by a grant from the Hermann und Lilly Schilling Foundation, Federal Republic of Germany

Table 1. Effect of autologous serum (As) on the number of colonies formed by human breast cancers in MC cultures

Patient	Age	Histology	Viability (%)	Colonies at 21 days	
				FCS ($\bar{x} \pm SD$) PE	AS ($\bar{x} \pm SD$) PE
BE	75	Ductal	25	6.5	8
NL	58	Ductal	5	5	–
RM	54	Ductal	60	16.5	1
AP	44	Ductal	20	3	–
CV	80	Ductal	18	13.5	6.5
CP	71	Ductal	20	3	2
LM	82	Ductal	11	20	–
CF	49	Ductal	5	3.5	11.5
PG	53	Ductal	9	10	32
FN	42	Ductal	27	15	15
BE	61	Ductal	38	8	2
NC	63	Ductal	30	28	18
PA	39	Ductal	35	28	16.5
ZL	56	Ductal	15	5	–
PP	61	Ductal	23	14	18.5
IA	63	Lobular	12	27.5	78.5
DA	56	Lobular	25	3.5	4.5
CS	48	Ductal	12	3	7
FM	72	Ductal	49	13	28
				(11.9 ± 8.8) 0.012%	(15.5 ± 20.5) 0.015%

MC, methylcellulose; *FCS*, fetal calf serum; *PE*, plating efficiency

receptor assay and for culturing. The sample to be cultivated was dipped in cold Iscove modified Dulbecco's medium (IMDM) supplemented with 10% fetal calf serum (FCS). Cell suspensions were obtained mechanically by finely mincing the tumor with scissors and then by teasing the cells with a loose-fitting Teflon pestle. The cells were filtered through sterile gauze, washed, counted and suspended to the desired concentration in IMDM. Eventual cell aggregates were gently dissociated by repeated passages in a syring equipped with needles of decreasing diameter until a single-cell suspension was obtained. Viability, as determined by trypan blue exclusion, ranged between 5% and 60% (Table 1). Each tumor sent for this study was accompanied by a sample of autologous peripheral blood, which provided the serum for cultures.

Cultures

Cultures were performed according to a modification of the method of Iscove and Schreier (1979): 10^5 viable tumor cells were suspended in 1 ml 0.8% methylcellulose (Fluka AG, Switzerland) in IMDM supplemented with 20% FCS or 10% AS and plated in 35-mm petri dishes. Both FCS and AS were inactivated at 56° C for 30 min before use. Cultures were always in duplicate. The plates were incubated at 37° C in a humidified atmosphere with

5% CO_2. Cell aggregates were counted as colonies provided that one of their axes exceeded 80 μm. The dimensions of the aggregates were evaluated by a micrometer inserted in the eyepiece (magnification × 12.5) of the microscope, equipped with different objectives (× 2.5; × 10; × 20).
At the time of plating (day 0), all dishes were screened for the possible presence of cell clumps or aggregates on the dish bottom. On the basis of this screening, colony counts were done after 7, 14, and 21 days of culture. Before and after culturing the cytological type of the tumor cells was identified by the Papanicolau staining of cytocentrifuge preparations. Collection of the colonies was obtained by diluting the MC with IMDM and the suspension pelleted at low speed (1,300 rpm, 10 min). Cells were then suspended in small volumes at the suitable concentration (10^6 cells per milliliter) of IMDM and cytocentrifuged.

Steroid Hormone Receptor Assay

Estrogen (ER) and progesterone (PR) receptors were assayed according to the charcoal-dextran method routinely used in many laboratories (Zava et al. 1982).

Results

Effect of Autologous Serum on Plating Efficiency

Nineteen tumors out of 37 grew in MC cultures with FCS and/or AS, i.e., 51% of the samples plated. This percentage and the PE of about 0.01% are in line with results using solid tumors cultured in agar semisolid media as reported by other authors (Hamburger and Salmon 1977; Von Hoff et al. 1980; Hamburger et al. 1978; Pavelic et al. 1980; Sandbach et al. 1982; Touzet et al. 1982). The influence of AS and of FCS on colony formation from breast cancers is compared in Table 1. AS did not significantly change the PE. The number of the colonies appeared independent both of the patients' age and of the viability of the single-cell suspension before plating. Furthermore, no selection of any histological tumor type was found in the group growing in MC cultures, i.e., the proportion of the ductal to the lobular type remained essentially that shown by the overall population of breast cancers diagnosed in our Institute (Table 1). An example of a human breast cancer colony obtained with our system is depicted in Fig. 1. In spite of the absence of bottom layers we never saw any fibroblast growth.

Growth and Steroid Hormone Receptors

The bioptic samples utilized for the steroid receptor assay and for culturing were always obtained from the same tumor mass. For both steroid receptor assay and culture, care was taken to collect samples representative of the entire biopsy. Among the 19 tumors growing in vitro, no correlation whatsoever was found between the amount of steroid receptors and PE. On the contrary, the receptor content of the colony-forming tumors was found to be significantly lower that that of the nongrowing group (Table 2). Figure 2 illustrates the relationship especially evident for progestinic receptors, where the difference between the two groups was highly significant ($p < 0.001$), as is shown by the analysis of variance (Table 2).

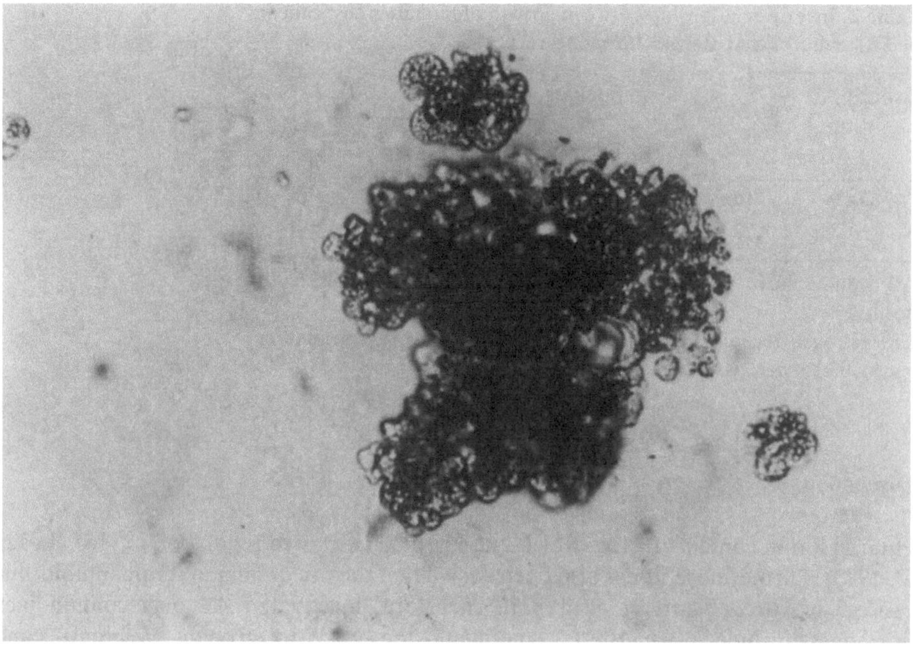

Fig. 1. Typical colony from a primary human breast cancer grown in methylcellulose. The two small aggregates should be regarded as clusters. × 500

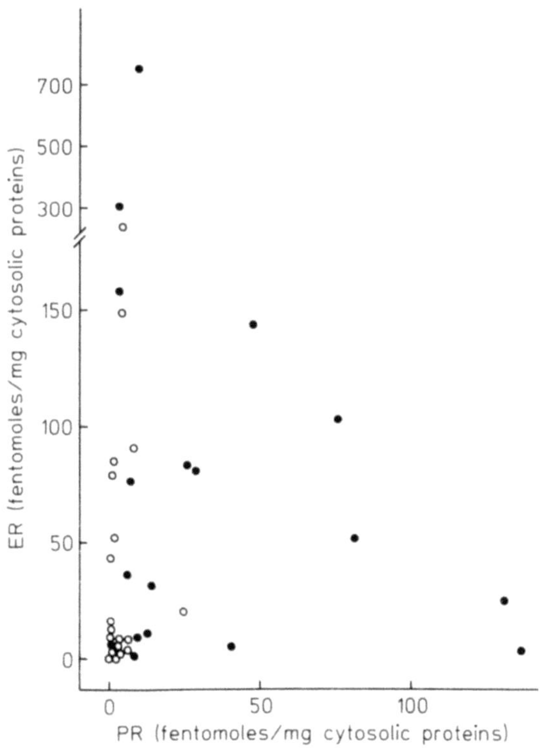

Fig. 2. Association between tumor content of steroid hormone receptors and ability to form colonies in methylcellulose. The values of estrogenic receptors *(ER)* were plotted against progestinic receptors *(PR)* for each tumor. ○, grown; ●, not grown

Table 2. Inverse relationship between growth of human breast cancer in MC cultures and steroid hormone receptors

Groups	*n*	fmol/mg protein	
		ER ± SE	PR ± SE
Growth	19	40.4 ± 13.1	3.3 ± 2.1
No growth	18	126.9 ± 49.9[a]	47.2 ± 12.8[b]

All values above 0.0 fmol/mg cytosolic protein were considered positive
ER, estrogen receptors; *PR*, progestinic receptors; *n*, total number
[a] $p < 0.1$; [b] $p < 0.001$

Discussion

Human serum contains factors which are involved in growth regulation in vivo (Lozzio et al. 1975). Furthermore, it has also been shown that the use of human serum inhibits human cancers in vitro (Pigott et al. 1982). Therefore, inactivated AS may contain factors, hormones included, which may modulate the in vitro growth of breast cancers. Nevertheless, replacement of FCS with AS in MC cultures of mammary tumors did not affect their PE. Both FCS and AS should contain, among other factors, variable amounts of steroid hormones, but in spite of this a negative relation emerged between high-receptor tumors and in vitro growth (Fig. 2, Table 2). This might in part reflect the limited predictive significance of the steroid receptor assay for the hormone dependence of breast cancers (McGuire 1975). Furthermore, the percentage (10%) of AS used might be inadequate for providing sufficient amounts of hormones and/or other growth-modulating factors. Unfortunately, after a few days of incubation higher percentages of AS produced a phase separation with the appearance of lipidous droplets which did not allow a dose-response investigation. The effect of a direct addition of steroid hormones to primary cultures of human breast cancer remains to be tested.

The PE of our cultures did not differ significantly from that recently reported for human breast cancers cultured in soft agar (Sandbach et al. 1982; Touzet et al. 1982). It should be noted that we plated 10^5 viable cells per dish, while in the majority of the studies using soft agar as semisolid medium, the number of cells plated was as much as five times higher (Hamburger and Salmon 1977; Von Hoff et al. 1980; Hamburger et al. 1978; Pavelic et al. 1980; Sandbach et al. 1982; Touzet et al. 1982). Therefore, in order to compare the absolute number of colonies between the various methods, one should theoretically multiply our results by a factor of 5. In this way, most of our samples would reach and overcome the empirical threshold of 20–30 colonies which, according to the literature, represents the minimal growth required for the specimen to be considered to have grown. On what biological basis this was established is presently unknown.

Another interesting observation concerns the low viability of the cell suspensions obtained by the mechanical dissociation of the tumors (see Table 1). In fact, the great and variable number of dead cells plated together with viable cells could influence the cloning. The difficulty of preparing viable single-cell suspensions from human solid tumors is widely recognized (Selby et al. 1982) and constitutes a major critical point in any culture technique.

With respect to ER and the ability of the tumors to grow in MC cultures, our results do not completely conflict with the findings reported by Sandbach et al. (1982), according to which ER status did not influence the percentage of tumors which formed colonies in soft agar. In fact, the association we found between low ER and growth is not really significant, although a certain trend toward significance does exist ($p < 0.1$, Table 2). This would allow our findings to be seen also as partially in line with those reported by Meyer et al. (1977) and Silvestrini et al. (1979), who found an inverse relationship between the amount of ER and a labeling index expressing the proliferative activity of mammary cancers from postmenopausal patients. As a matter of fact, our group of patients may also be considered as postmenopausal, the mean age being 63 ± 13 years. However, a direct relationship between labeling index and cloning capacity has not yet been the object of any investigation, and therefore we do not know whether labeling index and cloning capacity are two expressions of a unique property of the tumor.

Most striking was the association we found between PR and tumor growth; in fact, breast cancers with low content of PR showed a significant growth ($p < 0.001$) in our MC system, while specimens containing high amounts of PR did not (Fig. 2, Table 2). This may indicate that the latter tumors contain cells physiologically and ultrastructurally more differentiated, with a reduced ability to grow in semisolid media. In fact, one known effect of progesterone and other progestins is to promote differentiation of endometrial cells to a secretory state (Henderson et al. 1982). It therefore seems conceivable that also in breast epithelia malignant cells with high PR content may be more sensitive to the hormone and thus retain a degree of cellular differentiation which reduces their ability to form colonies in vitro. In conclusion, in our hands, the cloning of fresh specimens of human solid tumors seems difficult. This result does not conflict with the others obtained by use of different semisolid media.

Basic biological studies therefore seem to be essential prerequisites to an understanding of why some breast tumor samples do grow and form colonies while others do not. One can easily assume that some necessary growth factors are absent from FCS as well as from AS; on the other hand, the membranes and cytoplasm of growing tumor cells may have different quantitative and qualitative enzymatic characteristics from those of nongrowing cells or primary tumor cells.

Summary

Thirty-seven samples of human primary breast cancer were processed for direct cloning in methylcellulose (MC) cultures. Out of the 37 specimens plated, 19 (51%) tumors grew with a plating efficiency (PE) of 0.012%. Both growing and nongrowing tumors belonged mostly to the ductal histological type. Neither the use of autologous serum (AS) nor of fetal calf serum (FCS) affected the PE. Moreover, a negative correlation was found between the level of estrogens and especially of progestin receptors and the ability of tumors to grow in MC culture. These findings underline the difficulty of cloning fresh specimens of human solid tumors and indicate that malignant cells displaying high concentrations of progestinic receptors may also display a degree of differentiation which leads to a reduced clonogenic ability.

References

Hamburger A, Salmon SE (1977) Primary bioassay of human myeloma stem cells. J Clin Invest 60: 846–854

Hamburger AW, Salmon SE, Kim MB, Trent JM, Sohenlen B, David SA, Schmidt JH (1978) Direct cloning of human ovarian carcinoma in agar. Cancer Res 35: 3438–3443

Henderson BE, Ross RK, Pike MC, Casagrande JT (1982) Endogenous hormone as a major factor in human cancer. Cancer Res 42: 3232–3239

Hoang T, Iscove NN, Odartchenko N (1981) Agar extract induces release of granulocyte colony stimulating activity from human peripheral leukocytes. Exp Hematol 9: 499–504

Iscove NN, Schreier MH (1979) Clonal growth of cells in semisolid or viscous medium. In: Leskovits I, Pernis B (eds) Immunological methods. Academic, New York, pp 379–384

Lozzio BB, Lozzio CB, Bamberger EG, Lair SV (1975) Regulators of cell division: endogenous mitotic inhibitors of mammalian cells. J Int Rev Cytol 42: 1–47

McGuire WL (1975) Current status of estrogen receptors in human breast cancer. Cancer 36: 634–638

Meyer JS, Ramanath Raò B, Stevens SC, White WL (1977) Low incidence of estrogen receptors in breast carcinomas with rapid rates of cellular replication. Cancer 40: 2290–2298

Pavelic ZP, Slocum HK, Rustun YM, Creaven JP, Novak NJ, Karakousis C, Takita H, Mittelman A (1980) Growth of cell colonies in soft agar from biopsies of different human solid tumors. Cancer Res 40: 4151–4158

Pigott DA, Grimaldi MA, Dell'Aquila ML, Gaffney EV (1982) Growth inhibitors in plasma derived human serum. In Vitro 18: 617–625

Sandbach J, Von Hoff DD, Clark G, Cruz AB Jr, Obrien B (1982) South Texas humant umor cloning group, direct cloning of human breast cancer in soft agar culture. Cancer 50: 1315–1321

Selby P, Buick RN, Tannock I (1982) A critical appraisal of the "human stem cell assay". N Engl J Med 308: 129–134

Silvestrini R, Daidone MG, Di Fronzo G (1979) Relationship between proliferative activity and estrogen receptors in breast cancer. Cancer 44: 665–670

Touzet C, Ruse F, Chassagne J, Ferriere JP, Chollet P, Plagne R, Fonck Y, DeLatour M (1982) In vitro cloning of human breast tumor stem cells: influence of histological grade on the success of cultures. Br J Cancer 46: 668–669

Von Hoff DD, Casper J, Bradley E, Trent JM, Hodach A, Reichert C, Makuch R, Altman A (1980) Direct cloning of human neuroblastoma cells in soft agar culture. Cancer Res 40: 3591–3597

Zava DT, Wyler-Von Ballmoos A, Goldhirsch A, Roos W, Takahashi A, Eppemberger U, Arrenbrecht S, Martz G, Losa G, Gomez F, Guelpa C (1982) A quality control study to assess the inter-laboratory variability of routine estrogen and progesterone receptor assays. Eur J Clin Oncol 18: 713–721

Subject Index

Recent Results in Cancer Research

Managing Editors:
C. Herfarth, H.J. Senn

Volume 93
Leukemia
Recent Developments and Therapy

Editors: **E. Thiel, S. Thierfelder**
1984. 35 figures, 52 tables. Approx. 340 pages. ISBN 3-540-13289-9

Volume 92
Lung Cancer
Editor: **W. Duncan**
1984. 23 figures, 42 tables. IX, 132 pages. ISBN 3-540-13116-7

Volume 91
Clinical Interest of Steroid Hormone Receptors in Breast Cancer
Editors: **G. Leclercq, S. Toma, R. Paridaens, J.C. Heuson**
1984. 74 figures, 122 tables. XIV, 351 pages. ISBN 3-540-13042-X

Volume 90
Early Detection of Breast Cancer
Editors: **S. Brünner, B. Langfeldt, P.E. Andersen**
1984. 94 figures, 91 tables. XI, 214 pages. ISBN 3-540-12348-2

Volume 89
Pain in the Cancer Patient
Pathogenesis, Diagnosis and Therapy

Editors: **M. Zimmermann, P. Drings, G. Wagner**
1984. 67 figures, 57 tables. IX, 238 pages. ISBN 3-540-12347-4

Volume 88
Paediatric Oncology
Editor: **W. Duncan**
1983. 28 figures, 38 tables. X, 116 pages. ISBN 3-540-12349-0

Volume 87
F.F. Holmes
Aging and Cancer
1983. 58 figures. VII, 75 pages. ISBN 3-540-12656-2

Volume 86
Vascular Perfusion in Cancer Therapy
Editors: **K. Schwemmle, K. Aigner**
1983. 136 figures, 79 tables. XII, 295 pages. ISBN 3-540-12346-6

Springer-Verlag
Berlin
Heidelberg
New York
Tokyo

MIX
Papier aus verantwortungsvollen Quellen
Paper from responsible sources
FSC® C105338

If you have any concerns about our products,
you can contact us on
ProductSafety@springernature.com

In case Publisher is established outside the EU,
the EU authorized representative is:
Springer Nature Customer Service Center GmbH
Europaplatz 3, 69115 Heidelberg, Germany

Printed by Libri Plureos GmbH
in Hamburg, Germany